Indoor Geolocation
Science and Technology
At the Emergence of Smart World and IoT

RIVER PUBLISHERS SERIES IN COMMUNICATIONS

Series Editors:

ABBAS JAMALIPOUR
The University of Sydney
Australia

MARINA RUGGIERI
University of Rome Tor Vergata
Italy

JUNSHAN ZHANG
Arizona State University
USA

Indexing: All books published in this series are submitted to the Web of Science Book Citation Index (BkCI), to CrossRef and to Google Scholar.

The "River Publishers Series in Communications" is a series of comprehensive academic and professional books which focus on communication and network systems. Topics range from the theory and use of systems involving all terminals, computers, and information processors to wired and wireless networks and network layouts, protocols, architectures, and implementations. Also covered are developments stemming from new market demands in systems, products, and technologies such as personal communications services, multimedia systems, enterprise networks, and optical communications.

The series includes research monographs, edited volumes, handbooks and textbooks, providing professionals, researchers, educators, and advanced students in the field with an invaluable insight into the latest research and developments.

For a list of other books in this series, visit www.riverpublishers.com

Indoor Geolocation Science and Technology
At the Emergence of Smart World and IoT

Kaveh Pahlavan

Worcester Polytechnic Institute
USA

Routledge
Taylor & Francis Group

LONDON AND NEW YORK

Published 2019 by River Publishers
River Publishers
Alsbjergvej 10, 9260 Gistrup, Denmark
www.riverpublishers.com

Distributed exclusively by Routledge
4 Park Square, Milton Park, Abingdon, Oxon OX14 4RN
605 Third Avenue, New York, NY 10017, USA

First issued in paperback 2023

Indoor Geolocation Science and Technology At the Emergence of Smart World and IoT / by Kaveh Pahlavan.

Routledge is an imprint of the Taylor & Francis Group, an informa business

Publisher's Note
The publisher has gone to great lengths to ensure the quality of this reprint but points out that some imperfections in the original copies may be apparent.

While every effort is made to provide dependable information, the publisher, authors, and editors cannot be held responsible for any errors or omissions.

ISBN 13: 978-87-7022-967-8 (pbk)
ISBN 13: 978-87-7022-051-4 (hbk)
ISBN 13: 978-1-003-33853-6 (ebk)

*To my lovely wife, Farzaneh, my children, Nima, Nasim, and Shek,
and my grandchildren, Roya, and Navid*

Contents

Preface

The evolution of wireless localization technologies for indoor and urban area applications, where the Global Positioning System (GPS) does not work properly, has been an active area of research for commercial and public safety applications since the mid-1990s. At the time of this writing, received signal strength (RSS)-based Wi-Fi localization has been dominating the commercial market in smart devices complementing cell tower localization and GPS using the time of arrival (TOA) technology. These smart devices include smart phones, note pads, book readers, notebooks, and laptops, and in many of these devices, there is no cell phone or GPS. Wi-Fi localization technology takes advantage of the random deployment of Wi-Fi devices worldwide to support indoor and urban area localization with reasonable accuracies for hundreds of thousands of applications on smart devices, from modern location-based services for finding businesses, such as Yelp, to traditional turn by turn navigation. Public safety and military applications demand more precise localization for first responders, and they are discovering more sophisticated techniques for such precise indoor geolocation using hybrid techniques. Radio frequency (RF) localization techniques for these hybrid technologies are intent on more precision through TOA-based localization. Hybrid algorithms use a variety of sensors to measure the speed and direction of movement and integrate them with the absolute RF localization. In this book, we address the practical aspects of these technologies used for geolocation in indoor and urban areas for commercial and military applications. In addition, we will examine challenges in localization inside the human body that is emerging as an area of research for the future of the wireless health industry.

Science and engineering disciplines are going through a "transformation" from their traditional focused curriculum to a "multi-disciplinary" curriculum and "inter-disciplinary" research directed towards innovation and entrepreneurship. This situation demands more frequent updates and adjustments in the curriculum, project-oriented delivery of educational content, and the ability to form inter-disciplinary cooperation in research programs. A successful transformation of this form demands entrepreneurship and vision-ary talents to adapt to these frequent changes, and industrial experiences to direct the transformation towards emerging inter-disciplinary industries. Indoor and urban area opportunistic localization is an excellent example of a multi-disciplinary area of research and scholarship, which has emerged in the past few decades. Material needed for teaching opportunistic localization for indoor and urban areas includes several disciplines such as radio propagation analysis, signal processing, detection and estimation theory, and positioning algorithms. The content of courses on opportunistic localization is useful for traditional ECE and CS students as well as students in emerging multi-disciplinary programs such as Robotics and Biomedical Engineering as well as traditional Mechanical and Civil Engineering programs, which are similar to ECE, shifting towards inter-disciplinary curriculums. Therefore, there is a need for academic courses and a comprehensive textbook to address prin-ciples of opportunistic localization to be taught in these multi-disciplinary programs.

To prepare a textbook to be taught in academic courses in a multi-disciplinary area of technology, we need to provide selected details of practical aspects of a number of disciplines to give to the readers an intuitive feeling of how these disciplines operate and interact with one another. To achieve this goal in this book, we describe important positioning systems technologies, classify their underlying science and engineering in a logical manner, and give detailed examples of successful science and engineering that has turned into popular applications. Selection of detailed technical material for teaching courses in a multi-disciplinary area with a large and diversified set of technical disciplinary is very challenging, and this challenge becomes more difficult because people with different background skills are using the results.

This book provides a comprehensive treatment of the opportunistic local-ization science and technology applied to opportunistic positioning in indoor and urban areas. The novelty of the book is that it places emphasis on radio propagation and physical layer issues related to how the received wireless communication signals can be used for opportunistic RF positioning in a variety of wireless communication networks. The structure and sequence of material for this book was first formed in a lecture series by the author at the graduate school of the Worcester Polytechnic Institute (WPI), Worcester, MA, entitled "Wireless Access and Localization".

The book begins with an introduction chapter underlying the importance of localization in the emergence of the smart world, followed by nine chapters and an appendix chapter. The nine technical chapters of the book are divided into three parts. Part I consists of two chapters on RSS-based positioning. Part II consists of three chapters on TOA-based positioning. The Direction of Arrival (DOA) systems are treated as a part of TOA systems, because they rely on TOA estimation. In Parts I and II, the tool for analytical performance evaluations is the Cramer-Rao Lower Bound (CRLB) that is compared with the empirical results of performance evaluations. Part III presents the details of applied algorithms for positioning in four chapters on: basic positioning algorithms, pattern recognition algorithms used in RSS-based systems, TOA-based algorithms used for refining the TOA estimation, and hybrid algorithms to integrate relative and absolute localization techniques. The appendix chapter of the book is an independent chapter devoted to the review of the classical estimation theory applied to ranging and positioning applications. This chapter covers principles of estimation theory and derivations for CRLB applied to RSS- and TOA-based positioning techniques.

The emerging multi-disciplinary fields such as wireless networking, Robotics, or wireless positioning involves a variety of disciplines. Researchers with different skills work on different aspects of a unique project, but the experience that they gain in that endeavor is different and focused around their individual skills. Therefore, details of the research work, contents of the courses that emerge from the results of the research, and the books that are written on the same topic are based on the author's technical experiences. The author of this book has pioneering research experience and industrial exposure in design and performance evaluation of indoor geolocation based on empirical measurement and modeling of the behavior of the radio propagation in indoor areas and inside the human body. Presentation of the material is based on examples of research and development that his students have performed, his teaching experiences as a professors, and his experiences as a technical consultants to startup companies. As a result, this book organizes the result of research papers by his students, patents with the industrial colleagues, and his course notes to present the results of the research findings over more than a couple of decades to his graduate students beginning their education in this emerging area of research.

Since much of the material in indoor geolocation and in localization inside the human body are extracted from research work of the students at the center for wireless information network studies (CWINS), at the Worcester Polytechnic Institute (WPI), Worcester, Massachusetts, we are pleased to acknowledge the students' and colleagues' contributions to advancing the understanding of wireless channels and its application in opportunistic positioning using wireless communications signals. In particular, the author thanks Prof. Prashant Krishnamurthy of the University of Pittsburgh, also his first PhD student in this field, for careful editing of the material in the first few chapters of the book and weekly conference calls during preparation of the book. Without him, the author would have not completed the manuscript on time and with its current quality of presentation. Most of the MATLAB codes presented in the book and the solution book for the problems at the end of the chapters are prepared by Julang Ying, the author's current lead PhD student. He also helped in formatting the references and proof reading the equations. In addition, the author thanks Prof. Jacque Beneat, Prof. Xinrong Li, Dr. Ahmad Hatami, Dr. Robert Tingley, Dr. Bardia Alavi, Dr. Nayef Alsindi, Dr. Mohammad Heidari, Dr. Ferit Akgul, Prof. Muzzafer Kanaan, Dr. Yunxing Ye, Dr. G. Bao, Dr. Yishuang Geng, Dr. Umair Khan, Dr. Fardad Askarzadeh, and Dr. Nader Bargshady, and many other students of the CWINS; Prof. Mika Ylianttila and Dr. Juha-Pekka Makela of CWC, University of Oulu, Finland ; Prof. Sergey Makarov of WPI; Prof. Pratap Misra of Tufts University; Mr. Ted Morgan and Dr. Farshid Alizadeh of Skyhook Wireless; Prof. Jie He and Dr. Liyuan Xu of University of Science and Technology of Beijing (USTB), China; and Prof. Yongtao Ma of Tianjin University of China, who have directly or indirectly helped the author to extend his knowledge in this field and shape his thoughts for preparation of the new material in this book in a variety of research projects and associated publications. In particular, the author acknowledges Liyuan's contributions in preparing sections on sensor behavior and classification in Chapter 10. The author also acknowledges the Defense Advanced Research Projects Agency (DARPA), National Institute of Standards and Technology (NIST), National Science Foundation (NSF), Department of Defense (DoD), Department of Homeland Security (DHS), United Technology and Skyhook in the United States as well as Finnish Founding Agency for Technology and Research (TEKES), Nokia, Elektrobit, Sonera, and the University of Oulu in Finland, whose support of the CWINS research program at WPI enabled graduate students and the staff of the CWINS to pursue continuing research in this important field. A substantial part of the new material in this book has flowed out of these sponsored research efforts. The author also appreciates Dr. Allen H. Levesque for his contributions in other books with the author, which has indirectly impacted in formation of thoughts and the details of material presented in this book.

Much of the writing of the author in this book was accomplished during his sabbatical leave from WPI during spring and summer semester of 2018.

He expresses his deep appreciations to the Worcester Polytechnic Institute, in particular to Prof. Yehia Massoud, Head of the ECE Department, Prof. Winston Soboyejo, Dean of Engineering, and Prof. Bruce E. Bursten, the Provost of WPI, for their approval of his sabbatical leave for work on this book-writing project. Author dedicates the income of this book to the educational fund of his grandchildren, Roya and Navid Kablan, for the love and energy they gave him to complete this challenging mission, a large part of this writing occurred when they were around him giving him the needed energy to continue!

The author also thanks Mark de Jongh of River Publishers for his assistance and useful comments during various stages of production of the book, and Junko Nakajima and her team for help during manuscript proofreading and publication.

1

Introduction: Localization in Smart World

1.1 Introduction

Location and time are the prominent underlying features for any scientific and engineering observation. Every experience and every observation that is formed by an intelligent mind in a scientific and engineering sense is fundamentally associated with location and time. So, ever since the human being, a social animal equipped with linguistics to communicate with others, began to analyze and record its experiences intelligently, measuring the time of the experience and the location of the events has been an essential part for gathering intelligence for individual humans and the human society at large. In the same way as we humans transformed our civilizations from the farming economy to the industrialized world during the first industrial revolution in the late eighteenth century, over the past few decades we have transformed our civilization to the cyber physical world. In this modern industrial revolution, smartness and intelligence have become more important and so also is the importance of knowing the time and location of events, to analyze the behavior of objects, individuals, crowds, and society itself.

Historically, we began measuring time using the sun and the moon a few thousands of years ago at the dawn of civilization, eventually to have an *absolute* value of time, needed for intelligent agricultural processing and associated festivities and administrates. The granularity of time here was over longer horizons. Around the same time, we began measuring *relative* time with devices such as sand clocks for other important events with shorter durations. In the past few centuries, since the industrial revolution, human beings have discovered how to measure time precisely and the corresponding industry has learned how to make this information about time available and accessible to everyone, everything, and everywhere (Figure 1.1(a)). Today, every "Person" carries a device such as a smart watch or a smart phone, capable of measuring time with near-atomic clock accuracy and almost every

1

"Thing" has a clock with similar accuracies. However, this is not necessarily true for the location information.

The "location industry" also began in the early days of civilization using the sun, moon, and stars for navigating travelers on the ground and over the seas with a coarse precision. Later, these coarse technologies were complemented by compass and other devices using the Earth's magnetic field. Popular and more precise location technology is going through a similar discovery and metamorphosis as those of the time measurement industry, only in the past few decades.

Figure 1.1(b) illustrates the chronology of the evolution of localization technologies in recent years. Today's most popular localization system, Global Position System (GPS) with accuracy of tens of meters (and better under ideal conditions) was introduced in the 1970s for military applications and became available to commercial applications in the early 1990s. In the past decade, the dramatic fall of the costs of GPS chips have made them popular in smart phones and many other devices, to the extent that even some wearable fitness devices now have GPS. Since GPS signals do not work properly in indoor areas and the indoor applications normally need accuracy on the order of meters or less, research efforts in indoor geolocation science and technology began in the late 1990s and became prolific in the late 2000s

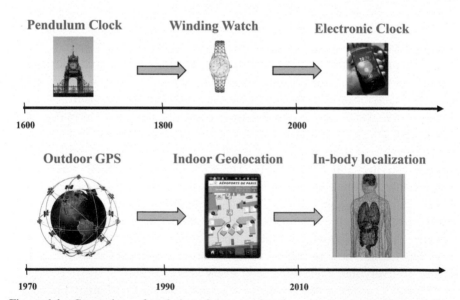

Figure 1.1   Comparison of evolution of time and location measurement methods (a) time measurement techniques in past few centuries, (b) positioning technologies in past few decades.

[Pah02, Pah10]. More recently, localization science has been progressing towards positioning medical devices inside the human body, where an accuracy on the order of a sub-centimeter is required [Pah12, Dav13] to locate the area where a medical problem may exist. These localization technologies have their own challenges, which are resolved in completely different scientific domains, resulting in significant applications-related useful scientific and engineering research. Because of complexity and diversity of science and technology involved in indoor geolocation, this area has emerged as its own discipline over the past two decades.

This multidisciplinary book presents the fundamentals of positioning and navigation science and technology used for the Smart World and the Internet of Things (IoT) applications in different platforms such as: smart devices, unmanned ground and flying vehicles, and existing cars operating as a part of intelligent transportation systems. Material taught in the book are beneficial for graduate students in the Electrical and Computer Engineering, Computer Science, Robotics Engineering, Biomedical Engineering, or other disciplines who are interested in integration of navigation into their multidisciplinary projects. The book provides for modeling the behavior of radio frequency and mechanical sensors used for localization applications. Then, it introduces the Cramer–Rao Lower Bound (CRLB), which uses these models for comparative performance evaluation of different localization techniques. Finally, it introduces variety of algorithms used for ranging, localization, navigation, and fusion of different sensor readings. We classify these algorithms into logical categories and provide practical example of specific algorithms. The examples include Wi-Fi localization algorithms, Ultra-Wideband (UWB) localization algorithms, and multi-sensor localization algorithms integrating mechanical sensor readings with radio frequency localization. We will provide examples with supporting MATLAB codes and hands-on projects throughout the book to improve the ability of the readers to understand and implement variety of algorithms. The book can be beneficial for both the academia, as a textbook with problem sets and projects, and the industry professionals, as a practical reference book.

1.2 Elements of Localization Science

The complexity of localization technology arises from the fact that the variety of applications require different precisions; different smart devices hosting these applications carry different sets of sensors, behavioral characteristics of the different sensors in different environments are complex;

and the availability of maps and the need for the visualization platform for different applications are quite diversified [Pah02, Bir11, Tar11]. As a result, the behavior analysis of the characteristics of the sensors and selection of the suitable algorithms to provide the needed precision for an application implemented on a platform has become a scientific area of research in the past couple of decades and the industry is in need of highly educated professionals in localization science and engineering [Pah13, Gu15].

Figure 1.2 illustrates the functional block diagram of a wireless geolocation system, which focuses on the functionality for positioning. The main elements of the system are as follows: (*i*) a number of location-sensing devices that provide metrics related to the relative position of a mobile station with respect to a known landmark, called reference point, (*ii*) a positioning algorithm that processes metrics reported by location sensing elements to estimate the location coordinates of the target moving object, (*iii*) a display system that illustrates the location of the target on a map, and (*iv*) a location intelligence engine that extracts useful features of location data tailored to generic applications.

The most popular location metrics for wireless localization are extracted from radio frequency (RF)-based localization systems [Zha90, Bar03]. The RF location metrics may indicate the approximated received signal strength (RSS), direction of arrival (DOA), or the time of arrival (TOA) of the signal or it can be a packet of information with the identity of an object read through RF signals [Col13, Col15]. Other localization metrics include imaging cameras used to extract visual information by direct image processing techniques [Hil08, Mul09, Dav07] such as the plate number of a car or by comparing features of consecutive images to analyze the motion of an object [Bis12, Moh14]. There are a number of mechanical sensors such as accelerometer, magnetometer, and barometer, which are used to determine the speed, direction, and height of a device such as a smart phone or a

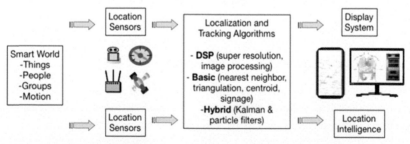

Figure 1.2   Elements of localization science and technology in the Smart World.

robotic platform. The positioning algorithm processes the received metrics to determine the coordinates of the target object. These algorithms may use signal processing algorithms such as super-resolution algorithms [Li04], to refine the sensor data and prepare better metrics, algorithms to process metrics and come up with a location estimate. As the measurement of metrics becomes less reliable or exact, the complexity of the position algorithm increases. In navigation applications, when we have some information regarding the movement of the target object, we combine the position estimates with the information on the pattern of movements of the object to refine the location estimates [Gen13] or algorithms such as Kalman filters [Au13, He14, Che12] or Particle filters [Bar15] to take advantage of the history of other location estimates as well as combining metrics obtained from different sensors [Kel11, Cas12a].

The display system pins the estimated location coordinates to a map of an environment. In the most traditional localization application, in direction finding for car driving, we refer to a geographic map with absolute coordinates. In other contexts, such as indoor areas or within a human body, it may be better to employ local or relative coordinates and an alternative "map" of the indoor area or body [Bao15]. In indoor areas, we have multiple maps associated with different floors of a building, when a smart device carried indoor and outdoor, we need different maps and a smart display system needs to sense the environment for selection of the appropriate map [Yin15]. In any case, the display system could be a service or an application residing in a server or a mobile locating unit, locally accessible software in a local area network, or a universally accessible service on the web such as Google maps. Obviously, as the horizon of the accessibility of the information increases, the design of the display system becomes increasingly complex.

Location intelligence is extracting information for a specific intelligent application using location and track information. Location information uses pattern of movement of a device to analyze the behavior of the person or a mechanical platform carrying the device [Wan11]. Such location intelligence includes location-time traffic analysis, Geo fencing (for elderly people, animals, prisoners, suspicious people, etc.), real-world consumer behavior, location certification for security, positioning IP addresses, and customizing contents and experiences. Video Clip 1.1 illustrates the customer behavior of Wi-Fi localization requests in New York City in different times of a day, collected by Skyhook, Boston, MA. Marketing agents of different organizations for targeted marketing use this type of location intelligence.

Video Clip 1.1 Customer behavior for Wi-Fi localization requests in NYC for different hours of a day. (Source: Skyhook Wireless).

 https://www.riverpublishers.com/book_details.php?book_id=687#video

1.3 Localization Technology and Applications

The word "smart world" became popular because a number of *smart applications* began to emerge to solve problems using time, location, sensor data, and general context. Figure 1.3 provides an overview of several popular

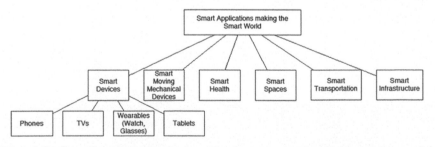

Figure 1.3 Overview of popular smart applications associated with the emergence of the Smart World.

and important smart applications leading to evolution of the smart world. As we integrated extensive computing abilities to mobile phones, starting with the iPhone, we began to call it a "smart phone". As we increased the artificial intelligence of moving mechanical devices using computer programs and electronic circuits, we called them Robots, to connote the fact that they benefit from computing intelligence. More recently, we have used the words "smart health" for better processing of a patient's data to improve the health outcomes and to reduce the cost of improving health services. We refer to buildings with extensive pervasive programmability and intelligence, such as knowing when to increase the thermostat, as smart spaces [Hel05] and we expect these environments to use RFID tags, iBeacon, and other technologies to enable intelligent location-aware robotic applications [Bae07]. We use the words "smart transportation" in the context of adding substantial computational intelligence to traffic monitoring and management. The electric grid that can adapt itself to changing weather and diverse sources of energy as well as intelligently handling electric usage in homes and buildings is now called the "smart grid". In essence, we are realizing a vision of the "smart world" through the variety of smart applications with intelligent capabilities to solve problems. All of these applications demand location information. In this section, we describe examples of some of these applications to show their need for localization information.

1.3.1 Localization for Smart Devices

Smart devices are the most popular platforms used for the creation of smart environment and the smart world. Billions of smart devices such as smart phones, tablets, smart watches, smart glasses, and smart TVs use location information for hundreds of thousands of applications either directly in applications, such as turn-by-turn direction finding, recommendations through Yelp, flight fares through Kayak, or indirectly as part of applications for gaming or customer behavior analysis. The accuracy and precision requirements for these applications are quite diversified and range from centimeters in gaming to meters in indoor geolocation, tens of meters in turn-by-turn direction, and hundreds of meters for broadcasting advertisements in targeted areas.

Smart phones are perhaps the most popular smart devices, and the overvaluing majority of those hundreds of thousands of applications have been designed for mobile application in smart phones. Smart phones also carry a number of location sensors. These sensors include the most popular RF-based location sensors such as GPS, Wi-Fi signal, cell phone signals, iBeacon (low-energy Bluetooth) as well as mechanical location sensors such as magnetometer, barometer and accelerometer, and of course high-quality cameras and microphones. Because of diversity of applications and their requirements, availability of multiple sensors on the platform, and the inherent mobility of the device, which demand operation in diversified environments, most of the research and developments for localization in smart devices have been initiated for smart phone applications.

1.3.2 Localization for the Robots

Robotic platforms are another essential components of the evolving smart world. A variety of ground rolling and flying robots of different sizes are emerging to facilitate operation in the smart environment of the smart world in a variety of applications in warehouse management, manufacturing, military missions, security, commercial delivery, aerial photography, and health. Every robot moves and we need to know where it is, so it should have a location and mapping system. The visualization platform for land robots is usually a 2D map, while flying robots need the more complex 3D localization and mapping. Robot localization systems need to know a landmark as origin or destination and a method to track the movement of the robot. Traditionally, robots benefit from 2D simultaneous localization and mapping (SLAM) algorithms for navigation [Hua11, Bru13] and, more recently, 3D localization for flying robots and for micro-robots inside the human being are emerging [Pah12, Hen12, Fue15].

Location sensors in robotic platforms include camera, speedometer, RF communication devices, RFID readers, and optical measuring meters. A land robot may carry all of these sensors while a flying robot or a robot with medical mission inside the human body may only carry a camera and an RF device to communicate between the controller and the robot [Idd00].

The accuracy and precision needs of moving mechanical devices vary by the application, size, and environment of operation of these devices and it may range from millimeters for robots operating inside the human body to a few meters for robots operating in indoor areas and tens of meters for outdoor robots. The mission of the robot also affects these values. For example, a robot operating in a laboratory in an indoor area or a flying miniature robot may need precisions on the order of centimeter or millimeter for certain tasks rather than a few meters accuracy commonly perceived for indoor operations [Gen15a].

1.3.3 Localization in Smart Health

Since we cannot see inside the human body directly and the inside of the human body comprises a number of complicated organs, each treated by a separate specialist doctor, the term "localization" in medicine is commonly used for locating a lesion, tumors, bleeding, or pain inside the human body. The term is also used for locating an external device such as intrusive surgery equipment or an endoscopy capsule inside the human body [Pah12]. In addition, in hospitals, there are numerous localization applications ranging from locating, doctors, patients, nurses, and visitors to locating commonly used equipment such as wheelchairs and specialized tools such as surgery equipment inside the surgery room. With the current wave of elderly health monitoring research using sensor networks [Hac14, Gen15b], localization technology for patients is needed everywhere. Technologies used for locating people or objects inside the hospitals are one of the major applications for the emerging indoor geolocation science and technologies [Pah02]. Other commonly used technologies for indoor and outdoor localizations are used for health monitoring whenever tracking and location of the patients are essential.

The accuracy and precision needed for localization inside the human body may vary from a fraction of a millimeter for neurons inside the brain to centimeters inside the GI-tract. Localization precision of the people and equipment in health-related applications follows the same guidelines as smart devices and robots.

For in-body localization and mapping, traditional 2D and, more recently, 3D X-ray [Mar14], 3D ultra-sound imaging [Yim13], 3D magnetic resonance imaging (MRI) [Tha12], and computer-aided tomography scan (CAT-scan) [Kut06] are used. More recently, RF signals [Pah12] as well as hybrid RF and imaging techniques have been used for locating micro robots inside the human body [Bao15]. An animated video clip demonstrating the RF and visual elements of hybrid localization inside the human body is available at [Bao12]. Since there is no map in vivo for tracking inside the gastro-internal tract (GI-tract) and the vascular tree, mapping inside the organs has been another current area of research.

1.3.4 Smart Spaces and Localization Using RFID

Radio-frequency identification (RFID) tags are another essential elements of the emerging smart worlds. Currently, passive RFID tags are used in numerous popular applications such as tagging newly born babies or patients in the hospitals, checking cars in the highway pay tolls, checking progress in long assembly lines, tracking small items in warehouses, implanting microchip IDs in animals and the human being, and tracking robots in indoor areas [Moa11]. It is expected that RFID tags become widely deployed in the smart buildings [Ma14, Spi10] and smart phones of the future carry RFID readers [Liu14, Hol09]. We expect trillions of RFID tags to be used for numerous applications in the smart world of the future. As important as it is to connect these tags through the Internet of Things (IoT), a need for localizing RFID tags is becoming essential. Passive RFID readers are simple proximity check localization systems communicating the tag information with the cyber domain. The real-time location system (RTLS) technology addresses indoor localization for RFIDs tags using Wi-Fi, iBeacon (Bluetooth), and UWB Technologies [Mil08]. As important as is to know the location of an RFID, if the RFID is installed in a fixed location, it can be used for opportunistic location and smart evacuations in challenging environments for RF localization such as inside the tunnels [Dea12] or they can be used to improve the accuracy of indoor geolocation systems as a component of a more complex hybrid localization system [Liu14].

1.3.5 Localization for Smart Transportation Systems

Indeed, smart transportation is an important part of smart cities and the smart world. As a result, smart transportation systems have received considerable attention in the literature [Gan13]. We know absolute location of the ground and air transportation systems operating in outdoor environments. If properly networked, this information revolutionizes efficiency of the transportation systems resulting in huge saving in fuel cost and transportation costs. The vehicular navigation systems primarily rely on GPS and mechanical sensors measuring the speed and direction of movement of the car [Pah13]. The weakness of current vehicular systems is lack of networking infrastructure to coordinate these movements in an intelligent manner. This is a long process involving networking standardizations. Another emerging localization application for vehicles is finding the relative location with respect to surrounding mobile and fixed objects. Such application is used to assist drivers in maneuvering the vehicle movements with respect to other moving vehicles as well as the fixed infrastructure surrounding the move. With the investments on the thousands of sensors on those moving vehicles and surrounding infrastructures, ideally we expect to know their locations as well as the trace. Such information carries mobility patterns of dynamic individuals and it enables further data mining-based applications such as smart itinerary guidance, trace-based social event analysis [Kar12], urban traffic planning [Cas12b], land usage monitoring [Hag12], smart public security [Gan13], and even traffic flow-based smart advertising [Lee12]. All these applications need a smart transportation system as the backbone; therefore, localization science and technology has a wide variety of applications in the emerging smart transportation systems as well.

1.3.6 Localization for Smart Infrastructure

Several critical national infrastructures are becoming smart – in particular, the electric grid, intelligent transportation, water systems, etc. [HLS]. Managing the resources of the critical infrastructures efficiently and for enabling their resilience and smooth operation, localization is essential. Although there is limited work in this area (see, for example, [He11, Fan14] for localizing faults in the smart grid), identifying where a resource is being constrained or added becomes important. When sensors are used for sampling water quality, or to assess the load in the electric grid, the locations of these sensors are important as they relate to the measured quantities and the infrastructure that is in place near the sensor. In terms of accuracy and precision, there are a variety of requirements – at a macroscopic level, it may be sufficient to know the locations of various components at the granularity of several tens of meters on a map, but within a specific component (e.g., an electric sub-station), the localization accuracy may have to be on the order of centimeters.

1.4 Some Existing Challenges in Localization

Localization science and engineering has made considerable fundamental growth in the past few decades, which has enabled a spurt in the number of smart applications [Pah13]. As the smart applications in the world expand, the need for localization in challenging environments grows. At the time of this writing, a number of challenges are facing the location science and technology as a fundamental enabling technology for the evolution of the so-called smart world.

Localization in crowded environments during the events in outdoor, such as stadiums, or indoors, such as large lecture halls, is an existing challenge for a number of smart applications tailored for these environments. These applications involve 3D map of the environment and an accuracy of less than a meter to locate targeted seats for delivery of a smart service. To support the audience with wireless access for smart phone as well as localization for delivery of physical services, most probably we need temporary infrastructure deployment using Balloons and Drones that can cover the areas of interest during the events attracting the crowd. To guide people intelligently to their seats, we may need to resort to hybrid and cooperative localization using RF signals, RFID-based signs, as well as mechanical sensors commonly available in smart phones. The map could be a 3D map layered into 2D map of individual floor or step levels or an interface with direction of movement pointers guiding the target smart phone to the seat location.

Finding *cost-efficient* locating systems for smart item finding in shelves of a department store is another long-standing problem for localization systems. Smart shopping of the future needs that type of technology to direct customers to the location of the desired products. Similar challenges may exist for localizing specific entities in the various critical infrastructures to enable smart applications. Such challenges are created in layers – as an example, first we have to find which floor the human being is located to select the associated map for the floor [Ye12, Yin15]. Next, we have to identify which shelf in which aisle contains the item of interest. The accuracy needed for this type of localization intelligence is around few centimeters to differentiate small items from one another. The most commonly used indoor localization of today is Wi-Fi localization with accuracy of a few meters, which is not adequate for this type of application. We need additional RF infrastructure and more complex hybrid and cooperative localization algorithms to achieve the needed accuracy. iBeacon technology, using mechanical sensors and Bluetooth in smart phones, or adding UWB infrastructure seem to be useful for such applications, but the technology has not yet stabilized, nor are the costs associated with it.

Another challenge for localization in the emerging smart world is locating flying smart tiny robots in indoor areas. 3D localization needed for these flying robots demands accuracy on the order of the size of these robots to navigate them intelligently without any crash incident. The display map for these applications needs to be 3D and certain amount of details for the furniture and other large items and people in the room is needed to control the movements of the robots. This is a very complex technology and very far away from the current states of the art. Similar to smart finding of items in the shelves, here accuracy of existing Wi-Fi infrastructure is not adequate and we may need additional RF infrastructure and integration of visual and mechanical sensors.

Application of microbots in the emerging smart medicine delivery and health monitoring systems has opened another horizon for localization of microbots inside the human body [Li12, Mar12]. Challenges for localization inside the human body begin with issues related to the map. We do not have any map for the path of movements of these microbots inside the specific human GI-tract or vascular tree in vivo. All we have is the general anatomic cartons of the shape of the organs and the paths, not the real 3D map needed for navigation inside the human body. Most microbots carry a camera, if the path of movement is reconstructed, the camera pictures can be used to reconstruct the inside of the individual organs in vivo. The infrastructure for localization in these scenarios is body-mounted sensors, which are always moving with respect to one another with the human body moves. Intelligent localization in an infrastructure that is in motion needs new algorithms to be discovered. Intelligent navigation of these microbots needs hybrid localization using RF signal and the images taken by the cameras [Bao15, Bao14]. RF propagation inside the human body is very complex because it is a non-homogeneous and liquid immersive environment, which opens a new horizon for scientific discoveries [For06, Mak11].

Finally, associated with the localization of objects and humans is the challenge of privacy and securing the information so that it is available to only authorized entities (e.g., the human whose location is being captured). While research work on location privacy has been ongoing in recent years, there are open questions on how location privacy can be maintained, while providing the utility of this information in a smart world.

1.5 Overview of the Book

This book is prepared as a textbook to guide the reader to learn the fundamental technical aspects of localization science and technology for indoor and urban areas. There are three fundamental technical skills needed for understanding the design and performance evaluation of these systems, modeling the behavior of the location sensors, performance evaluation using CRLB, and design of algorithm for optimize the precision of a positioning system. In the first chapter, we first explained technical complexity, applications, and open challenges for localization science and engineering in the emergence of the Smart World and IoT. We explain why localization science is a complex multi-disciplinary area of research and technology and identify the technical aspects associated with this field of research. We categorized different applications, the precision they need, and technologies they use; we introduced a few technical challenges and explained what is needed to overcome these challenges; and finally we provided an overview of the rest of the book.

The remainder of the book is divided into three parts and an appendix chapter. Part I cosists of two chapters on received signal strength (RSS)-based systems. It begins with Chapter 2 on behavior of RSS and range estimation followed by Chapter 3, describing popular RSS systems. Part II consists of three chapters on Time-of-Arrival (TOA)-based positioning. It begins with Chapter 4 on principles of TOA ranging, followed by Chapter 4 on description of opportunistic TOA positioning techniques using existing communication system infrastructure and technologies. The Direction of Arrival (DOA) systems are treated as a part of TOA sytems, because they rely of TOA estimation. Since the analysis of the effects of multipath in indoor and urban areas on the performance of TOA systems is vital, Chapter 6 is devoted to the analysis of the effects of multipath. In Parts I and II, the tool for analytical performance evaluations is the CRLB that is compared with the empirical results of performance evaluaitons. Part III presents the details of applied algorithms for positioning. It begins by Chapter 7 to ppresent the basic positioning algorithms. Then, it presents applied algorithms used for RSS- and TOA-based positioning Chapters 8 and 9, respectively. Since RSS is an unreliabe measure of rangings, pattern recognition algorithms are the center of discussion. The TOA is much more reliabe, but it is very sensitive to multipath. The algorithms introduced for TOA are signal processing algorithms to mitigate the effect of multipath. The final Chapter 10 of this part is devoted to hybrid positioning algorithms for integration of relative and absolute positioning. We begin by introducing different sensors for measurement of speed and direction of movement for reltive localization. Then, we provide examples of Particle and Kalman filtering applications for integration of absolute and relative localization. Since understanding of certain basic concepts in the classic estimation theory is an important analytical background for understanding of the book, Appendix A is devoted to the review of the parameter estimation theory and derivations of the CRLB. In classical estimation theory, ranging and positioning are referred to as single and multiple parameter estimation and CRLB is a mean for calculation of variance of these estimates. We review all these specific parts of the estimation theory in Appendix A.

The emerging multi-disciplinary fields such as wireless networking, robotics, or wireless positioning involve a variety of disciplines. Therefore, people with different skills work on different aspects of a unique project. But the experience that they learn in that endeavor is different and focused around their individual skills. Therefore, details of the research work, contents of the courses that are taught, and the books that are written on the same topic are based on the author's experiences. The author of this book has pioneering research experience and industrial exposure in design and performance evaluation of indoor geolocation based on empirical measurement and modeling of the behavior of radio propagation. The presentation of the material is based on examples of research and development that he has been engaging with as a professor or as technical consultant to startup companies. As a result, this book organizes results of research papers, patents, and course notes of the author to present the results of his research findings and educational experiences over a couple of decades to his students.

Assignments for Chapter One

Answer the following questions:

(you can use all information available at the Internet found by search engines such as Google or Wikipedia)

1. Why localization is becoming more important for the information networks?
2. What are the elements of the Smart World and what is the role of localization in formation of these elements?
3. Explain the meaning of "location intelligence".
4. Why smart phones apps are of interest in knowing the user's location information?
5. Why do we need special skills to get into the field of opportunistic positioning, and what are the types of skills needed to understand this field?
6. What is SLAM and how it is used in robotics?
7. What is the role of positioning systems in emerging "smart transportation technologies"?
8. Why localization with millimeter accuracy is needed for inside the human body and what are the challenges in developing that technology?
9. Identify the name, location (City and Country), and the indoor navigation technology of three companies providing indoor geolocation.
10. What are the challenges for the future of the positioning systems?
11. Why cellular service providers are interested in location-based services? Give some examples of location-based services.
12. What are the basic elements of a wireless geolocation system?
13. Name the three major distance-based techniques used for location finding and explain how are they different from each other?

14. How can nodes in an ad hoc or sensor network determine their locations even if they are not directly connected to nodes that are location aware?
15. What is Wi-Fi localization and how does Wi-Fi localization complement GPS technology?

Project 1.1:

Search Chapter 1 and the Internet (IEEE Explore, Wikipedia, Google Scholar, ACM Digital) to identify one area of research and one area in business development which you think are the most important for the future of the positioning industry. Give your reasoning why you think the area is important and cite at least one paper or a website to support your statement.

Project 1.2:

Use Google Scholar to find the two most cited popular interesting articles related to indoor positioning and navigation in the last three years. Read these articles and write a half-page summary for each of them.

2

Fundamentals of RSS Ranging

2.1 Introduction

In Chapter 1, we explained that the fundamental elements of localization are: a geographical map of an area identifying recognizable reference points, and a method to measure the distance of a device to these reference points. One of the most popular methods to measure the distance of a device from a reference point is to use the received signal strength (RSS) of a signal transmitted from the reference point with the target device or equivalently measure the RSS of the signal transmitted from the device at the reference points. Many off-the-shelf RF devices measure averaged values of the RSS. Since radio channels are reciprocal, both measurements are the same and we can use either of them for measuring the distance between the device and the reference point. The reference point could be the location of a Wi-Fi access point, a cell tower, or an iBeacon (Figure 2.1) and the device could be a smart phone, a laptop, or a robot or an autonomous mobile vehicle. The fundamental questions related to the technologies that enable RSS-based ranging are: "Given the RSS value, how should we calculate the distance from the source and how good is such a calculation?" and "How does it relate to the radiating source and environmental characteristics?"

Understanding of the RSS-based ranging is essential for learning how RSS-based localization systems work. RSS-based localization is the dominant technique used for Wi-Fi localization of the smart mobile devices such as smart phones and laptops. To understand the popularity of this technology, one can imagine that today the number of smart phones is close to the population of the planet, and millions of applications have been developed for the smart phones that extensively use the location of the user for a variety of technical and marketing objectives. These smart phone applications prominently use the simple and inexpensive RSS-based Wi-Fi localization for their needs. The device reads the RSS of its surrounding Wi-Fi access points

17

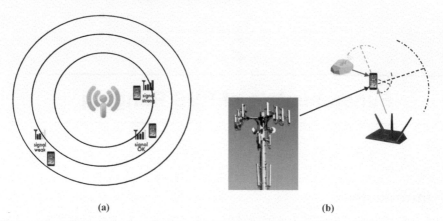

(a)   (b)

Figure 2.1   Fundamental questions for RSS based ranging, (a) "Given the RSS value, how can I measure the range of the device from the source and how accurate is that measurement?". (b) "How does my measurement relate to the radiating source location, transmitted power, and environmental characteristics?".

and sends them to a centralized Wi-Fi location center holding a Wi-Fi location database and localization algorithms, such as the ones owned by companies like Skyhook, Google, or Apple, and the center sends the location estimate back to the device. At the time of this writing, the database at Skyhook received around 1 billion location estimation inquiry messages per day! This chapter is devoted to the RSS-based ranging, which describes how we model the behavior of the RSS, how we can estimate the distance from the RSS readings, and how we can measure the accuracy of the distance estimates using the RSS. Chapter 3 will address the question: "how do RSS-based localization systems work?".

2.2 Modeling RSS Behavior for Ranging

Multipath arrival of the signal in indoor and urban areas, where the applications discussed in this book operate, causes extensive fluctuations of the amplitude of the received signal in time [Pah13]. Figure 2.2 illustrates the variation of the amplitude in dB as a function of the logarithmic distance between the transmitter and the receiver as a receiver is moved away from a transmitter. This figure also shows how we approach the modeling of these variations of the signal for different localization applications. The instantaneous received amplitude in a multipath environment always varies with time and with small local changes in distance. The Fourier transform

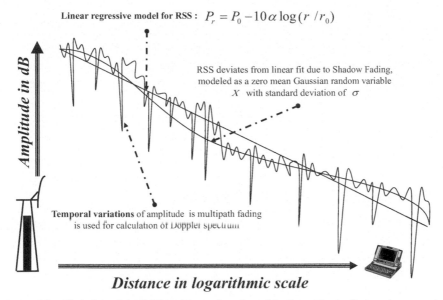

Figure 2.2 Variation of the RSS in dB as a function of the logarithmic distance between the transmitter and the receiver and how we approach to model them for different purposes.

of these changes is referred to as Doppler spectrum, and in Section 10.6, we explain how it is used for calculation of speed of movement of a device. The *average* received signal strength (RSS) in dBm is what we use for RSS-based localization to determine the distance between an antenna and a device. The traditional method to model how the RSS is related to the distance from the transmitter is to use linear regression and least-square (LS) estimation to calculate the parameters of the model using empirical data [Pah13]. The general statistical regression model for the RSS behavior in dBm is given by:

$$P_r = P_0 - 10\alpha \log_{10}(r/r_0) + X(\sigma), \tag{2.1}$$

where the various parameters are defined as:

P_r: RSS (the average received power)

r: Distance between the transmitter and the receiver

r_0: An arbitrary known range as a reference distance

P_0: RSS at the reference distance r_0

α: Distance-power gradient

X: A Gaussian random variable representing shadow fading

σ: Standard deviation of zero mean shadow fading

Intuitively, this model assumes that the RSS in Watts or mW (not in dB) is inversely proportional to the distance raised to the power indicated by the distance-power gradient, α. The reference range r_0 is an arbitrary location close to the transmitter. Shadow fading represents variations of the RSS from the linear regression line in dB caused by objects shadowing the path between the transmitter and the receiver.

This model is very simple, and we can determine its parameters with a simple empirical measurement. The following subsection clarifies this point and shows how we can measure the model parameters.

2.2.1 LS Estimation of RSS Model Parameters

We can use the traditional least square (LS) algorithm in statistical modeling to estimate the path-loss model parameters, (P_0, α, σ), given by Equation (2.1), from a set of measurements of the RSS at specific known locations. Let us assume that we have measured N samples of the RSS, $\{P_i; \ 0 < i \leq N\}$ in *dBm*, from a radiating antenna at distances, $\{r_i; \ 0 < i \leq N\}$. Further, let us assume that we are operating in an indoor area, where we let the reference distance r_0 to be 1 m. To solve the problem of estimating parameters using the LS approach, we define a cost function, which is the average of the difference between expected received power at a distance, $\widehat{P}_i = P_0 - 10\,\alpha \log(r_i)$, and the measured power, P_i, at that location:

$$\varepsilon(P_0, \alpha) = \frac{1}{N} \sum_{i=1}^{N} (\widehat{P}_i - P_i)^2 = \frac{1}{N} \sum_{i=1}^{N} (P_0 - 10\alpha \log r_i - P_i)^2. \quad (2.2a)$$

By taking the derivatives of this cost function, we will have two sets of linear equations with two unknowns:

$$\begin{cases} \frac{\partial \varepsilon}{\partial \alpha} = \sum_{i=1}^{N} -2 \times 10 \log r_i (P_0 - 10\alpha \log r_i - P_i) = 0 \\ \frac{\partial \varepsilon}{\partial P_0} = \sum_{i=1}^{N} 2(P_0 - 10\alpha \log r_i - P_i) = 0 \end{cases}, \quad (2.2b)$$

and we can calculate P_0 and α by solving these two equations with two unknowns. Substituting the estimated values of (P_0, α) in Equation (2.2a), we can obtain an estimate for the variance of the shadow fading:

$$\sigma^2 = \varepsilon(P_0, \alpha) = \frac{1}{N} \sum_{i=1}^{N} (P_0 - 10\alpha \log r_i - P_i)^2. \qquad (2.2c)$$

This formulation in Equation (2.2) allows us to compute the necessary parameters for the RSS linear regressive model.

Example 2.1 (MATLAB code for calculation of parameters)

In this example, we provide MATLAB code for solving the above linear regression problem to calculate the parameters for RSS behavior model, $(P_0,\ \alpha,\ \sigma)$. Figure 2.3(a) shows the MATLAB code to fit the data. In this code, the vector r represents the range of measurement locations and vector Pr represents the associated RSS measurement in dBm. Figure 2.3(b) shows the results of curve fitting and location of a sample measurement point and its associated estimated value with the RSS model. The program also prints out the estimated model and the standard deviation of shadow fading as:

```
The Estimated Path Loss Model is:
Pr=-23.5916-17.1571*log(r)
Mean value of shadow fading is:
     -0.6353

Standard Deviation of shadow fading is:
3.2478
```

Fitting this output to the model in Equation (2.1), we have $P_0 = -23.5916$, $\alpha = 1.71571$, and $\sigma = 3.2478$. The estimation of shadow fading has a bias of -0.6353.

The above method for modeling the average RSS has been used by a variety of standardization organizations to model the behavior of the RSS for different wireless communication systems. These models are used for calculation of coverage and the expected data rate of a wireless device at a given distance from an access point or a base station. A comprehensive coverage of these models is available in various wireless communication textbooks [Pah05, Rap02, Mol12]. Here, we provide a summary of three popular models for inside the human body applications, indoor areas, and urban areas, which are useful for the analysis of RSS-based localization applications described in this book.

In the original models for wireless communication applications, the received power is replaced by the relative difference between the RSS and the

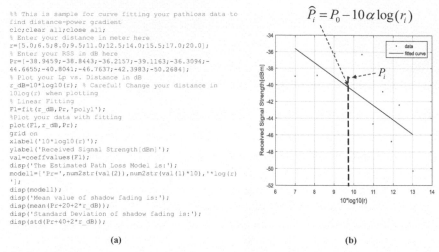

(a)   (b)

Figure 2.3   Empirical measurement of parameters of the linear regression model for behavior of the average RSS (a) MATLAB code to fit the data to a regression line, (b) resulting plot with locations of a sample measurement and its estimated location using the RSS behavior model.

transmit power, which is the path-loss, $L_p = P_t - P_r$, of the channel. In our presentation, we present the results in terms of the received signal strength, P_r, in the format of Equation (2.1).

2.2.2 NIST Model for RSS inside the Human Body

Researchers at the National Institute of Standards and Technology (NIST) developed a path-loss model [Say09, Say10] for understanding how the RSS attenuates from inside of the human body to the surface of the human body. This model has been adopted by the IEEE 802.15.6 standards group for body area networking. Since empirical measurements of radio propagation inside the human body is not practical or feasible, NIST researchers used the High Frequency Structure Simulation (HFSS) software to simulate radio propagation and calculate radio signal strength attenuation inside the human body using numerical solutions to Maxwell's equations[1]. The commercially available HFSS software is one of the most popular tools designed for this purpose and offers a 3D graphical user interface for emulation of radio wave propagation inside the human body with its organs. Figure 2.4 shows a sample

---

[1] An overview of software simulation tools for RF propagation in and around the human body is available in Section 6.9.

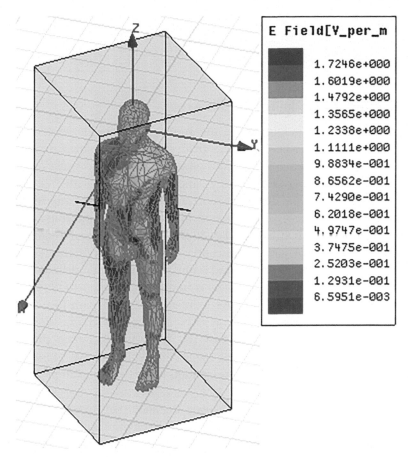

E Field[V_per_m

	1.7246e+000
	1.6019e+000
	1.4792e+000
	1.3565e+000
	1.2338e+000
	1.1111e+000
	9.8834e-001
	8.6562e-001
	7.4290e-001
	6.2018e-001
	4.9747e-001
	3.7475e-001
	2.5203e-001
	1.2931e-001
	6.5951e-003

Figure 2.4 Graphical user interface of the HFSS for radio propagation study inside the human body.

result of the graphical user interface for HFSS for simulation of RF signal propagation inside the human body. Using the HFSS, NIST researchers have defined two scenarios of radio propagation for deep and near-surface tissues. For each scenario, they have made multiple measurements to determine the parameters for the RSS behavior model given by Equation $(2.1)^2$. Table 2.1 shows the parameters of the model for the two scenarios. These simulations are made for MedRadio bands at 402–405 MHz. Note that $L_0 = P_t - P_0$ is the path-loss at the reference distance r_0.

---

[2]Method for estimation of path-loss model parameters is described in Example 2.1.

Table 2.1   Parameters of the NIST model for RSS inside the human body

Implant to Body Surface	r_0 (mm)	L_0(dB)	α	σ(dB)
Deep tissue	5	47.14	4.26	7.85
Near surface	5	49.81	4.22	6.81

2.2.3 IEEE 802.11 Model for Indoor Areas

For small distances in indoor areas, the transmitter and the receiver are often in the same room, where the RSS from the direct line-of-sight (LOS) paths dominates the power arriving from other paths. In these situations, the distance power gradient α is close to 2, and is associated with free space propagation [Pah13]. As the distances separating the transmitter and receiver increases such that the receiver is beyond the room where the transmitting antenna resides in, walls obstruct the direct LOS signal and reduce the received power from that path significantly. The reception of the received power from other reflected or diffracted paths increasingly contribute to the overall RSS. The distance power gradient α in these obstructed-LOS (OLOS) conditions increases substantially. As a result, popular path-loss model standards designed for WLAN and WPAN applications define *distance-partitioned* models for the behavior of the RSS, which define different distance power gradients for different ranges of distances. We present the distant-partitioned IEEE 802.11 recommended model as an example model for RSS behavior in indoor areas.

Figure 2.5 shows the general two-piece distance partitioned model for RSS behavior, recommended by the IEEE 802.11 standardization committee for indoor areas. The RSS is modeled as two piecewise linear segments separated by a break point at r_{bp} by:

$$P_r = X(\sigma) + P_0 - \begin{cases} 10\alpha_1 \log_{10}(r) & ; r < r_{bp} \\ 10\alpha_1 \log_{10}(r_{bp}) + 10\alpha_2 \log_{10}(r/r_{bp}) & ; r > r_{bp} \end{cases} . \quad (2.3)$$

The distance-power gradient in the two segments are $\alpha_1 = 2$ and $\alpha_2 = 3.5$. Note that $\alpha_1 = 2$ associates with open LOS areas and $\alpha_2 > \alpha_1$ associates with OLOS situations. The top row of Equation (2.3) for distances less than the breakpoint distance is the same as Equation (2.1) with a distance power gradient of 2 and $r_0 = 1$ m used as the reference distance. The bottom row is also the same as Equation (2.1) with the break point used as the reference distance.

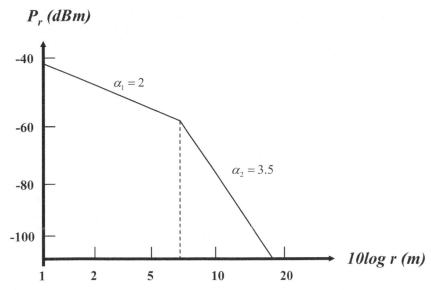

Figure 2.5 The general two-piece distance partitioned model for RSS behavior, recommended by the IEEE 802.11 standardization committee for indoor areas.

The IEEE 802.11 standard defines six different indoor scenarios for path-loss models with four different break points. Table 2.2 shows the parameters associated with these models. Model A is a *flat fading* model with a single path between the transmitter and the receiver. The breakpoint distance for this model is at 5 m and the standard deviation of the shadow fading ibreak 5 dB. Model B is recommended for a typical residential environment with LOS conditions and more than one effective path between the transmitter and the receiver. The path-loss parameters of this model are the same as those of Model A. Model C is recommended for a typical residential or small office environment with LOS and OLOS conditions between the transmitter and the receiver. The breakpoint for this model is still at 5 m, but the standard deviation of the shadow fading is increased to 8 dB. Model D is recommended for a typical office environment with OLOS conditions with a 10 m breakpoint distance and an 8 dB standard deviation of shadow fading. Model E is recommended for a typical large open space and office environments in areas with OLOS conditions and it has a breakpoint distance of 20 m and a shadow fading standard deviation of 10 dB. Model F is recommended for a large open space with indoor and outdoor environments in the areas with OLOS conditions.

Table 2.2   Parameters for different IEEE 802.11 recommended path-loss models

Environments	d_{bp} (m)	α_1	α_2	Shadow Fading Std. Dev. (dB)
A	5	2	3.5	5
B	5	2	3.5	5
C	5	2	3.5	8
D	10	2	3.5	8
E	20	2	3.5	10
F	30	2	3.5	10

2.2.4 Okumura-Hata Model for Urban Areas

Outdoor models for the RSS behavior are mostly designed for cellular tele-
phony applications and they include details of antenna height and frequency
of operation. This is because in outdoor areas the height of the antennas
varies substantial. It could be on top of a hill or deep in a canyon, while
in WLAN indoor applications antennas are commonly installed on the same
level ceilings. Variations of the height of antennas affects the coverage. In
addition, the range of frequencies used in the traditional cellular networks
changes over an order of magnitude from hundred MHz up to a few GHz and
that has to be included in the model as well. Traditional WLANS operate at
2.4 and 5.2GHz which have much closer propagation characteristics.

  One of the most popular models for RSS in outdoor areas is the Okumura-
Hata model, which describes the RSS by Equation (2.1) with distances in *km*.
The reference distance $d_0 = 1\,Km$. The other two parameters of the model
are given by:

$$\begin{cases} P_0 = P_t - 69.55 + 26.16 \log f_c - 13.82 \log h_b - a(h_m) \\ \alpha = [4.49 - 0.655 \log h_b] \end{cases}, \qquad (2.4)$$

where f_c is the frequency in MHz and h_b and h_m are the heights of base
station antenna and mobile antenna, respectively. The function $a(h_m)$ is a
correction factor that depends on the frequency of operation and the environment. Table 2.3 provides functions for calculation of the correction factor
and the range of operation of this model. More elaborate channel models for
wider range of frequencies for outdoor operations are available in [Mol12].
The variance of shadow fading in urban areas varies between 3 and 10 dB
[Gud91].

Table 2.3 Correction factor and the range of operation for Okumura-Hata model

			Range of Values	
Center frequency f_c in MHz			150–1500 MHz	
h_b, h_m in meters			30–200 m, 1–10 m	
$a(h_m)$ in dB	Large city	$f_c \leq 200$ MHz	$8.29 \ [\log \ (1.54 \ h_m)]^2 - 1.1$	
		$f_c \geq 400$ MHz	$3.2 \ [\log \ (11.75 \ h_m)]^2 - 4.97$	
	Medium-small city	$150 \geq f_c \geq 1500$ MHz	$1.1 \ [\log f_c - 0.7] \ h_m - (1.56 \log f_c - 0.8)$	

2.2.5 Behavior of Shadow Fading and Localization Applications

The RSS models for wireless communications explained earlier assume we have a fixed shadow fading for a given environment. For example, the IEEE 802.11 model (Table 2.2) assumes three separate values of 5, 8, and 10 dB for the standard deviation of the shadow fading in five different environments. However, as we get closer to a radiating antenna, independent RSS measurements at the same short distance become more similar (have lesser variations), which means the standard deviation of shadow fading becomes smaller. Hypothetically, when the distance from an antenna approaches zero, we always expect to measure the same value of the RSS and standard deviation of the shadow fading approaches zero. Therefore, the standard deviation of shadow fading begins at zero close to the transmitter and increases as we move away from the antenna, and after certain distances, it settles to a fixed value. In communication applications, we use the RSS model for calculation of coverage for deployment of the antennas and we are not keen on standard deviation of shadow fading at the close distances to the antenna. In these applications, the shadow fading at the cell edge affects the deployment approaches. In indoor geolocation applications such as Wi-Fi localization, we are interested in all distances and if the RSS at shorter distances is less variable, we can benefit from that to improve the precision of the localization algorithm. Therefore, having a model for the standard deviation of the shadow fading in close vicinity of the transmit antenna is useful for localization applications.

A group of students at Worcester Polytechnic Institute measured the RSS in the close vicinity of a radiating antenna and modeled the standard deviation of the shadow fading with an exponential function. Figure 2.6

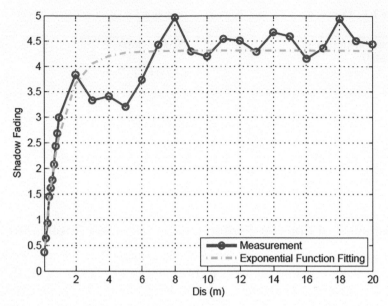

Figure 2.6   Standard Deviation of shadow fading near the transmitter antenna in an indoor area [Liu15].

shows the results of this experimental measurements and exponential curve fitting for the data [Liu15]. The best fit curve for the standard deviation in this experiment is:

$$\sigma = \lambda \times e^{-\beta \times d} + \gamma,$$

where $\lambda = -4.28$, $\beta = 0.9372$ and $\gamma = 4.31$. This empirical model explains that the standard deviation of the shadow fading begins at zero and settles to a value of 4.31 at around a 5 m distance from the transmit antenna. With a lower variance of shadow fading closer to the antenna, we can expect a more accurate estimate of the range at shorter distances.

2.3 RSS-Based Ranging and Distance Measurement Error

The general statistical model for the RSS in Equation (2.6) can be used for measurement of the distance between a transmitter and a receiver as well as analysis of the anticipated error in the calculation of the distance or range using the RSS. In the previous section, we introduced methods for empirical measurement and computation of the model parameters in an environment and we provided three popular standard models for inside human bodies, in indoor areas, and in outdoor areas. In wireless communications, these models are used for calculation of coverage for deployment of the infrastructure and to calculate the data rate of a link. In this chapter, we use these models for RSS-based ranging. We want to find optimal methods for finding the range of a device that is measuring the RSS from an RF radiating antenna and to analyze the pattern of errors in measurement of the distance when we use the RSS. We begin by addressing the problem with a common sense solution and then we extend the solution using classical estimation theory formulations.

2.3.1 Measurement of Distance using the RSS

Let us say we have measured a sample of RSS, P_r, at an unknown distance, r, from a radiating RF source and we want to estimate the distance or range from the source, $\hat{r}$. Then, we can argue that the shadow fading, $X(\sigma)$, in the model presented by (2.1) is a zero mean Gaussian random variable and *on average* it does not affect the received power. Therefore, we can determine the distance by solving the equation:

$$P_r = P_0 - 10\alpha \log\left(\hat{r}/r_0\right), \qquad (2.5a)$$

which leads to the estimate of the range as:

$$\hat{r} = r_0 \times 10^{-\frac{P_r - P_0}{10\alpha}}. \qquad (2.5b)$$

Our estimation algorithm (2.5b) calculates the distance using the measured or observed received power, P_r, and two of the characteristic parameters of the channel, P_0 and α, which are estimated previously by the deployer of the localization system, a standards organization, or a combination of the two.

Example 2.1: (Calculating the range using RSS)

In this example, we use the IEEE 802.11 Model D to calculate the distance of a device located at a 15 m distance from a Wi-Fi access point measuring an RSS value of $P_r = -50$ dBm. Model D for the relation between RSS and range is given by:

$$P_r = P_0 - 20 \log 10 - 35 \log\left(\frac{r}{10}\right).$$

Assuming $P_0 = -21$ dBm, the estimate of range from Equation (2.5b) is:

$$\hat{r} = 10 \times 10^{\frac{50 - 21 - 20 \log(10)}{35}} = 18.1m.$$

At a 15 m distance, we were expecting an RSS of:

$$P_r = -21 - 20 \log 10 - 35 \log\left(\frac{15}{10}\right) = -47.2\text{dBm},$$

but we have measured -50 dBm, and that 2.8 dB of error in measurement of RSS resulted in a 3.1 m error in measurement of the range.

As we demonstrated in the above example, assuming we have a perfect estimate of the parameters of the model, (α, P_0), still the estimated value of the range, $\hat{r}$, obtained from Equation (2.5b) is not the same as the actual range, r, because the measurement of the RSS, P_r, involves shadow fading randomness, $X(\sigma)$. The difference between the estimated and the actual value of the error is the distance measurement error (DME) and it is given by:

$$DME = \varepsilon = \hat{r} - r. \tag{2.5c}$$

The value of this error is a random variable because each time we measure the RSS at a different shadowing condition but at the same distance, due to shadow fading, we measure a different value of the RSS. The statistics of ε reflect the *quality* of our approach to estimate the range by observing the RSS. We can use the standard deviation of ε as a measure for accuracy of our method used for estimating the range:

$$\text{Measure for Quality of Estimation}: \sqrt{E\left(\varepsilon^2\right)} = \sqrt{E\left\{(r-r)^2\right\}} \tag{2.5d}$$

The simplest method to measure these statistics of the error is to simulate the measurement with the shadow fading by using Equation (2.1) and a random number generator, create many samples of received power, P_r, at a given range, and use those values in Equation (2.5b) to estimate the distance associated with the simulated values of power. Then, it is possible to use the above definition to calculate the DME and its statistics as a measure of the quality of our estimation algorithm. The following simple example further clarifies this concept.

Example 2.2 (DME statistics for RSS ranging)

In this example, we use the IEEE 802.11 Model D to simulate the RSS with shadow fading, when the distance between the transmitter and the receiver is 15 m. Then, we use the simulated RSS values to calculate the range and the associated distance measurement error. We repeat the process 1000 times, plot the histogram of the distance measurement error, compute its mean and variance.

The general path loss model for the IEEE 802.11 is described in Equation (2.3), and the parameters of Model D are given in Table 2.2. For this model, the breakpoint distance is 10 m and standard deviation of shadow fading is

8 dB. Using these parameters, the RSS is calculated from:

$$P_r = P_0 - 20 \log 10 - 35 \log \left(\frac{r}{10}\right) + X \quad (8)$$

Assuming $P_0 = -21$ dBm, the MATLAB code provided in Figure 2.7(a) generates a thousand RSS values from the above equation and uses the following:

$$\hat{r} = r_{bp} \times 10^{-\frac{P_r - 21 - 20 \log(10)}{35}}$$

to estimate the distances associated with the thousand power readings. The code then forms the DME, plots its histogram (Figure 2.7(b)), and calculates the mean and standard deviation of the error, shown at the bottom of Figure 2.7(b). Note that since the distance is 15 m, the DME cannot be less than -15 m, but on the positive scale, it can be higher than 15 m. Also if we repeat the experiment with different seeds for the random number generator, we will find slightly different values for the statistics and the histogram.

2.3.2 An Analytical Method to Calculate the Variance of DME

If we repeat Example 2.2 for calculation of the statistics of the DME, each time the answers are slightly different. By increasing the number of samples used for simulation, the difference reduces, but it never vanishes. This is an

```
clear all; close all;
r=15;
r_bp=10;
sig=8;
shadow=normrnd(0,sig,[1,1000]);
Pr(1:1000)=-21-20*log10(r_bp)-35*log10(r/r_bp)+shadow;
r_hat=10.^(-(Pr+21+20*log10(r_bp))/35)*r_bp;
dme=r_hat-15*ones(1,1000);
dme_mean=mean(dme);
dme_std=std(dme);
figure
hist(dme,50);
disp([mean(dme) dme_std]);
```

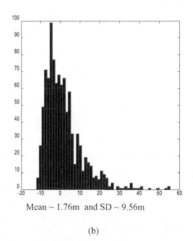

Mean ~ 1.76m and SD ~ 9.56m

(a)   (b)

Figure 2.7 (a) MATLAB code for simulation of the DME of RSS-based ranging for a shadow fading with standard deviation of 8, (b) histogram of the results of DME simulations and the mean and variance of error in meters.

incentive for the search for an analytical solution for this statistical problem. Analytical solutions are often more mathematically complex and we cannot find tractable solutions all the time. However, the statistical analysis of the DME for RSS-based ranging has a relatively simple solution that we introduce next.

Consider the general RSS model that we use to calculate the range, r, from the measured RSS, P_r:

$$P_r = P_0 - 10\alpha \log \frac{r}{r_0}. \tag{2.6a}$$

The derivative of this equation shows the relation between small variations of measured RSS and small variations of the range:

$$dP_r = -\frac{10\alpha}{\ln 10} \cdot \frac{1}{r} dr. \tag{2.6b}$$

Both the differential value of the RSS measurement and its associated estimated distances are random variables, and the variances of the two are related by:

$$E\left\{[dP_r]^2\right\} = \left[\frac{10\alpha}{\ln 10} \cdot \frac{1}{r}\right]^2 E\left\{[dr]^2\right\}. \tag{2.6c}$$

The variance of the received power is due to the shadow fading and it has the same value as variance of the shadow fading:

$$E\left\{[dP_r]^2\right\} = \sigma^2 \tag{2.6d}$$

Therefore, the variance of DME is given by:

$$\sigma^2 = \left[\frac{10\alpha}{\ln 10} \cdot \frac{1}{r}\right]^2 E\left\{[dr]^2\right\} \Rightarrow E\left\{[dr]^2\right\} = \left[\frac{\ln 10}{10\alpha}\right]^2 \sigma^2 r^2 \tag{2.6e}$$

The measure for quality of this estimate defined by Equation (2.5d) is square root of this value given by:

$$\text{Measure of Quality of Estimate: } \sqrt{E(\varepsilon^2)} = \sigma_r = \left[\frac{\ln 10}{10\alpha}\right] \times \sigma \times r \tag{2.6f}$$

In the following section, we confirm this value with classical Maximum Likelihood estimation theory.

2.4 Classical Estimation Theory and RSS-Based Ranging

The estimate of distance for an object that is at 15 m from a transmitter when the standard deviation of the RSS is close to 10 may not be satisfactory for some applications. To reduce the standard deviation of the DME, one intuitive solution which comes to mind immediately is to measure additional samples of the RSS. As we will see in the next chapter, RSS-based localization systems often resort to multiple measurements at a location to improve the reliability of the location estimates. Assuming we have multiple RSS measurements at the same distance, either we can average the various estimated distances obtained from the various measurements of RSS values or we can average the RSS values and use them in Equation (2.5b) to have a better estimate of the distance. But, how many samples we should take? Which one of the two techniques provides a better estimate of the distance? How can we find the accuracy of these estimates? Classical estimation theory, which has evolved over the past half a century to solve radar and other estimation problems, can be used to answer these questions systematically. Appendix A provides a review of classical estimation theory applied to positioning applications. In the remainder of this subsection we apply that to RSS-based range measurement that is the topic of this chapter.

2.4.1 ML and MMSE Estimation for RSS-Based Ranging

Classical estimation theory provides methods for modeling, estimation, and calculation of the bounds of the performance of an estimator [Van04, Kay13, Poo13]. In the classical estimation theory terminology, estimation of the range using the RSS is referred to as a single parameter (the range) estimation using observation of the function of the parameter (the RSS) in additive Gaussian noise (the shadow fading variations). Figure 2.8 shows the basic formulation of single parameter estimation using observation of a function of a parameter in zero mean additive Gaussian noise in classical estimation theory. The analysis begins with a desire to estimate the parameter, α, when we observed, O, its function, $g(\alpha)$, in zero mean Gaussian noise, η with standard deviation of, σ:

$$O = g(\alpha) + \eta(\sigma). \tag{2.7a}$$

In classical estimation theory, there are two popular methods to calculate the estimate of the parameter $\hat{\alpha}$: Maximum Likelihood (ML) estimation and the

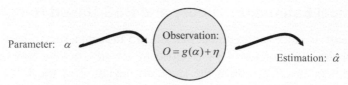

Figure 2.8   Basics formulation of classical single parameter estimation process for observation of function of a parameter in additive noise.

Minimum Mean Square Error (MMSE) estimation. In ML estimation, we form a likelihood function

$$f(O/\alpha) = \frac{1}{\sqrt{2\pi}\sigma}e^{-\frac{[O-g(\alpha)]^2}{2\sigma^2}}, \tag{2.7b}$$

and then we determine the value of α that maximizes this function. We declare that value as the ML estimate of the parameter, $\hat{\alpha}_{ML}$. Equivalently, since the likelihood function is an exponential function, we can maximize the log likelihood function:

$$(O/\alpha) = \log\left[f(O/\alpha)\right] = \log\left(\frac{1}{\sqrt{2\pi}\sigma}\right) - \frac{[O-g(\alpha)]^2}{2\sigma^2}. \tag{2.7c}$$

In MMSE estimation, we define an error cost function:

$$\varepsilon(\alpha) = E\left\{[O-g(\alpha)]^2\right\}, \tag{2.7d}$$

and we find the value of α that minimizes this cost function and we declare that value as the MMSE estimate of the parameter, $\hat{\alpha}_{MMSE}$.

As shown in Appendix Sections A.1.2 and A.2.2, both ML and MMSE estimations provide the same answer:

$$\hat{\alpha} = \hat{\alpha}_{ML} = \hat{\alpha}_{MMSE} = g^{-1}(O) \tag{2.7e}$$

This means, when we measure or observe the function of a parameter to find the ML or MMSE estimate, we apply the inverse of the function of the observation to obtain the estimate of the parameter. It is interesting to note that both types of classical estimators are independent of the variance of the observation noise. Considering this formulation, now we can apply classical estimation theory to our RSS-based range estimation problem.

In range estimation using the RSS, we measure the RSS in shadow fading (see Equation (2.1)), our parameter is the range, *r*, and our observation is the measured RSS, P_r, and we can formulate the problem in terms of classic estimation theory as:

$$
\begin{cases}
O = P_r = P_0 - 10\alpha \log_{10}(r/r_0) + X(\sigma) \\
g(r) = P_0 - 10\alpha \log_{10}(r/r_0) \\
\eta(\sigma) = X(\sigma)
\end{cases}
\quad (2.8a)
$$

Therefore, from Equation (2.7e), the ML and MMSE estimates of the range are given by:

$$
\hat{r} = \hat{r}_{ML} = \hat{r}_{MMSE} = g^{-1}(O) = g^{-1}(P_r) = r_0 \times 10^{-\frac{P_r - P_0}{10\alpha}}, \quad (2.8b)
$$

which is the same as our simple common sense estimator in Equation (2.5b). The practical benefit of the classical estimation theory formulation is that we can extend the results to more complex situations, calculate the *best* estimate for multiple observations of the RSS at the same location, and calculate the Cramer–Rao Lower Bound (CRLB), which provides a quality of the estimation.

2.4.2 Range Estimation with Multiple RSS Measurements

As explained in detail in Chapter 3, RSS-based localization systems often use RSS fingerprints of reference radiating antennas during the localization process. In indoor areas, the collection of the fingerprint is performed manually and it is traditional to measure multiple RSS values in each location, while the person holding the measurement device turns around to create different samples of shadowing effects. In this section, we show the classical estimation theory methods for estimation of the range using multiple RSS measurements.

In classical estimation theory, for N-observations of function of a parameter in noise, we have:

$$
O_i = g(\alpha) + \eta_i; \quad i = 1, 2, \dots, N. \quad (2.9a)
$$

As shown in Appendix Sections A.1.4 and A.2.3, the classical ML and MMSE estimations for this problem are given by:

$$
\hat{\alpha} = \hat{\alpha}_{ML} = \hat{\alpha}_{MMSE} = g^{-1}\left[\frac{1}{N} \sum_{i=1}^{N} O_i \right]. \quad (2.9b)
$$

Casting multiple observations of the RSS in the same location in this formulation, we have:

$$\begin{cases} \hat{r} = g^{-1} \left[\frac{1}{N} \sum_{i=1}^{N} O_i \right] = g^{-1} \left[\frac{1}{N} \sum_{i=1}^{N} P_{r-i} \right] \\ g(r) = P_0 - 10\alpha \log_{10}(r/r_0) \end{cases} \qquad (2.10a)$$

If we define the average RSS over all samples as:

$$\bar{P}_r = \frac{1}{N} \sum_{i=1}^{N} P_{r-i}, \qquad (2.10b)$$

then the estimate is given by:

$$\hat{r} = r_0 \times 10^{-\frac{\bar{P}_r - P_0}{10\alpha}}. \qquad (2.10c)$$

Equation (2.10c) is the same as Equation (2.8b) except that we have replaced the single value of the RSS by the average RSS. For multiple measurements of the RSS, we have two choices: (2.1) to use individual RSS measurements to estimate their associated ranges and then use the average of all ranges as the final estimate or (2) to average all the measured RSS and calculate the associated range for average RSS as the estimate of the range. Equation (2.10c) obtained from the classical estimation theory shows that the second approach results in the optimum ML and MMSE estimation of the range. Selection of the right answer in this choice was only possible through the classical estimation theory formulation and we could not reach that with intuitive deductions.

2.4.3 CRLB for Ranging with RSS Measurement

In Example 2.2, we present the DME and its statistics using a simple simulation of the RSS. The CRLB calculates the bound on the estimation of the variance of the DME. The ML estimate provides the best estimate, and the CRLB provides an analytical tool for calculation of the bounds on the variance of the ML estimate. Therefore, if we use any other algorithm to estimate the distance, the CRLB provides a universal lower bound to our estimation and we can use it as a benchmark for evaluation of the performance of any estimation algorithm that we may use, either for simplicity or practicality.

In the design of an RF localization system, we need to compare the performance of different alternatives for localization. The CRLB allows for comparing the precision of location estimations by alternative approaches for localization. The smaller the variance, the smaller is the chance that the error in location estimate is large. In the same way that different information transmission applications have different error rate requirements, different localization applications have different precision requirements. For a conceptual system design, a positioning engineer may compare the CRLB for different metrics used for localization to select the appropriate technology or decide on the density for installation of the infrastructure to meet certain positioning accuracy requirement.

In classical estimation theory terminology, the *smallest* variance of the estimate of a parameter based on noisy Gaussian observation is the CRLB, and it can be calculated by inverting the Fisher Information Matrix (FIM) [Van04]:

$$CRLB = Var\left[\hat{\alpha}(O) - \alpha\right] \geq F^{-1}. \tag{2.11a}$$

The FIM matrix, which is actually scalar in single parameter estimation, is given by

$$\mathbf{F} = E\left[\frac{\partial\ (\mathbf{O}/\alpha)}{\partial\alpha}\right]^2 = -E\left[\frac{\partial^2\ (\mathbf{O}/\alpha)}{\partial\alpha^2}\right], \tag{2.11b}$$

in which $(\mathbf{O}/\alpha)$ is the log likelihood function in ML estimation process, defined by Equation (2.7b).

To calculate the CRLB for range estimation using the measurement of the RSS, we need to calculate Equation (2.11) for the log likelihood function of observation of the function of a parameter in noise given by Equation (2.7c). Appendix Section A.3.3 provides the details of this derivation and shows that for the observation of the function of a parameter, $g(\alpha)$, in zero mean Gaussian noise with standard deviation of, σ, the FIM and the CRLB are given by:

$$\begin{cases} \mathbf{F} = -E\left[\frac{\partial^2 \ln f(O/x)}{\partial x^2}\right] = E\left[\frac{\partial \ln f(O/x)}{\partial x}\right]^2 = \frac{[g'(\alpha)]^2}{\sigma^2} \\ CRLB \geq F^{-1} = \frac{\sigma^2}{[g'(\alpha)]^2} \end{cases} \tag{2.12}$$

For RSS-based ranging, $g(r) = P_0 - 10\alpha \log_{10}(r/r_0)$ and the FIM becomes:

$$\begin{cases} \mathbf{F} = \frac{[g'(r)]^2}{\sigma^2} \\ g'(r) = -\frac{10\alpha}{(\ln 10)r} \end{cases} \Rightarrow \mathbf{F} = \frac{(10)^2\alpha^2}{(\ln 10)^2\sigma^2 r^2}, \tag{2.13a}$$

resulting in a CRLB of:

$$CRLB \geq \mathbf{F}^{-1} = \frac{(\ln 10)^2}{100} \frac{\sigma^2}{\alpha^2} r^2 . \tag{2.13b}$$

Since the CRLB is a bound on variance of the DME, the standard deviation of DME is given by:

$$\sigma_r = \sqrt{CRLB} \geq \frac{\ln 10}{10} \frac{\sigma}{\alpha} r. \tag{2.13c}$$

This powerful result analytically relates the DME to the distance and shows that in RSS-based ranging, the accuracy of measurement is directly proportional to the distance, r, and the standard deviation of shadow fading, σ, while it is inversely proportional to the distance-power gradient, α. Since the standard deviation of shadow fading and the distance-power gradient are fixed values related to the environment, we can conclude that in RSS-based ranging, precision of the ranging is proportional to the distance. Since the coverage of an antenna limits the range of operation, if we use an RFID with a coverage of 1 m, the accuracy of the RSS ranging would be approximately 1 m and if we use RSS ranging for a Wi-Fi with a coverage of around 30 m, the accuracy of ranging may become around 10–20 m. A quantitative example further clarifies the situation.

Example 2.3 (Standard Deviation of DME using CRLB)

For the indoor environment described in Example 2.2, the distance power gradient was $\alpha = 3.5$, the standard deviation of shadow fading was $\sigma = 8$ dB, and the distance was $r = 15$ m. Using Equation (2.13c), we can calculate the standard deviation as:

$$\sigma_r = \sqrt{CRLB} \geq \frac{\ln 10}{10} \frac{\sigma}{\alpha} r = 0.53 \times r = 7.95.$$

Comparing the result of Example 2.3 with the results of simulation in Figure 2.7, we can recognize the importance and usefulness of using CRLB. The CRLB provides an analytical expression for calculation of the DME statistics, which eliminates the need for time consuming simulations. In addition, CRLB sheds light on the relation between the ranging error and characteristics of the environment, which adds to the scientific understanding of the problem.

Equation (2.13c) is the same as Equation (2.7e), which we calculated intuitively in Section 2.3.2. Intuitive calculation of the standard deviation of RSS-ranging error when we have multiple observations of RSS becomes difficult, because we need to answer the famous question that we should average the estimate of distance from RSS values or average the RSS values and calculate the distance. This ambiguity is resolved in classical estimation theory solution to the problem.

2.4.4 CRLB for Ranging with Multiple RSS Measurements

Appendix Section A.3.4 provides a detailed derivation of CRLB for multiple observations of function of a parameter in noise:

$$O_i = g(x) + \eta_i(\sigma), \qquad (2.14a)$$

where σ is the standard deviation of observation noise. Results of these derivations reveal that for N observations of the function of a parameter in independent zero mean Gaussian noise, the CRLB is:

$$CRLB \geq \mathbf{F}^{-1} = \frac{\sigma^2}{N\left[g'(x)\right]^2} \qquad (2.14b)$$

Then, the standard deviation of the estimation error is bounded by:

$$\sigma_r = \sqrt{CRLB} \geq \frac{\ln 10}{\sqrt{N}10}\frac{\sigma}{\alpha}r. \qquad (2.14c)$$

This result reveals that an increase in the number of samples reduces the standard deviation of the DME by the square root of the number of RSS measurements.

Example 2.4 (CRLB for Multiple RSS Measurements)

For the indoor environment described in Examples 2.2 and 2.3, the distance power gradient was $\alpha = 3.5$, standard deviation of shadow fading was $\sigma = 8$ dB, and the distance was $r = 15$ m. Using Equation (2.14c), we can calculate the standard deviation for 10 independent samples in one location:

$$\sigma_r = \sqrt{CRLB} \geq \frac{\ln 10}{\sqrt{10}10}\frac{\sigma}{\alpha}r = 0.17 \times r = 2.52.$$

Having 10 samples increases the precision slightly over three times.

2.5 Confidence Regions for RSS-Based Ranging

Figure 2.9 shows a radiating antenna and a device measuring an RSS value of P_M from the radiating antenna at a distance, r, which we do not know

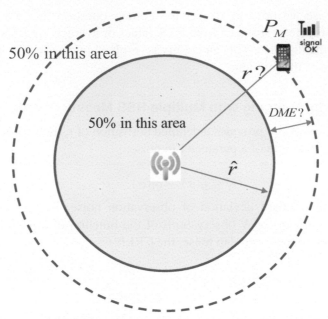

Figure 2.9   Range, RSS measurement, range estimation and confidence on having a device in an area.

and we want to estimate. We use the measured RSS and the model for RSS behavior to estimate the distance, $\hat{r}$, using Equation (2.8b). The estimate, $\hat{r}$, is different from actual distance, r, and we have a distance measurement error of $DME = \hat{r} - r$. We do not know the specific value of DME because we do not know the value of, r, but we can calculate the variance of this error using CRLB. The circle with a radius of, $\hat{r}$, divides the space into two regions inside and outside of the circle, and based on our estimate of the distance, there is a 50% chance that the device is localized inside the circle and a 50% chance that it is outside the circle. Figure 2.10 shows the same problem from a different angle, relating the RSS model, probability distribution function of the zero mean Gaussian shadow fading, $f_{SF}(x)$, and confidence level for the device to be in one of the two regions separated by our estimate of the range. The confidence level for an estimate to be in an area is the probability that the device measuring the RSS, P_M, is in that area. We have three possibilities for the location of the terminal, inside, on, or outside the circle. As shown in

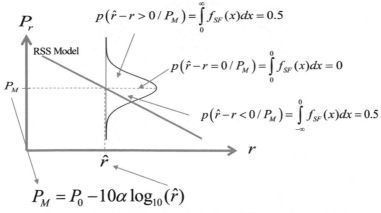

Figure 2.10   Calculation of confidence on estimation of distance from a RSS measurement.

Figure 2.10, we can calculate these probabilities using the fact that shadow fading is a zero mean Gaussian random variable. These probabilities are given by:

$$
\left\{
\begin{aligned}
p\left(\hat{r} - r > 0/P_M\right) &= \int_{-\infty}^{0} f_{SF}(x)dx = 0.5 \\
p\left(\hat{r} - r = 0/P_M\right) &= \int_{0}^{0} f_{SF}(x)dx = 0.0 \\
p\left(\hat{r} - r < 0/P_M\right) &= \int_{0}^{\infty} f_{SF}(x)dx = 0.5
\end{aligned}
\right. \qquad (2.15)
$$

In the remainder of this section, we use the relationship between the measurement of the RSS by a device, the estimate of the distance of the device from the antenna, and the calculation of confidence on having the device in a specific region to introduce a few practically interesting concepts related to RSS-based ranging.

2.5.1 Circular Confidence Regions for an RSS-Based Ranging

Figure 2.9 explains how we can use an RSS measurement, P_M, to divide the area into two regions separated by a circle with radius $\hat{r}$ and assigning a confidence value to each region. Intuitively, as shown in Figure 2.11, if we

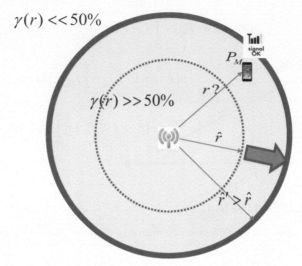

Figure 2.11   Range, RSS measurement, range estimation and certainty regions.

increase the radius, $\hat{r}$, to a larger value of $\hat{r}' > \hat{r}$, the probability of having the terminal inside the circle increases and the probability of having the terminal outside the circle decreases. To increase the probability of being inside the circle from 50% to a higher value, as shown in the fiure, we need to reduce or mark down the measured RSS used for calculation of distance estimate. In the wireless communication literature, the value of mark down on the RSS in dB to increase the probability of coverage is referred to as the *fade margin*, F_σ.

Figure 2.12 shows the relationship among fade margin, F_σ, the radius of certainty region, $\hat{r}'$, and the probability of having the device measuring the RSS, P_M, inside the certainty circle, γ:

$$\gamma = p\left(\hat{r}' - r > 0 / P_M - F_\sigma\right)$$
$$= \int_{-\infty}^{F_\sigma} f_{SF}(x)dx = 1 - \frac{1}{2}erfc\left(\frac{F_\sigma}{\sqrt{2}\sigma}\right) \qquad (2.16a)$$

The relation between marked down measured power, $P_M - F_\sigma$, and radius of the circle dividing the regions, $\hat{r}'$, shown in Figure 2.12 is found by using the RSS behavior model:

$$P_M - F_\sigma = P_0 - 10\alpha \log_{10}(\hat{r}'). \qquad (2.16b)$$

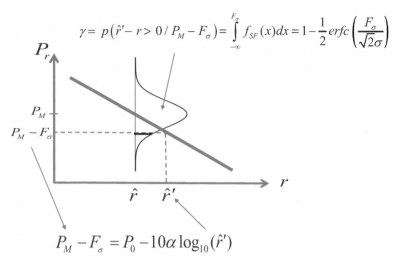

$$\gamma = p\left(\hat{r}'- r > 0 \,/\, P_M - F_\sigma\right) = \int_{-\infty}^{F_\sigma} f_{SF}(x)dx = 1 - \frac{1}{2}erfc\left(\frac{F_\sigma}{\sqrt{2}\sigma}\right)$$

$$P_M - F_\sigma = P_0 - 10\alpha \log_{10}(\hat{r}')$$

Figure 2.12 Calculation of probability of coverage or confidence on ability to measure power in a given distance.

Therefore, for a given desired certainty, γ, we can calculate the fade margin, F_σ, from Equation (2.16a):

$$F_\sigma = \sqrt{2}\sigma \times erfcinv\left[2(1 - \gamma)\right], \tag{2.17a}$$

Where *erfcinv* is the invert of the *erfc* function and it is available at MATLAB. The radius of certainty region is then calculated from:

$$\hat{r}' = 10^{\frac{P_0 - P_M + F_\sigma}{10\alpha}}. \tag{2.17b}$$

Equations (2.17a,b) allows us to calculate the certainty range for any RSS reading, given the parameters of the RSS behavior in an environment.

Example 2.5: (Location display design for an iBeacon)

The iBeacon application development manufacturers often design a graphical user interface on smart phones that uses the RSS value of the iBeacon to calculate its range and they display the location as a dot in a circle. Assume the transmitted power of an iBeacon is fixed at a value that results in an RSS reading of -50 dBm at distance of 1 m and a smart phone reads an RSS value of -60 dBm from the device. Calculate the radius of the circle for the display of the location with 90% confidence.

For short distances in an open area, where we generally use iBeacon devices, we can assume a distance power gradient of $\alpha = 2$, and standard deviation of shadow fading of $\sigma = 5dB$ and using the *erfcinv* function of MATLAB, the fade margin for 90% certainty in coverage is:

$$F_\sigma = \sqrt{2} \times 5 \times erfcinv[2(1 - 0.9)] = 6.4dB$$

and the radius of 90% coverage becomes:

$$\hat{r}' = 10^{\frac{-50+60+6.4}{10 \times 2}} = 6.6m.$$

The ML estimate of the location to map the dot on the display is:

$$\hat{r} = 10^{\frac{-50+60}{10 \times 2}} = 3.2m.$$

2.5.2 Rim-Shaped Confidence Regions for RSS-Based Ranging

We can use a similar method as the one used in Section 2.5.1 to calculate a rim-shaped region with an arbitrary confidence (Figure 2.13). In this case,

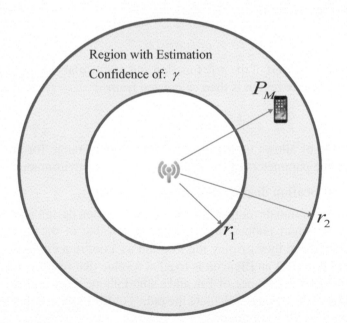

Figure 2.13   A rim region with an arbitrary confidence for location of a device using RSS reading from an antenna.

we calculate two ranging values, r_1, r_2, so that the confidence in having the device with RSS measurement of P_M inside the rim is an arbitrary value of γ. Figure 2.14 explains the method used to relate confidence to the auxiliary parameter F_σ and variance of shadow fading:

$$\gamma = p\left(r_1 < r < r_2/P_M, F_\sigma\right) = 1 - erfc\left(\frac{F_\sigma}{\sqrt{2}\sigma}\right) \qquad (2.18a)$$

The value of r_1, r_2 is calculated from the RSS model using the measured power, P_M, and the auxiliary parameter, F_σ:

$$P_M \pm F_\sigma = P_0 - 10\alpha \log_{10}(r). \qquad (2.18b)$$

From Equation (2.18a), we calculate fade margin:

$$F_\sigma = \sqrt{2}\sigma \times erfcinv(1 - \gamma), \qquad (2.19a)$$

and from Equation (2.18b), we have:

$$r_1, r_2 = 10^{\frac{P_0 - P_M \pm F_\sigma}{10\alpha}}. \qquad (2.19b)$$

Using Equations (2.19), we can calculate the rim boundaries for an arbitrary confidence value.

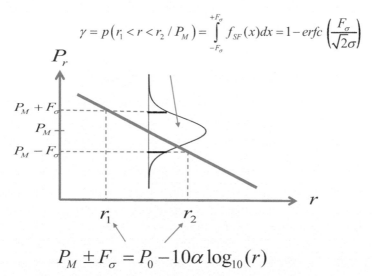

Figure 2.14 $\quad$ Fade margin and calculation of a rim region with an arbitrary confidence factor.

Example 2.6 (Rim region with arbitrary confidence of estimation)

Following Example 2.5, for 90% confidence area, using Equation (2.18a), the fade margin is:

$$F_\sigma = \sqrt{2} \times 5 \times erfcinv(0.1) = 8.2 dB.$$

If a device reads the RSS as -50 dBm from an access point with $P_0 = -20$ dBm in an environment with a distance power gradient of $\alpha = 3.5$, we have $-50 = -20 - 35 \log(d) \pm 13.2$. Then, using Equation (2.18b), the radiuses for 90% confidence become:

$$\left\{ \begin{array}{l} r_1 = 10^{\frac{-50+60-8.2}{10 \times 2}} = 1.2m \\ r_2 = 10^{\frac{-50+60+8.2}{10 \times 2}} = 8.1m \end{array} \right. .$$

The ML estimate of the location is:

$$\hat{r} = 10^{\frac{-20+50}{10 \times 3.5}} = 7.2m.$$

Our confidence that the device is at the range of 7.2 is zero because we know that we always have noise and we can never measure the exact distance using the RSS. But we know with 90% confidence that the device is in the range of 1.2–8.1 m from the access point. As we increase the confidence level, the difference between the radius of the inner and outer circles of the region increases. For 100% confidence, the device is in the range of zero to infinity and the device can be anywhere.

2.5.3 Confidence and Probability of Coverage

The RSS, P_r, in a given range from an antenna is a random variable calculated from the statistical model for behavior of RSS, given by Equation (2.1). Since the shadow fading noise is modeled by a Gaussian distribution, theoretically, we can measure an RSS in any location, no matter how far it is. But the probability of measuring any RSS from an antenna reduces with distance. In practice on another hand, a given receiver has a sensitivity, P_s, and it cannot measure the RSS if it is less than that sensitivity value. If the device that has a measurement sensitivity of P_s, measures the RSS at a location with a distance r from an antenna, the probability of having a measurement P_r is $p(P_r - P_s > 0)$. This probability is our confidence on the ability to measure the RSS at a given location and we refer to that as the probability of coverage, $p_c(r)$.

Figure 2.15 shows the general concept of the probability of coverage and the method for calculation of its value at a given distance r. Since $P_r = P_0 - 10\alpha \log r + X(\sigma)$, we can calculate the probability of coverage, P_c, as:

$$p_c(r) = p(P_r - P_s > 0) = p[P_0 - 10\alpha \log r - P_s > X(\sigma)]$$
$$= \int_{-\infty}^{P_0 - 10\alpha \log r - P_s} f_{SF}(x)dx = 1 - \tfrac{1}{2}erfc\left(\frac{P_0 - 10\alpha \log r - P_s}{\sqrt{2}\sigma}\right) . \quad (2.20)$$

The practical importance of this analysis is that it explains how we may measure any RSS at any location. However, the probability of measuring a value of RSS at a given distance, following Equation (2.20) and it is higher for higher-power devices. This observation will help us in performance evaluation of RSS-based localization systems with non-homogeneous transmission power in emerging IoT localization applications [Yin17].

Example 2.7 (Probability of coverage power)

Figure 2.16(a) shows the probability of coverage for distances from 1 to 120 m from a radiating antenna with the received power at 1 m, P_0, of -20, -30, -40, and -50 dBm and with a receiver sensitivity, P_s, of -90 dBm operating in an area with a distance power gradient of 3.5, when $\sqrt{2}\sigma$ is normalised. Figure 2.16(b) shows the MATLAB code used for generating these plots. The probability of coverage follows a similar pattern. Up to certain distances, the probability is close to one and after that, it falls rapidly

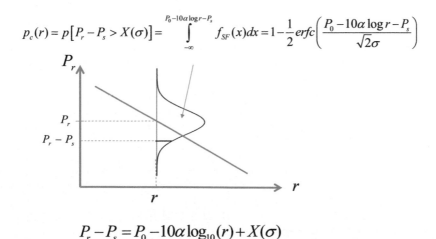

$$P_r - P_s = P_0 - 10\alpha \log_{10}(r) + X(\sigma)$$

Figure 2.15 Calculation of probability of coverage or confidence on ability to measure power in a given distance.

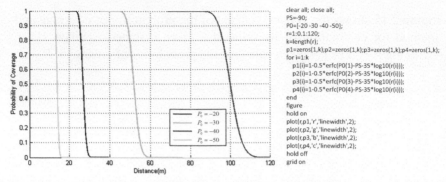

Figure 2.16   Probability of coverage as a function of distance, (a) plots for different transmitted power levels, (b) MATLAB code to produce the plots.

to a value close to zero. Obviously, higher transmitted powers provide a larger coverage.

Assignments for Chapter Two

Questions

1. What are the key parameters for modeling of the RSS behavior for RSS-based ranging application? Explain how we can measure these parameters empirically.
2. Name three models for RSS behavior inside the human body, indoors, and outdoors applications. What are the differences among the modeling parameters in these models?
3. What do DME, FIM and CRLB stand for and how they are mathematically related to one another?
4. What is fading margin? Explain its meaning in terms of the fraction of coverage at the edge of a cell.
5. What are the situations where the Okumura-Hata model is applicable?
6. Explain the meaning of confidence on an estimated range. In localization applications, how can we benefit from confidence of an estimate?
7. Explain the meaning of probability of coverage and explain how it can help in localization?
8. What is the difference between ML and MMSE estimation when they are applied to RSS-based ranging?

Problems

Problem 2.1:
The transmitted power of an LTE base station in a large city is 10 W, the minimum signal strength for the receiver is -100 dBm, and it operates at 800 MHz bands. The base station antenna height is 50 m, and mobile station antenna height is 1 m.

a) Use Okumura-Hata model to determine the coverage of the base station.
b) Determine the fade margin needed to increase the probability of coverage at the fringe of the coverage to 95%. Assume that the variance of shadow fading in the area of coverage is 10 dB.

Problem 2.2:
The transmitted power of an IEEE 802.11g Access Point (AP) operating at 2.4 GHz is 100 mW. The minimum required power for the operation of the AP is -90 dBm.

a) Determine the coverage of the AP in a residential or a small office using the IEEE 802.11 channel model D.
b) For the received signal strength of -78 dBm, calculate the ML distance rage with 95% certainty.

Problem 2.3:
An iBeacon development system uses low-energy Bluetooth (BLE) technology and provides the user with the RSS reading of the iBeacon, from another iBeacon, in dBm, and an ML distance estimate, d, in meters between the two iBeacons. However, the manufacturer does not release the path loss model they have used to relate RSS to the distance. A user curious to know the model sets the transmitted power to 0 dBm and goes to a location with distance reading of 1 m and another location with distance reading of 10 m and records the associated RSSs of -30 dBm and -55 dBm, respectively.

a) What is the path-loss model used by the manufacturer, if the user assumes a single gradient model was used by the manufacturer?
b) Assuming the sensitivity of the receiver is -80 dBm, plot the probability of coverage of the device as a function of distance. Assume the standard deviation of shadow fading is 8.

Problem 2.4:

For a single observation of a single parameter in zero mean Gaussian noise with variance σ^2, we have:

$$\begin{cases} O = \alpha + \eta \\ f(O/\alpha) = \frac{1}{\sqrt{2\pi}\sigma} e^{-\frac{(O-\alpha)^2}{2\sigma^2}} \end{cases}$$

Calculate the FIM and CRLB from:

$$F = -E\left[\frac{\partial^2 \ln f(O/\alpha)}{\partial \alpha^2}\right]$$

Problem 2.5:

Repeat Problem 4 for single observation of function of a parameter in zero mean Gaussian noise with variance σ^2: $O = g(\alpha) + \eta(\sigma)$

Problem 2.6:

Repeat problem 5 for N-observations of function of a parameter in zero mean Gaussian noise with variance σ^2:

$$\begin{cases} O_i = g(\alpha) + \eta_i; \quad i = 1, 2, ..., N \\ f(\mathbf{O}/\alpha) = \prod_{i=1}^{N} \frac{1}{\sqrt{(2\pi)}\sigma_i} e^{-\frac{[O_i - g(\alpha)]^2}{2\sigma_i^2}} = \frac{1}{\sqrt{(2\pi)^N} \prod_{i=1}^{N} \sigma_i} e^{-\sum_{i=1}^{N} \frac{[O_i - g(\alpha)]^2}{2\sigma_i^2}} \end{cases}$$

Calculate the FIM and the CRLB from:

$$F = E\left[\frac{\partial \ln f(O/\alpha)}{\partial \alpha}\right]^2$$

Problem 2.7:

a) Determine the CRLB of RSS-based ranging using IEEE 802.11 model B
b) Plot the CRLB as a function of distance for distances between 1 and 15 m
c) Repeat (b) if the variance of shadow fading is given by the following equations from [Liu15]

$$\sigma = \lambda \times e^{-\beta \times d} + \gamma$$

λ	β	γ
-4.28	0.9372	4.31

Problem 2.8:

a) Write the MATLAB code, using IEEE 802.11 channel model D, to generate 1000 random samples of the received signal strength (RSS) with different random values of the shadow fading when the distance between the transmitter and the receiver is 15 m.

b) Use each sample of the RSS for ML/MMSE estimation of the distance and determine the distance measurement error (DME) associated with that sample measurement of the RSS. Repeat the process 1000 times and plot the histogram of the distance measurement error.

c) Determine the mean and standard deviation of the DME

d) What is the ML estimate of the distance for all 1000 observation and its associated standard deviation of DME? How does it relate to the results of (c)?

Projects

Project 2.1 (Wi-Fi RSS data collection and analysis)

In this project, we measure the RSS of Access Points (APs) to map the Wi-Fi infrastructure deployed in a typical office building (e.g. Atwater Kent Laboratory at WPI). Then, we collect a data base of the Wi-Fi RSS readings in a typical floor of the building in LOS and OLOS conditions to model the behavior of the RSS in typical indoor areas. We collect the data using a laptop with a commercially available data collection software (e.g. WirelessMon platform)[3].

Measurement system and scenario:

a) Prepare a map of a typical floor of the building.

b) Do war driving in the building to mark the location of all APs in that floors of the building on the floor maps. Note that you might have several APs in one physical location with different MAC addresses. Measure and record the RSSs at approximately 1 m distance under each AP location.

c) Select at least 5 different locations in the selected floor in LOS and 5 locations in OLOS situations in the surrounding areas of a specific AP, where you can read the RSS from that AP. When choosing the points, spread them over the entire area as much as you can so that you have maximum diversity of distances between the AP and the laptop.

---

[3]If you can acquire or write code on an Android device to read the RSS of APs, using a smart phone you can conduct this project.

d) Measure the physical distance between all selected points and the selected AP in that floor.

e) Go to the actual locations to measure samples of RSS from the selected AP. Measure at least 10 samples of the RSS at each location, while you are keeping the laptop in front of and you are turning around in the same spot. This should provide you various measured values caused by shadow fading from different positions of your body.

f) Apply the MATLAB code in Example 2.1 to calculate the parameters, (P_0, α, σ), of the RSS behavior model for LOS and OLOS data separately.

g) Using parameters of part (f) develop a partition RSS behavior model for the environment of your selected floor.

h) Compare your RSS behavior model with the IEEE 802.11 model that fits your selected building.

i) Calculate standard deviation of the DME for each of the LOS/OLOS measurement locations using the ten RSS measurements in that location. Compare that standard deviation with the square root of the CRLB obtained by using your channel measurement parameters.

3

RSS Positioning Systems

3.1 Introduction

In Chapter 2, we presented models for the RSS behavior, explained how to measure the range using these models, and also explained how we can measure the quality of estimate using the CRLB. In this chapter, we address RSS-based positioning systems, where we utilize the RSS from multiple reference point (RP) antennas to determine the location coordinates of a device. We begin with the calculation of CRLB for RSS-based positioning and then describe popular RSS-based positioning systems. From the classical estimation theory point of view, the ranging process is a single-parameter, r, estimation and positioning is a multiple-parameter, (x, y, z), estimation of the coordinates of a location. In ranging, we were using a single antenna as a RP, but in positioning, we measure the range from multiple radiating RPs. Channel models for ranging and positioning are the same as we described them in Chapter 2. However, positioning algorithms are more complex and we discuss them later in Chapter 8. In this chapter, we provide the derivation of the CRLB for RSS-based localization and describe popular RSS-based positioning systems. We divide the popular RSS positioning systems into two categories: systems using RSS directly and systems using a fingerprint database for positioning. For each class of systems, first we derive the CRLB for performance evaluation and then introduce examples of popular system in that category. The first category of RSS-based localization technologies are those used for Wireless Video Capsule Endoscope (WVCE), Radio Frequency Identification (RFID), and low-energy Bluetooth (LEB) or iBeacon. These three technologies directly use the RSS for localization. The second category consists of Wi-Fi localization and Cell Tower localizations, which use a fingerprint database of the RSS measurements for localization.

The example systems we describe reveal a variety of practical issues related to RSS-based localization. In the WVCE application, the RF signal is designed to carry the frames of video taken by a camera from inside the gastrointestinal (GI) tract of a human being, to a belt mounted data acquisition device. The localization system opportunistically measures the RSS of the transmitted signal using body mounted sensors deployed on the torso of a patient to position the location of the capsule in the GI tract. The WVCE is an expensive medical procedure – each capsule costs approximately US $500 and the coverage of the RF signal transmitted from the capsule is a few tens of centemeter. Traditional RFID devices are passive and they are built to transfer short messages read by an RFID reader in proximity of the device. The cost of a passive RFID tag is a few cents, while the RFID reader may cost over several hundreds of dollars. Typical RFID devices have a range of around 1.5 m. More recently, active RFID technologies have gained momentum and iBeacon has been the leading technology in that domain. The iBeacon technology is made for proximity broadcasting and communication with smart devices in the proximity, eliminating the cost of expensive RFID readers. iBeacon devices are sold for a few dollars and they have a range of coverage with adjustable transmitted power and an average coverage of 10 m. In WVCE, RFID, and iBeacon technologies, the infrastructure for localization is deployed by the user of the location data and the measured RSS is directly used for localization.

Wi-Fi access points have a coverage of approximately 50 m, cost approximately US $50, and they are typically deployed indoors. Two industries have emerged around Wi-Fi localization concept, the Real-Time Location Services (RTLS), and Wireless Positioning Systems (WPS), both using RSS fingerprinting for positioning. The RTLS technology relies on relatively labor-expensive site surveys for RSS fingerprinting in specific buildings to provide accuracies on the orders of a few meters. WPS is a citywide Wi-Fi localization system with reasonable coverage in urban and sub-urban areas, where most smart devices operate. This technology uses a GPS-assisted RSS fingerprint database which is collected by war driving in the streets. Cell-towers on the average cover a few kilometers, cost a few tens of thousand dollars, and the antennas are installed outdoors. RSS-based Cell-tower Positioning Systems (CPS) provide citywide comprehensive coverage, an accuracy of a few hundred meters, and they use GPS-assisted finger-printing. Figure 3.1 summarizes the cost-coverage characteristics of popular technologies that we study in this chapter.

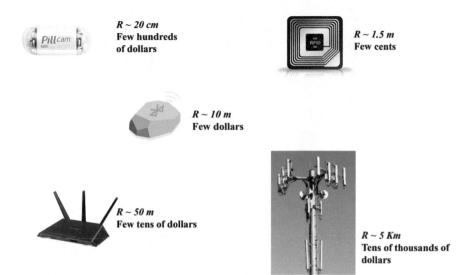

Figure 3.1   Coverage and cost comparison of technologies using RSS based localization.

3.2 Performance of RSS Positioning Methods

Figure 3.2 provides the elements of RSS-based positioning in relation to the infrastructure. We have a grid deployment pattern with radiating antennas used as the RPs for localization that are installed in the corners of the grid. These antennas have a coverage of R and they are deployed with a minimum distance of D from each other. To ensure that all four antennas are covering the entire area, we need to have $R \geq \sqrt{2}D$. A smart device at a location (x, y) is reading RSS values from the surrounding RPs and uses that to estimate the location of the device, $(\hat{x}, \hat{y})$. We are interested in the statistics of the distance measurement error (DME) defined by:

$$DME = \varepsilon = \sqrt{(\hat{x} - x)^2 + (\hat{y} - y)^2}. \tag{3.1}$$

The precision of location estimates depends on the range of the device from the RPs and it changes from a location to another. In Section 2.4.3, we use CRLB to analyze the precision of RSS-based ranging. From Equation (2.13c), we see that precision depends on the distance between the antenna and the device, the distance-power gradient, and the variance of shadow fading of the RSS in the environment. Therefore, we expect that the precision of the positioning system also depends on these parameters. Usually, for positioning purposes, we have multiple antennas and so, the precision would be related to distances from all of these antennas.

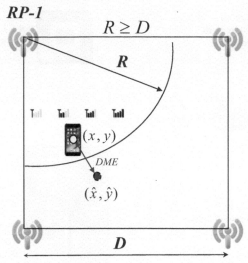

Figure 3.2   The fundamental elements of localization, deployment density, D, coverage, R and precision, *DME*. A device located at (x, y), measures RSS from four access points and wants to estimate the location of the device, $(\hat{x}, \hat{y})$, "with what precision, *DME*, device can be located?"

If the model for calculation of the received power is accurate, the error in localization is caused by the effects of shadow fading that is modeled as a Gaussian random variable with variance σ^2. The formulation of this problem is very similar to the formulation for calculation of the statistics of DME for range estimation using CRLB that we discussed in Sections 2.3 and 2.4. In classical estimation theory terminology, the range estimation problem was developed for observation of a function of a single parameter in zero mean Gaussian noise. Here, we have N-observations of a function of two parameters, (x, y), in zero mean Gaussian noise with variance, σ^2.

3.2.1 Positioning Using RSS Directly

Consider the case with N-RPs used as landmarks to locate a device using RSS. From Equation (2.1), the relation between RSS and the distances from the RPs is given by:

$$P_i = P_0 - 10\alpha \log r_i + X_i(\sigma), \quad i = 1, 2, \ldots, N, \qquad (3.2a)$$

in which:

$$r_i = \sqrt{(x - x_i)^2 + (y - y_i)^2}, \tag{3.2b}$$

where X represents the shadow fading with variance, σ^2, (x, y) is the location of the device, and (x_i, y_i) are locations of RPs.

Formulating this problem in classical estimation theory in vector notation, we have N-RSS observations $\mathbf{O} = \begin{bmatrix} P_1 & P_2 & . & P_N \end{bmatrix}^T$ in zero mean Gaussian noise $\mathbf{X} = \begin{bmatrix} X_1 & X_2 & . & X_N \end{bmatrix}^T$:

$$\mathbf{O} = \mathbf{G}(x.y) + \mathbf{X}, \tag{3.2c}$$

where:

$$\mathbf{G}(x, y) = \begin{bmatrix} P_0 - 10\alpha \log \sqrt{(x - x_1)^2 + (y - y_1)^2} \\ P_0 - 10\alpha \log \sqrt{(x - x_2)^2 + (y - y_2)^2} \\ . \\ P_0 - 10\alpha \log \sqrt{(x - x_N)^2 + (y - y_N)^2} \end{bmatrix}. \tag{3.2d}$$

The common sense and the ML estimate for positioning the location of the device are both obtained when $\mathbf{O} = \mathbf{G}(x, y)$ or

$$\begin{bmatrix} P_1 \\ P_2 \\ . \\ P_N \end{bmatrix} = \begin{bmatrix} P_0 - 10\alpha \log \sqrt{(x - x_1)^2 + (y - y_1)^2} \\ P_0 - 10\alpha \log \sqrt{(x - x_2)^2 + (y - y_2)^2} \\ . \\ P_0 - 10\alpha \log \sqrt{(x - x_N)^2 + (y - y_N)^2} \end{bmatrix}. \tag{3.2}$$

We have a set of quadratic equations with two unknowns, (x, y). Figure 3.3 shows a graphical interpretation of this problem. Each RSS measurement results in a range estimate and a circle around the source. The location of the device is somewhere in the intersection area of all the circles and finding that location needs a localization algorithm. Localization algorithms have different alternatives, and we will discuss them in Chapter 7. We continue with the calculation of the CRLB for this problem as a bound for the precision of any RSS-based localization algorithm.

3.2.2 CRLB for Positioning Using RSS Directly

The distinct feature of CRLB is that it provides a unique value as a bound for the performance of numerous localization algorithms. As a result, we

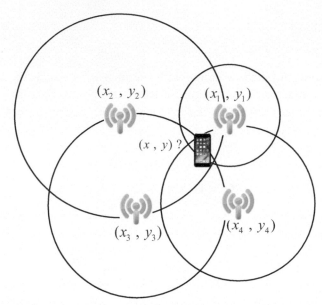

Figure 3.3   Basic concept of RSS based localization, a terminal measures RSS from several reference points. Each RSS measurement results in a range estimate and a circle with the radius of range around the source. Location of the device is somewhere in the intersect of all circles.

can analyze the fundamental characteristics of a techniques, for example, RSS-based positioning, to relate the performance to transmitted power, the deployment scenario, and environmental parameters, such as the distance power gradient and the variance of shadow fading. Based on the discussion of the bounds for ranging in Section 2.4.3, one may next think of the relation between ranging error and positioning error. To examine this issue, we provide the analysis of the positioning estimation error for RSS-based localization systems, following [CHE02].

Shadow fading causes variations in the received power, *d***P**, and the variation in this power causes a variation in the range estimate, *d***r**. Since the variation in power from a given RP is given by:

$$dP_i(x, y) = -\frac{10\alpha_i}{\ln 10}\left(\frac{x - x_i}{r_i^2}dx + \frac{y - y_i}{r_i^2}dy\right); \quad i = 1, ...N \qquad (3.3a)$$

In vector form:

$$d\mathbf{P} = \begin{bmatrix} dP_1 \\ dP_2 \\ \cdot \\ dP_N \end{bmatrix}, \quad d\mathbf{r} = \begin{bmatrix} dx \\ dy \end{bmatrix}. \qquad (3.3b)$$

$$\begin{cases} d\mathbf{P} = \mathbf{H} \times d\mathbf{r} \\ \mathbf{H} = \nabla_{x,y}\left[\mathbf{G}(x,y)\right] \end{cases}. \qquad (3.3c)$$

In other words, differential changes in power, $d\mathbf{P}$ and $d\mathbf{r}$, are related by the Jacobian matrix, $\nabla_{x,y}\left[\mathbf{G}(x,y)\right]$.

Using the values of the vector function, $\mathbf{G}(x,y)$, from Equation (3.2d), we have:

$$\mathbf{H} = \nabla_{x,y}\mathbf{P} = -\frac{10}{\ln 10}\mathbf{I}_N[\alpha_1........\alpha_N] \begin{bmatrix} \frac{x-x_1}{r_1^2} & \frac{y-y_1}{r_1^2} \\ \cdot & \cdot \\ \cdot & \cdot \\ \frac{x-x_N}{r_N^2} & \frac{y-y_N}{r_N^2} \end{bmatrix}. \qquad (3.3d)$$

We need relations between $d\mathbf{r}$ and $d\mathbf{P}$, since $d\mathbf{P} = \mathbf{H}d\mathbf{r}$ and $\mathbf{H}$ is not a symmetric matrix:

$$d\mathbf{r} = \left(\mathbf{H}^T\mathbf{H}\right)^{-1}\mathbf{H}^T d\mathbf{P}. \qquad (3.3e)$$

The shadow fading is a zero mean Gaussian random variable, therefore:

$$E\left\{|d\mathbf{P}|^2\right\} = cov\,(d\mathbf{P}) = \begin{bmatrix} \sigma^2 & 0 \\ 0 & \sigma^2 \end{bmatrix} = \sigma^2\mathbf{I} \qquad (3.4a)$$

We are interested in variance of distance, $cov(d\mathbf{r}) = E\left\{|d\mathbf{r}|^2\right\}$, and using Equation (3.3e), we have:

$$cov(d\mathbf{r}) = E\left\{(\mathbf{H}^T\mathbf{H})^{-1}\mathbf{H}^T d\mathbf{P}\left[(\mathbf{H}^T\mathbf{H})^{-1}\mathbf{H}^T d\mathbf{P}\right]^T\right\}$$
$$= \sigma^2(\mathbf{H}^T\mathbf{H})^{-1} \qquad (3.4b)$$

Therefore, the covariance of the location estimate is:

$$E\left\{|d\mathbf{r}|^2\right\} = cov\,(d\mathbf{r}) = \sigma^2\left(\mathbf{H}^T\mathbf{H}\right)^{-1} = \begin{bmatrix} \sigma_x^2 & \sigma_{xy}^2 \\ \sigma_{xy}^2 & \sigma_y^2 \end{bmatrix} \qquad (3.4c)$$

The CRLB is variance of the positioning error caused by shadow fading given by:

$$CRLB \geq \text{Tr}(\mathbf{F}^{-1}) = \sigma_r^2 = \sigma_x^2 + \sigma_y^2. \tag{3.4d}$$

The FIM matrix for this problem is given by:

$$\mathbf{F} = E\left\{|d\mathbf{r}|^2\right\}^{-1} = \frac{\mathbf{H}^T\mathbf{H}}{\sigma^2}. \tag{3.4e}$$

We could also arrive to these results by calculating the FIM directly and inverting that to obtain the CRLB.

Example 3.1: (MATLAB for contour of DME in a room)

Figure 3.4 shows the results of the above analysis to determine the contour of RSS-based localization errors in a 30 m × 30 m room. The MATLAB code for generating the plot is provided below. The distance power gradient for the RSS model is $\alpha = 2$, and the standard deviation of the shadow fading is assumed to be $\sigma = 2.5\ dB$. We have five RPs, four in the corners and one in the middle of the room. In the plot of the contour, red represents higher values.

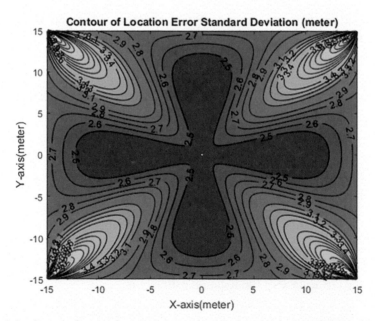

Figure 3.4 Contour for standard deviation of positioning error using CRLB in a 30 × 30 meters room.

The standard deviation of position error is higher along the sidelines of the area. In the central areas, we have lower errors. In general, the positioning error is on the order of the ranging error. However, the distribution of errors in the area is different and they fluctuate around the values of the ranging error from different APs. When we are in the central area, we get equally accurate ranges from all RPs, which gives us a better estimate of location.

```
%%
% The simulation is based on Kobayashi's paper 'Signal Strength Based
Indoor Geolocation'
% The result can be validated by Fig. 3 in the paper
% Calculation of CRLB is based on Equation (19), (22) and (23)
%%
close all;clear all;clc;warning off;
%% Initialization
% Locations of Access Points
APx(1)=15;APy(1)=15;
APx(2)=15;APy(2)=-15;
APx(3)=-15;APy(3)=-15;
APx(4)=-15;APy(4)=15;
APx(5)=0;APy(5)=0;
SD=2.5; % Standard Deviation of Shadow Fading
NUM=5; % Number of Access Points
% Locations of Receivers
pace=0.1;
mx=-15:pace:15;
my=-15:pace:15;
nxy=length(mx);
for yi=1:nxy
    for xi=1:nxy
        for i1=1:NUM
        alpha=3.5;
        r(i1,xi,yi)=sqrt((mx(xi)-APx(i1))^2+(my(yi)-APy(i1))^2);
% Distance Between Transmitter and Receiver
                H1(i1,xi,yi)=-10*alpha/log(10)*(mx(xi)-
APx(i1))/(r(i1,xi,yi))^2; % First Column of H Matrix
                H2(i1,xi,yi)=-10*alpha/log(10)*(mx(yi)-
APy(i1))/(r(i1,xi,yi))^2; % Second Column of H Matrix
        end
        H(:,:,xi,yi)=[H1(:,xi,yi),H2(:,xi,yi)];
        Covv(:,:,xi,yi)=SD^2*((H(:,:,xi,yi))'*H(:,:,xi,yi))^(-1);
% Covariance Matrix of Error Estimate
            SDr(xi,yi)=sqrt(Covv(1,1,xi,yi)+Covv(2,2,xi,yi));
% Standard Deviation of Location Error
    end
end
SDr=SDr';
figure (1)
```

```
[C h]=contourf(mx,my,SDr,20);
h.LevelList=round(h.LevelList,1) %rounds levels to 3rd decimal place
clabel(C,h);
xlabel('X-axis(meter)');
ylabel('Y-axis(meter)');
title('Contour of Location Error Standard Deviation (meter)')
```

The above derivation of CRLB assumes that in all locations, we read the RSS from all APs and this does not include the probability of coverage (see Section 2.5.3), which is a function of the transmitted power. If the transmitted power from the APs and consequently their coverage are different, we will need to include the probability of coverage into the computation. Situations like that are expected to occur in IoT applications, where many radiating devices with non-homogeneous transmission powers are involved [Yin17]. Another issue that is not considered in this calculation is the variations of the standard deviation of shadow fading at close proximity to the AP.

3.2.3 RSS-Based Ranging Using a Fingerprint

In RSS fingerprinting techniques, we take advantage of the RSS measurements in a known location to sense the location of a device in an unknown location from its measured RSS. Figure 3.5 shows the basic concept behind

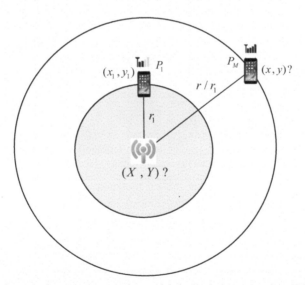

Figure 3.5   How can we estimate the range of a device using a previous RSS reading in a known range? What is precision of this estimate?

the RSS fingerprinting. We have an RF radiating device, for example, a Wi-Fi access point, at an unknown location, (x , y). Assuming a device at a known location, (x_1 , y_1), has measured an RSS fingerprint value of P_1 and we know the RSS value, we can use the RSS behavior model and calculate the distance of the fingerprint location from the access point using:

$$P_1 = P_0 - 10\alpha \log r_1 + X_1(\sigma). \tag{3.5a}$$

The same or another device in an unknown location, (x , y), measures an RSS value of P_M from the same access point:

$$P_M = P_0 - 10\alpha \log r + X_M(\sigma) \tag{3.5b}$$

The fact that we have measured power from the same antenna in two locations gives us an idea that the unknown location of the device measuring P_M must be somewhere around the known location, (x_1 , y_1), where we measured, P_1. The difference between the two powers is:

$$
\begin{aligned}
P_M - P_1 &= -10\alpha \log \frac{r}{r_1} + [X_M(\sigma) - X_1(\sigma)] \\
&= -10 \log \frac{r}{r_1} + X\left(\sqrt{2}\sigma\right),
\end{aligned}
\tag{3.5c}
$$

since the summation of two zero mean Gaussian random variables, X_M, X_1, with identical variances, σ^2, is a zero mean random variable, X, with variance $2\sigma^2$. If we pull this equation into classical estimation theory terminologies, we have an observation of the difference between two powers as a function of the range in additive Gaussian noise:

$$O = P_M - P_1 = g(r) + X(\sqrt{2}\sigma), \tag{3.6a}$$

where the function is defined as:

$$g(r) = -10\alpha \log \frac{r}{r_1}, \tag{3.6b}$$

and the variance of the noise is $2\sigma^2$. As shown in Appendix section A.1.2, the ML estimate of the observation of a function of a parameter in noise is found by solving:

$$O = g(r) \Rightarrow -10\alpha \log \frac{r}{r_1} = P_T - P_1. \tag{3.6c}$$

Therefore, the ML estimate of the range of the device from the reference antenna is:

$$\hat{r}_{ML} = r_1 \times 10^{-\frac{P_M - P_1}{10\alpha}}. \tag{3.6d}$$

A simple algorithm to solve this problem is to use the measured RSS in the known location to calculate r_1 and substitute it into Equation (3.6d). But we do not know the location of the antenna to calculate, r_1, and we need other algorithms that we describe in Part III, when we address algorithms. However, using the above formulation we can calculate the CRLB to compare this approach with other localization methods. To calculate the CRLB for this problem, we can use classical information theory. As shown in Appendix section A.3.2, the Fisher information matrix for observation of the function of a parameter in noise is:

$$\mathbf{F} = -E\left[\frac{\partial^2 \ln f(O \mid r)}{\partial r^2}\right] = E\left[\frac{\partial \ln f(O \mid r)}{\partial r}\right]^2 = \frac{[g'(r)]^2}{2\sigma^2}.$$

Since $g'(r) = -\frac{10\alpha}{(\ln 10)r}$,

$$\mathbf{F} = \frac{(10)^2 \alpha^2}{(\ln 10)^2 2\sigma^2 r^2}, \tag{3.7a}$$

and the CRLB is given by:

$$CRLB \geq \mathbf{F}^{-1} = \frac{(\ln 10)^2}{100} \frac{2\sigma^2}{\alpha^2} r^2 \quad \Rightarrow \quad \sigma_P \geq \frac{\ln 10}{10} \frac{\sqrt{2}\sigma}{\alpha} r. \tag{3.7b}$$

The discussion presented in this section showed that if we do not know the location of an antenna, but we have an RSS fingerprint at a known location, we can still estimate the range of the device from the access point. The variance of the estimation error is double of the case when we know the location of the AP.

3.2.4 CRLB for Positioning Using an RSS Fingerprint Database

Figure 3.6 shows the basic idea behind RSS-based positioning using fingerprinting. An access point antenna located at (X, Y) radiates RF signal

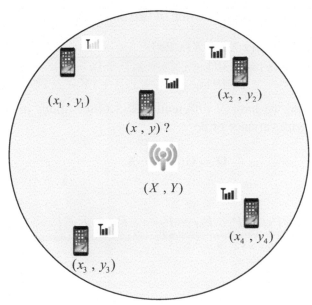

Figure 3.6 Basic idea of RSS fingerprinting for positioning, we have measured the RSS in several locations from an access point antenna to create a fingerprint database to be used as virtual reference point for positioning the device using its RSS in an unknown location.

that can be measured with a device as long as the device is in the area of coverage of the access point. We move around with the device and measure N-samples of RSS values, $\{P_i \; ; \; i = 1, 2, ..., N\}$, in specific locations, $\{(x_i, y_i), \; i = 1, 2, ..., N\}$, which we map on a coordinate system. Now, we can locate the device at any other location, (x, y), using the RSS measurement in that location, P_M. This formulation of the problem is the same as the ranging formulation of Equations (3.6a,b,c) in vector notation. Our observations are the differences between the measured RSS in the unknown location and RSS in known locations:

$$O_i = P_M - P_i = -10\alpha \log \frac{r}{r_i} + X\left(\sqrt{2}\sigma\right)$$

$$= -10\alpha \log r + 10\alpha \log r_i + X\left(\sqrt{2}\sigma\right). \tag{3.8a}$$

These are N-observations of a function of two parameters (x, y):

$$\begin{cases} O_i = g_i(x,y) + X(\sqrt{2}\sigma) \\ g_i(x,y) = -10\log\sqrt{(x-x_i)^2 + (y-y_i)^2} \\ \qquad +10\log\sqrt{(X-x_i)^2 + (Y-y_i)^2} \end{cases} \quad . \qquad (3.8b)$$

In vector notation, we have N differential RSS observations in zero mean Gaussian noise with variance of $2\sigma^2$:

$$\mathbf{O} = \mathbf{G}(x.y) + \mathbf{X}, \qquad (3.9a)$$

where

$$\begin{cases} \mathbf{O} = \begin{bmatrix} P_M - P_1 & P_M - P_2 & . & P_M - P_N \end{bmatrix}^T \\ \mathbf{X} = \begin{bmatrix} X_1 & X_2 & . & X_N \end{bmatrix}^T \end{cases} \qquad (3.9b)$$

and

$$\mathbf{G}(x,y) = \begin{bmatrix} -10\alpha\log\sqrt{(x-x_1)^2+(y-y_1)^2} + 10\alpha\log\sqrt{(X-x_1)^2+(Y-y_1)^2} \\ -10\alpha\log\sqrt{(x-x_2)^2+(y-y_2)^2} + 10\alpha\log\sqrt{(X-x_2)^2+(Y-y_2)^2} \\ . \\ -10\alpha\log\sqrt{(x-x_N)^2+(y-y_N)^2} + 10\alpha\log\sqrt{(X-x_N)^2+(Y-y_N)^2} \end{bmatrix} . \quad (3.9c)$$

The common sense and the ML estimate for positioning the location of the device is obtained when $\mathbf{O} = \mathbf{G}(x, y)$ or

$$\begin{bmatrix} P_M - P_1 \\ P_M - P_2 \\ . \\ P_M - P_N \end{bmatrix} = \begin{bmatrix} -10\alpha\log\sqrt{(x-x_1)^2+(y-y_1)^2} + 10\alpha\log\sqrt{(X-x_1)^2+(Y-y_1)^2} \\ -10\alpha\log\sqrt{(x-x_2)^2+(y-y_2)^2} + 10\alpha\log\sqrt{(X-x_2)^2+(Y-y_2)^2} \\ . \\ -10\alpha\log\sqrt{(x-x_N)^2+(y-y_N)^2} + 10\alpha\log\sqrt{(X-x_N)^2+(Y-y_N)^2} \end{bmatrix} . \quad (3.10)$$

This is a set of N-quadratic equations with two unknowns, (x, y), each equation representing a hyperbola with two canonic centers at, (x, y) and (X, Y). We need an algorithm to solve this problem, and Chapter 8 will discuss those algorithms. We continue by calculation of the CRLB for this problem as a bound for precision of any RSS-based algorithm using the difference of RSS values for fingerprinting-based positioning. The Jacobian matrix for the power difference is the same as Equation (3.3d).

The formulation of the problem in classical estimation theory given in Equations (3.8a and b) is very similar to the formulation when we use the RSS directly, as in Equation (3.2c). For the calculation of the CRLB, we need to form the Jacobian matrix of $G(x, y)$. Since the difference between $G(x, y)$ using RSS directly, Equation (3.2d), and the same matrix for fingerprinting, Equation (3.9c), is the last part of each column, $10\alpha \log \sqrt{(X - x_1)^2 + (Y - y_1)^2}$, and this part is a fixed number, when we take the derivative of the two matrix to form the Jacobian matrix, results become the same as Equation (3.3d). The only difference is that in fingerprinting, the variance of shadow fading and consequently variance of estimation error is doubled. Since the variance of the shadow fading doubled but the Jacobian matrix is the same, the FIM matrix of Equation (3.4e) becomes:

$$\mathbf{F} = E\left\{ |d\mathbf{r}|^2 \right\}^{-1} = \frac{\mathbf{H}^T \mathbf{H}}{2\sigma^2}, \tag{3.11a}$$

and the CRLB is:

$$CRLB \geq \mathrm{Tr}(\mathbf{F}^{-1}) = \sigma_r^2 = \sigma_x^2 + \sigma_y^2. \tag{3.11b}$$

An interesting observation from this derivation is that in fingerprinting it does not matter that we know or we do not know the location of the antenna – in both cases the performance bound using the CRLB is the same. The difference between the direct use of RSS and using the RSS fingerprint is doubling the variance or a 1.4 times increase in the standard deviation of statistics of the DME. However, in practice, there is a huge difference between positioning systems in Figures 3.3 and 3.6. In the system shown in Figure 3.3, RPs are actual transmitting antennas such as a Wi-Fi access point or a cell tower, which can be a very expensive piece of hardware with a considerable cost of deployment. In Figure 3.6, RPs are virtual and we have only made an inexpensive measurement in each location. The only cost is the design of an inexpensive software patch in the device to store the measurements and their locations. Using fingerprinting provides a means for inexpensive deployment of dense infrastructure because each measurement in a location acts as a RP for localization (although the labor of capturing such measurements may have a cost). As a result, as we discussed before, RSS localization systems using direct RSS are often used for smaller areas either inside the human body or inside a room and fingerprinting systems are used in the rest of popular indoor and urban area localization systems.

3.3 Positioning Systems Using RSS Directly

In RSS-based localization for inside the human body for WVCE, in localization using RFID tags, or in-room localization with iBeacon technology, the area of coverage for a device is small and deployment of infrastructure is easy and inexpensive. We may install tens of body mounted sensors on the surface of a human body to locate the WVCE inside the GI tract, we may install RFID tags in every floor tile of a manufacturing floor to locate a robot, or we may install iBeacons on each wall or several on the ceiling of a room to locate a device. The density of deployment is tailored to provide certain accuracy for specific applications using a positioning system. Dense deployment of infrastructure for positioning in a specific important area such as inside the human body or dense deployment of RFID on the floor of a warehouse for accurate localization is justifiable, and in such dense deployments, it is possible to directly use the RSS for positioning. In this section, we describe popular examples of this technology. Section 3.4 describes Wi-Fi and cell-tower opportunistic localization using existing wireless communication infrastructure. The density of deployment of the infrastructure of these communication systems is not adequate for accurate localization and so localization systems resort to RSS localization using fingerprinting.

3.3.1 RSS-Based Localization Inside the Human Body

Localization inside the human body has its own unique challenges, which has made this medium an interesting area of research for localization applications [Pah11]. The inside of the human body is mostly made of liquids, it has high conductivity, and besides that it is a non-homogeneous propagation environment because the speed of radio propagation changes in different organs with different compositions. The path of motion of the objects inside the human body is extremely complex and often unstable and we do not know the nature of multipath radio propagation conditions inside the body. Radio propagation experiments inside the human body in vivo are extremely complex, time-consuming and expensive. We cannot monitor the motion visually because the capsule or other devices are inside the body.

The example that we examine in this section is localization of the WVCE inside the GI tract in its three major organs: the stomach, the small intestine, and the large intestine. Figure 3.7 demonstrates the anatomy of these three

Figure 3.7 Anatomy of the three major organs inside the GI tract of the human body, (a) stomach, (b) small intestine, and (c) large intestine.

major organs. The most challenging of the three organs is the small intestine with a length of up to nine meters and a volume close to the size of a baseball and an unknown curly path of motion, which may also be hosting numerous diseases or medical conditions to be detected and localized. The WVCE is the size of a large capsule with a length of approximately 2.5 cm and a diameter of around 1 cm carrying one or more video cameras and a wireless transmitter that can transfer the pictures taken at an approximate rate of two per second to the surface of the body to be stored in a belt mounted data acquisition device. When a patient ingests the capsule, it takes approximately 8 hours to pass through the GI tract, and in that period, it transmits approximately 5500 pictures, half of them from the small intestine. For medical doctors, it is important to associate the pictures of bleeding or tumors to specific locations inside the tract, which is a very challenging problem. One simple approach is to use the RSS of the transmitted signal to localize the capsule. Figure 3.8 shows the architecture of this system. We install a number of body mounted sensors on the surface of the body to measure the RSS of the signal transmitted from the capsule, and using these data, we locate the capsule.

To determine the accuracy of the RSS-based localization for WVCE, a group of students at the Worcester Polytechnic Institute [Wan11] used a 3D autonomic picture of the human body and, in a manner similar to Example 3.1, calculated the CRLB for localization in the three different organs using different patterns of sensor deployment on the torso of the human body. The channel model used for this analysis was developed by NIST, which is presented in Table 2.1 in Section 2.2.2. Figure 3.9 compares the performance of the RSS-based localization using CRLB in the three major organs of the GI tract, the stomach, the small intestine, and the large-intestine for 16 and 32 body mounted sensors on the torso. The results shows that an

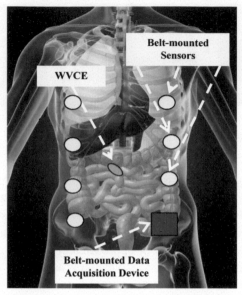

Figure 3.8   Overall architecture of a localization system for inside the human body identifying WVCE, body mounted sensors and the belt mounted data acquisition device on the human torso.

accuracy of 1–6 cm in absolute location can be achieved in 90% of locations in the GI tract. The best precision is obtained in the small intestine with 32 sensors.

3.3.2 RSS-Based Passive RFID Systems

RFID technology evolved from the barcode industry where a tag attached to an object enables an optical reader to automatically read the identification information from the tag. RFID uses electromagnetic fields to transfer the identity from the tag to the reader. The advantage of RFID over barcodes is that it can operate over longer ranges and in obstructed LOS conditions. Today, we have passive and active RFID technologies. The passive RFID collects the energy from the tag reader and transmits its memorized information by reflecting that energy. The coverage of typical passive RFID tags is approximately 1.5 m. Active RFID tags, such as the iBeacon using BLE with variable transmitted power options, carry batteries and cover distances of up

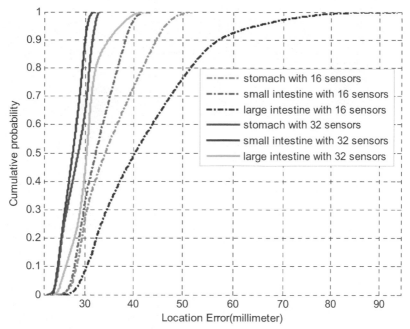

Figure 3.9 Cumulative probability distribution of the DME inside major organs of GI tract of the human body, the stomach, small- and large-intestine, using results of the CRLB for 16 and 32 body mounted-sensors on the torso.

to 10 m. RFID devices operate at different frequencies, animal tags at 120–150 kHz (LF), smart cards at 13.56 MHz (HF), traditional active RFIDs at 433 MHz (UHF), and iBeacon at 2.4 GHz in the ISM bands.

If we integrate the geolocation information as a part of the ID of a tag that is fixed at a location, or we register the location of such an RFID tag in an external database, an RFID device can be used for localization applications. Figure 3.10 shows the overall block diagram of an RFID localization system. Figure 3.11 [Ruf09] shows the application of RFID for localization inside a mine tunnel, where GPS or cell tower signals cannot penetrate. The RFID reader and RFID tag communicate to determine the location information, and then, the location information is transferred from the RFID readers to a server to position the location of the tag on a display of the map of the area of operation.

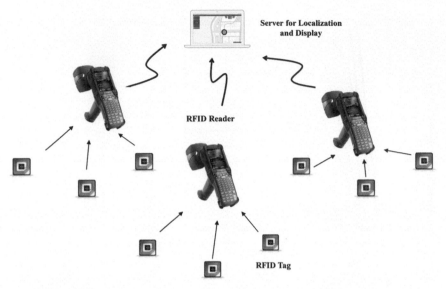

Figure 3.10   Overview of a typical RFID positioning system. RFID readers report the location information from the RFID tag reading, the server determines the location and add that to the layout of the area.

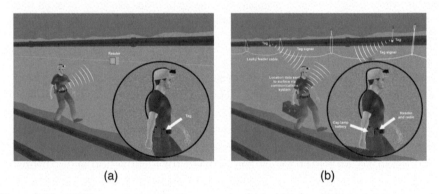

Figure 3.11   RFID sign post localization scenarios for inside of a mine tunnel, (a) user carries the RFID and readers are deployed on the wall, (b) user carries the reader and RFID tags are deployed in the environment.

There are two methods to install the infrastructure for an RFID localization system. In Figure 3.11(a), user is carrying the RFID tag and the readers are installed on the walls to check who is passing along their vicinity and report this information to the network. In Figure 3.11(b), the user carries the tag reader and using a wireless communication technology, sends the RFID's location to the system for display. This basic idea for RFID positioning systems have been used in some of the early successful commercial applications for indoor geolocation. For example, in the mid-1990's a few companies designed systems for monitoring new born babies inside nursing rooms. In this application, the newly born babies are tagged with an RFID tag and an RFID reader installed at the entrance of the doors to the nursing room monitors whether a specific baby is in inside or out of the room. Another very popular daily application of RFID tags is in highway monitoring of cars to pay tolls. Figure 3.12(a) shows a typical handheld RFID reader, and Figure 3.12(b) shows a few pictures of RFID tags for different aforementioned applications. Around 2010, Kiva Robotics used RFID tags installed at the center of floor tiles for precise localization and tracking of Robots in warehouses. A video clip showing the application of Kiva Robotics, Boston, for transportation of goods among different departments inside an Amazon warehouse is available at [V-IEE-08]. The straight movement paths of the robots to carry the merchandise among the shelves benefits significantly from reading the location of the RFID on the floor tiles with the RFID reader installed at the bottom of the robot. This way a localization accuracy of less than length of a title is achieved, which is very instrumental for navigating unmanned robots inside the warehouse without accidents.

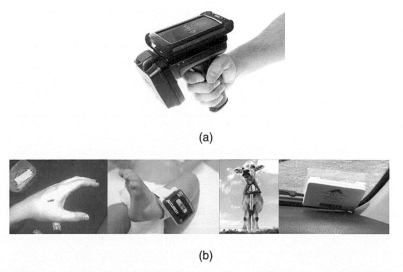

(a)

(b)

Figure 3.12   (a) a typical handheld RFID reader (b) typical RFID in different applications.

3.3.3 RSS-Based Active RFID Systems

The iBeacon protocol, released by Apple in 2013, based on Bluetooth Low Energy (BLE), also known as Bluetooth Smart technology opened another avenue for more precise indoor geolocation. iBeacon technology natively supports smart device operating systems (iOS or Android) simplifying application development for these devices. iBeacon features proximity context broadcasting for numerous applications, and BLE is tailored for IoT applications inside buildings. An overview video of the iBeacon technology and its applications is available at [V-Est16].

The iBeacon technology can be implemented for ad-hoc indoor localization, where we can quickly design a local ad-hoc localization environment. For example, we can use iBeacon for localization and tracking of the children in nursery rooms of the hospitals with more features than traditional RFID-based systems [Li16]. With the huge expectation on the influence of BLE and iBeacon technology in implementation of IoT localization, the use of these technologies is expected to receive a momentum in the near future and for the location aware IoT [Yin17]. The difference between localization using iBeacon and Wi-Fi localization is that iBeacon has a smaller coverage and a much smaller size making it easy to install in an ad-hoc manner with minimal cost. This feature supports in-room ad-hoc localization with accuracies of a few meters without the need for collecting a signature or fingerprint database. We simply stick a few iBeacons in a work space and create an ad-hoc localization system. A video description of this procedure is available at [V-Est15]. The application scenario would be similar to that of Example 3.1 and performance evaluations using CRLB follows the methods described in that example. The MATLAB code in Example 3.1 can be used to create coverage plots, similar to Figure 3.4, for performance evaluation of iBeacon with different deployments in a given work space. The IEEE 802.11 model B, described in Table 2.2, with a distance power gradient of 2 and a standard deviation of 5 dB for shadow fading fits this application well. We will leave that example as one of the assignments at the end of this chapter.

3.4 Positioning Systems Using RSS Fingerprint Database

Figure 3.6 shows the basic idea behind localization using fingerprinting. We go around in the coverage area of an RF radiating antenna and use a device to measure the RSS values from that antenna in known locations to form a fingerprint database. Then, we use that database to localize a device in an unknown location using its RSS reading. In Section 3.2.4, we used the CRLB to show that the location of an RSS fingerprint can be utilized as a *virtual* RP for localization using the difference of RSS readings. Using virtual RPs by fingerprinting doubles the variance of the lower bound on the location estimate as it is compared to localization using physical antennas. Installing an antenna involves the cost of the access point device as well as cost of wiring and labor for the installation. Measurement of signature of an antenna in a location only involves the labor to go to the specific location and storing the measurement of the RSS fingerprint of an antenna at that location. Indeed, when at a given location, we can measure the signature of *all* antennas covering that location, which increases the efficiency of the procedure for collecting an RSS fingerprint database. This discussion underlines the importance of RSS fingerprint databases in RSS-based localization to achieve high-precision positioning at low cost. RSS fingerprinting provides an opportunity for virtual dense deployment of RPs for localization. As a result, both popular Wi-Fi localization systems, RTLS and WPS, use RSS fingerprinting, and RSS fingerprinting cell tower positioning systems (CPS) are gaining in popularity. The process of driving around to collect an RSS fingerprint database is commonly referred to as war driving. War driving in indoor and in urban areas are different because in war driving in urban areas, we can automate the process by using GPS to tag the location of the fingerprint measurement. Since GPS does not provide a good coverage indoors, war driving in indoor areas is commonly performed manually. In this section, we describe RTLS, WPS, and CPS as example systems that use RSS fingerprinting. These examples use existing Wi-Fi and cellular network infrastructure for opportunistic localization of smart devices.

An RSS-based localization system relies on the density of infrastructure and pattern of deployment, which have substantially different characteristics for cellular and Wi-Fi networks. In cellular networks, the antennas are installed outdoors by service provider companies coordinating the location of antennas. The base stations are complex and expensive. Finding the land for deployment is complex, but coverage is comprehensive. A cellular network installs a variety of base stations with different coverage sizes in different areas based on the user traffic load of an area. Minimal infrastructure and taller cell towers with largest coverage, macro-cells, are deployed in wide open sub-urban areas to cover up to tens of kilometers on highways. In densely populated urban canyons, micro-cells are deployed on the walls of high-rise buildings or lampposts to cover a few hundred meters. Smaller pico-cells, with coverage of tens of meters, are deployed on the roof of populated areas such as airports or shopping malls.

Wi-Fi access points are mainly deployed indoor with more or less the same coverage of around 50 m. Access points are inexpensive and easy to deploy, and deployment in a larger area is often uncoordinated. Moreover, the coverage need not be comprehensive.

Wi-Fi deployment in residential areas is usually limited to one router per residence connecting to the Internet and distributing the access inside the residence. This is changing with some self-configuring mesh networks such as Google, Wi-Fi, and Plume[1], but we always have only a few routers per resident. In dense urban areas, people reside in apartment buildings with numerous access points installed in different residential apartments. The size of the land in residential houses in sub-urban North America are on average from a fractions up to a few acres, and this increases as we move away from downtown areas to the farming areas far away from the center of cities. Therefore, the highest density of deployment of Wi-Fi access points is in the downtown areas, and as we move away from the downtown, the density of access points usually reduces. When we go far from populated areas and drive on a highway, we often have no Wi-Fi coverage. While cell tower deployments follow a similar pattern, the density of Wi-Fi deployment in urban and suburban areas is higher, but cell towers cover even highways.

We have selected three RSS-based positioning systems, RTLS, WPS, and CPS, opportunistically using their communication infrastructures for localization. We explain how these technologies evolved and we review sample empirical measurements of the performance of these systems. We refer to algorithms just by name and we leave the details of the algorithms for Chapters 8–11.

3.4.1 RTLS Wi-Fi Positioning Using RSS Fingerprints

The term Real-Time Locating Systems (RTLS) is commonly used in the industry for the systems designed for automatic identification and tracking of tags in contained areas. The word has been used since the late 1990s when the first commercial indoor geolocation systems using TOA were introduced by PinPoint, Billerica, MA and WhereNet, Santa Clara, CA. In the early 2000s, pioneering indoor Wi-Fi localization companies, such as

---

[1]For example, see https://www.plume.com.

Ekahau, Helsinki, Finland and Newbury Networks, Boston, MA, introduced Wi-Fi RTLS technology using RSS fingerprinting of existing Wi-Fi networks deployed for wireless communication applications.

The most popular applications of RTLS have been asset tracking in warehouses, locating "in-demand" equipment and personnel inside hospitals, developing map-guides for visitors to public areas such as museums, locating the elderly and patients with special needs in nursing homes, and monitoring children or pets away from visual supervision. The first generation of Wi-Fi RTLS products were software programs running on laptops and palmtype computers equipped with Wi-Fi devices. The software operates in two modes: a *data collection mode* in which the user builds up the reference database inside a building for fingerprinting; and the *localization* mode in which the software locates the terminal based on the RSS readings from the surrounding access points. Later on, Wi-Fi chipsets were integrated in a small RFID localization tag to form an embedded system for RTLS applications. More recently, some of the manufacturers have integrated GPS chipsets within the RFID tag to provide continual tracking of the tags when the device is moved between two surveyed sites. The business model for generating income from the software solutions was focused on the site licensing and the cost of site survey for collection and maintenance of the database. The introduction of Wi-Fi RFID tags added a new source of revenue out of sales of the individual tags. This market was large enough to sustain a few small companies in this field worldwide.

Figure 3.13 shows the general architecture of a Wi-Fi RTLS system. To build up the fingerprint database, the tag or a laptop is moved to visually identified locations on the layout of a floor plan of a building and the location manually geo-tag the RSS reading from different APs. Since manual geotagging of the measurement locations is time-consuming, acquiring several sets of measurements in each tag positions is recommended by the manufacturers. To create independent sample of RSS, the operator taking samples of RSS in a location circulates around in the location to create different levels of shadow fading. When the RSS fingerprint database is ready, the tag or the laptop is switched to localization mode. In this mode, the tag reads the RSS of the APs in the surrounding area and reports that to the positioning engine. The positioning engine uses the fingerprinting database and its localization algorithm to calculate the location of the tag and forward that to a display unit to position the tag location in the building layout.

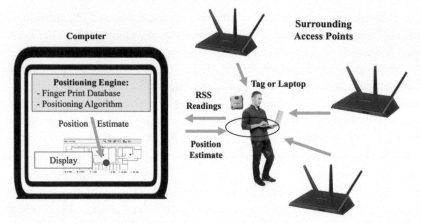

Figure 3.13 Overall architecture of a Wi-Fi RTLS system for two phases of operation: (1) war driving with a laptop to create the RSS fingerprint database, and (2) localization of a tag or the laptop using the fingerprint data base.

Example 3.2: (Typical Performance of RTLS)

Figure 3.14 [Hei07] shows the performance of Ekahau's software for 1, 2, and 3 access points and 4, 10, and 27 training points in a typical laboratory building in the third floor of the Atwater Kent Laboratories at the Worcester Polytechnic Institute. Figure 3.15 shows the location of the three access points and 27 training points. The performance evaluation curves in Figure 2.14(a), demonstrates that accuracies of around 1 m

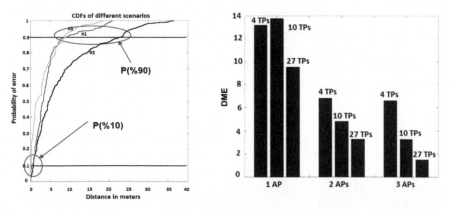

Figure 3.14 Performance of Ekahau Wi-Fi RTLS in the third floor of Atwater Kent Laboratory at the Worcester Polytechnic Institute.

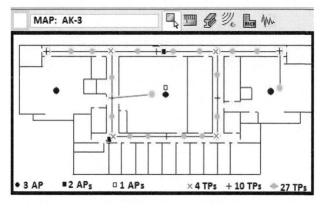

Figure 3.15 An indoor scenario for performance evaluation of the RTLS algorithms on the third floor of the Atwater Kent Laboratory, Worcester Polytechnic Institute, Worcester, MA.

are achieved in 90% of locations. These products use proprietary pattern recognition algorithms that we discuss in Chapter 9.

3.4.2 WPS Wi-Fi Positioning Using RSS Fingerprints

Wi-Fi Positioning System (WPS) was first introduced for commercial applications by Skyhook Wireless, Boston, MA in 2003. The basic technical concept behind WPS is similar to RTLS; it utilizes Wi-Fi localization technology with fingerprinting. However, WPS is aiming on a comprehensive metropolitan area coverage, similar to cell tower localization and the GPS. RTLS focuses on RSS fingerprints inside the buildings and WPS mainly focuses on fingerprinting in the streets of a metropolitan area. We can automatically tag the location with GPS and use a car to drive and collect a massive Wi-Fi fingerprint database. Since GPS does not work reliably in indoor, RTLS system collects the fingerprint database manually. Integration of WPS and GPS in the positioning process creates additional values for both of the technologies. WPS complements GPS for indoor coverage, reduces the time to fix, reduces power consumption, and offers resistance to interference. GPS complements WPS to provide outdoor coverage for Wi-Fi localization and WPS can use the universal GPS coordinates as a reference frame. In 2008, the leading wonder of emerging smart devices, the iPhone, started to use Wi-Fi technology inside a mobile device to complement cellular telephone connection. Wi-Fi, wherever it is available, provides high-speed wireless

Internet access at no cost. In addition, it allowed integration of Skyhook's emerging Wi-Fi localization technology for localization everywhere. The cell tower position was backed up for Wi-Fi localization wherever Wi-Fi is not available. WPS databases owned by Skyhook, Apple, and Google receive several billion hits per day, and this technology is the most popular localization technic for smart devices.

Figure 3.16 shows the general architecture of the WPS systems. In WPS, the fingerprinting database is collected through war driving in the streets of a metropolitan area, using GPS to tag the location and time of measurements. Figure 3.17 shows sample measurements obtained from a short driving in close vicinity of the Worcester Polytechnic Institute campus [Rob09]. Figure 3.17(a) shows the Google map of the campus with the green dots representing the location of access points in different buildings. Figure 3.17(b) illustrates the RSS readings from a single access point installed in the corner of the Atwater Kent Laboratory at the intersect of Salisbury and West Streets. Figure 3.17(c) shows the measurements taken from all surrounding APs at a rate of approximately one measurement set per second while we move in

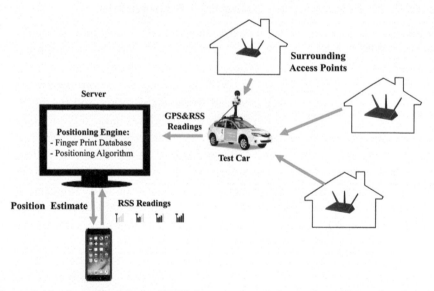

Figure 3.16   Overall architect of WPS for two phases of operation: (1) creating the RSS fingerprint database and (2) localization of the a device using the fingerprint data base.

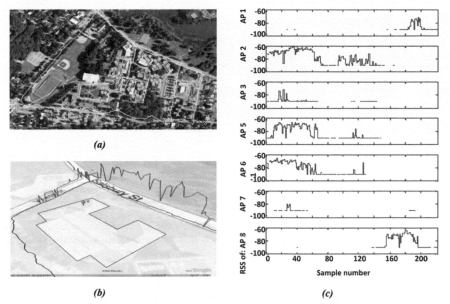

Figure 3.17 Wi-Fi fingerprint data from war driving next to the Worcester Polytechnic Institute campus, (a) Location of APs in different buildings, (b) typical measurement from one AP in Salisbury and West streets, (c) RSS of all APs from driving in Salisbury street [Rob09].

Salisbury St. Since Wi-Fi access points are commonly installed inside the buildings, the signal level read by war driving in the streets is always weak and includes locations without readable data. In Chapter 9, we will discuss how this affects design of algorithms for WPS systems. Access points are identified by their MAC addresses, and they are usually obtained by reading the beacon packets periodically transmitted to allow devices to join the Wi-Fi network.

From the example given in Figure 3.17 for a short drive in a street, one can imagine the huge size of the Wi-Fi fingerprint databases. WPS service providers use all readings with the same MAC address to estimate the location of the access points and keep the database of all these locations at a server. Figure 3.18 shows the map of access points in Skyhook's database in the Bay Area, Manhattan, and Seattle. At the time of this writing, the database of any WPS service providers has over a billion access points worldwide.

Client software in the smart devices, shown in Figure 3.16, reads the RSS of the surrounding access points and sends that with their MAC addresses

| Bay Area | Manhattan | Seattle |

Figure 3.18   The Map of Wi-Fi location database of the Skyhook in the Bay Area, Manhattan, and Seattle.
Source: Skyhook Wireless.

to the server. The server uses the service provider's proprietary localization algorithm, RSS readings of the smart device, and the location of the access points in the fingerprint database to estimate the location of the device and send it back to the client software in the device.

3.4.3 WPS versus GPS

The main body of the fingerprinting database for WPS systems is collected by war driving in streets, which relies on GPS technology to tag the location where we record RSS from the Wi-Fi access points. Therefore, a WPS position estimate is based on record of GPS position estimates in different locations. Looking from this viewpoint, a WPS system can be considered as a software system for memorizing and refining GPS locations at later times without the need for physical presence of GPS hardware. This software GPS system provides a low-cost, low-power, and fast localization technique. The most important feature of WPS is that it operates based on Wi-Fi readings, which extends the coverage of WPS to indoor areas, where GPS faces serious challenges. GPS provides a universal coverage outdoors but it does not work in a large fraction of locations in indoor and urban areas, where most location-based mobile computing applications are used. Integration of WPS software and GPS hardware provides for that comprehensive coverage, wherever a device can afford a GPS chip set.

To compare the performance of WPS with the GPS, we need to focus on outdoors areas where both systems work. In these areas, we expect WPS to perform worse than GPS because WPS estimates are driven by GPS estimates. However, the WPS industry has found ways to improve its performance over the GPS in dense urban areas. Since location of the Wi-Fi devices are fixed but the location of the satellites providing for GPS location estimates is changing, we can associate several GPS readings with different levels of accuracy to the same Wi-Fi access points by collecting the Wi-Fi fingerprints in different times when the location of the satellite is different. We can also use pattern recognition techniques to correct GPS readings with the actual driving map. These opportunities allow a WPS system the possibility of using more precise locations than the GPS to build up the fingerprinting database, in particular in dense urban areas where GPS loses its accuracy due to the multipath and shadowing of the line of sight to the satellites by high-rise buildings.

Example 3.3: (Performance of the WPS vs GPS)

Figure 3.19 shows results of performance comparison for Skyhook WPS and GPS on a test route in downtown San Francisco. Figure 3.19(a) shows the Google map of the area, and Figure 3.19(b) shows the path route taken for the test drive. Figure 3.19(c) shows the empirical performance of the WPS and GPS errors calculated on the specified route. The median error (50%) of

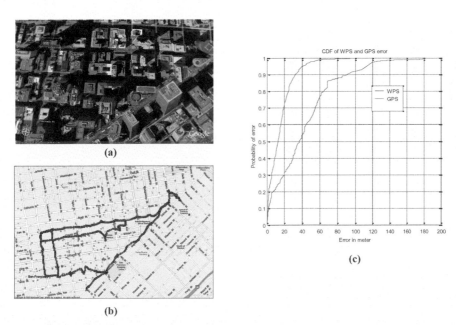

(a)

(b)

(c)

Figure 3.19   WPS performance in San Francisco's downtown area: (a) the Google map of the area, (b) driving route for measurements on the street map, and (c) performance of GPS against WPS.

this experiment for the GPS is close to 40 m, while WPS provides accuracies of slightly more than 10 m. This figure illustrates the fact that in dense urban areas, Wi-Fi localization can have a better performance than GPS in terms of the cumulative distribution of the error. This situation reverses as we go to suburban areas, where the density and coverage of Wi-Fi signals are restricted, and we can see GPS satellites directly that can now provide higher precision. Figures 3.20 provide a similar performance evaluation in a route in the Boston suburban area with more open space in which GPS performs better than WPS. Figure 3.20(c) shows that the median error for the GPS is reduced to 20 m, while WPS median error has increased to close to 30 m.

3.4.4 WPS and Organic Data

The major cost for implementation of a WPS system is the cost for war driving, which involves costs of driving route planning and management as well as operational costs for car and driver. In the early days of this industry, Skyhook Wireless began by hiring cab drivers in Boston, to carry their data acquisition devices around the town, to keep the cost low. Later

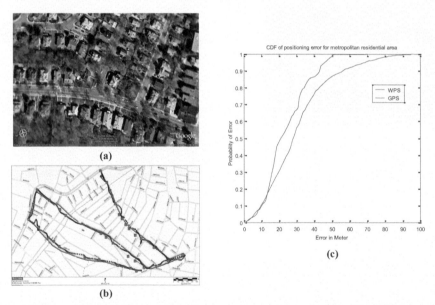

Figure 3.20   WPS performance in sub-urban residential Boston metro area: (a) The Google map of the area, (b) Path of measurements on the street map, and (c) Performance of GPS against WPS.

on, they managed their own cars. Google had its own cars driving in the streets for mapping and taking pictures of the buildings; they used them for RSS fingerprinting as well. Similar to cellular telephone industry, WPS industry began by scanning major cities and then expanding the covered to the smaller cities. Since we have continual change in pattern of access point deployments, as we install new access points or we change the location of access points, WPS service providers need to continually update their database by re-scanning to maintain the database.

An effective technique to reduce the cost of maintenance, expand the coverage, and increase the re-scanning intervals is to take advantage of user's *organic data* to update the database. The organic data is collected by the user device either at a time when the user starts a localization application or with an automated program installed in the device, collecting data periodically. Using this organic data to update the database will expand the size of the database and reduce the refreshment intervals, which results in a significant reduction of the maintenance cost and expansion of the size of the database to include new installations and re-locations.

Integration of the organic data into the systematically collected database with coordinated war driving procedures requires *data mining algorithms* to ensure that the additional organic data does not reduce the overall accuracy of localization. In a database collected for WPS, the geo-tag carries over GPS errors, the number of measurements spread over the area depends on the speed of the drivers, and the geographical coverage of the database depends on coordination plan among drivers. We need another set of algorithms for post processing of the collected database to minimize GPS geo-tagging errors and to make the spatial distribution of the database close to uniform. These algorithms are different from the actual localization algorithms used for WPS.

The localization algorithms designed for WPS have to cope with the uncertainties of the database caused by stochastic spatial and temporal characteristics of Wi-Fi APs and uneven distribution of the data associated with individual APs. These algorithms should process a huge database, for which direct use of nearest-neighbor-based algorithms may not be the optimum solution all the time. The radio propagation environment for WPS involves a variety of complex indoor to outdoor scenarios, which are more unpredictable than indoor to indoor radio propagation in the case of RTLS. These characteristics can be sometimes exploited to improve the accuracy of the algorithms.

These complexities for the design of WPS have opened a field for innovative engineering art and science for companies to engage in database collection and post processing for Wi-Fi in metropolitan areas. In the same way that many companies have search engines, but Google leads the others by having larger and better processed data, a better Wi-Fi localization in metropolitan areas is becoming an expertise for companies who may have the largest database, and the best algorithms for post-processing the data and locating a device with more accuracy and wider availability.

3.4.5 CPS Cell Tower Localization Using RSS Fingerprinting

We refer to Cell-tower Positioning Systems (CPS) as a localization system using RSS fingerprints of cell-tower to locate a device. CPS systems operate similar to the WPS, but they use the fingerprint of the cell towers instead of Wi-Fi access points. The deployment density of cell tower is smaller than the Wi-Fi access points; therefore, the accuracy of CPS is less than the WPS. Coverage of cell towers are more comprehensive than Wi-Fi; therefore, smart devices use CPS as a backup for WPS in areas where Wi-Fi signal is not available. The original iPhone was using only WPS and CPS and later version included the GPS as well. Figure 3.21 shows a comparison among accuracy of the three technologies in different areas of operation.

Figure 3.22 shows the overall architecture of a CPS system. It is exactly the same as the architecture of the WPS systems shown in Figure 3.16. The only difference is that in CPS, we use cell towers instead of Wi-Fi access points. Since cell towers are installed outdoors and we are war driving outdoors, we can potentially cover the entire area of coverage of the tower, observing a wide range of RSS values. In WPS, access points are deployed inside the buildings, and when we drive outside, we are often in the fringe of the coverage with very weak signals and a small range of RSS values.

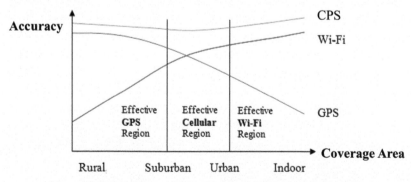

Figure 3.21   Accuracy of CPS, WPS and GPS in different areas.

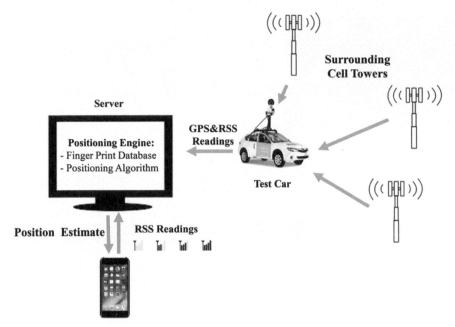

Figure 3.22 Overall architect of CPS for two phases of operation: (1) creating the RSS fingerprint database and, (2) localization of the a device using the fingerprint data base.

The other difference caused by the pattern of deployment of cell towers is that during war driving, we can visually identify the exact location of the cell towers and pin them on a map. When we drive outside, we have no visual means to see the exact location of an access point that may be inside a building. Knowing the location of the cell tower and having a wider range of variations in the RSS provide an edge for improving the accuracy of CPS systems to counteract the effects of lower density of deployment in CPS, as it is compared with the WPS. The overall accuracy of CPS is much less than WPS in densely populated areas, while in the open areas, such as on the highways, or in remote areas, such as farms, CPS provides a better solution. The following example provides some test results for performance of CPS.

Example 3.4: (Performance of the CPS)

Figure 3.23 shows the CPS performance in downtown Boston, MA [Akg09]. Figure 3.23(a) shows the Google map of the area with the test route and location of cell towers, and Figure 3.23(b) shows the colored map of the signal strengths on the route (c), performance of two algorithms, the range of variations of the RSS is over 60 dB. Figure 3.23(c) shows the performance

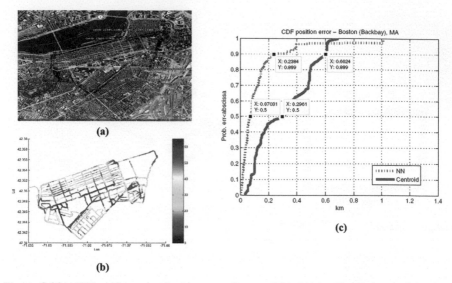

Figure 3.23 CPS performance in downtown Boston, MA (a) the Google map of the area with the test route, (b) colored map of the signal strengths on the route (c) Performance of two algorithms.

of CPS in downtown Boston using the centroid and the nearest-neighbor (NN) algorithms. The centroid algorithm uses only the location of cell towers, and the NN uses the signature database (Chapter 9 provides details of both algorithms). The median accuracy of the centroid is approximately 296 m, and using the signature database reduces it to 70 m. In the downtown areas, we have micro-cellular deployment with a separation distance of approximately 500 m, which is consistent with the estimate using centroid algorithm. Using the signature database improves the performance more than four times. The accuracy of the NN algorithm should be compared with that of WPS, which provided an accuracy close to 10 m in downtown areas (see Example 3.3). Figure 3.24 provides the CPS performance in Shrewsbury, MA, in a suburb of Worcester. Figure 3.24(a) illustrates the Google map of the area with the test route and location of the cell towers in the area, and Figure 3.24(b) shows the colored map of the signal strengths on the route. Figure 3.24(c) shows the result of performance evaluation using the centroid and NN algorithms. In this area, the centroid algorithm results in a median error of 1 km, which reflects the increase in separation of the cell towers in that area. The NN provides a median accuracy of 32 m, which is close to the performance of WPS in a suburb of the Boston (Example 3.3). The number of access points read

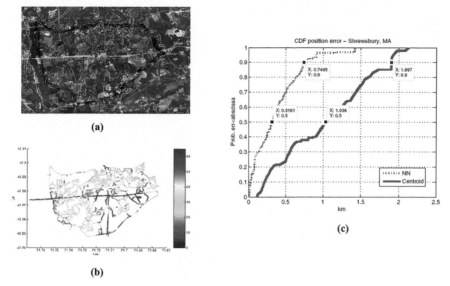

Figure 3.24 CPS performance Shrewsbury, MA area (a) the Google map of the area with the test route, (b) colored map of the signal strengths on the route (c) Performance of two algorithms.

in suburban areas is close to the number of cell towers read in these areas. In downtown areas, this balance changes significantly in favor of access points.

Assignments for Chapter Three

Questions

1. Explain the differences between GPS, wireless cellular-assisted GPS, and indoor geolocation systems.
2. Why does GPS not function adequately in indoor areas?
3. Why is the RSS not a very good measure of the distance between a transmitter and a receiver? How can distance estimates with RSS be improved?
4. What is the error relationship between ranging and position location?
5. What are the differences between RTLS and WPS technologies in terms of precision requirements, database collection technique, environments where they are used, and localization algorithms?
6. How can a WPS system (which uses GPS to tag the location) provide results that are more accurate than GPS?

7. Why is WPS referred to as software GPS?
8. In RSS-based localization, what are the advantages and disadvantages of using the exact location of APs and using RSS reading at arbitrary locations?
9. What is specific about RFID localization that is different from other localization schemes?

Problems

Problem 1:
Modify the MATLAB code in Example 3.1 to plot the contour of the 2D CRLB for RSS-based positioning in a 4 m × 5 m room with four iBeacon devices installed as RPs in the middle of each of the four walls.

Problem 2:
Assume in Problem 1 we have only one iBeacon in the center of the room, but we collect four fingerprints in the middle of each wall.

a) Apply Equation (3.3) to this problem
b) Calculate the Jacobian matrix H
c) Give equations for calculation of Fisher matrix and the CRLB

Problem 3:
Extend the 2D CRLB for RSS-based localization to 3D scenarios. Name an application where the 3D bounds help.

4

Fundamentals of TOA Positioning

4.1 Introduction

RF localization is the most popular method for localization, and it is implemented based on either the measurement of RSS or the TOA of the received signal. In Chapters 2 and 3, we described principles of RSS-based ranging and positioning systems. Chapters 4–6 are devoted to principle of TOA-based localization. Analysis of the DOA technologies is closely related to the TOA, and we include them in our discussions for TOA-based systems.

The TOA-based GPS technology is the central element of all traditional vehicles, vessels, and airplanes navigation systems. In addition, modern smart phones use GPS as an important part of their localization platform, and Unmanned Automotive Vehicles (UAV) and Drones rely heavily on GPS for their navigation. The TOA-based ZigBee and UWB sensors are expected to enter the smart world to support precise localization inside manufacturing floors, for first responders entering unknown indoor operational environments, for interactive electronic gaming, and other popular applications. The TOA-based localization is also popular in short distance ranging with low-density deployment for RFID and sensor network applications. The TOA-based technology is used for E-911 cell tower localization applications and massive multiple-input-multiple-output (MIMO) and millimeter wave (mmWave) at 60 GHz technologies for multiple streaming in 5G cellular networks, rely on precise localization using TOA technology. More precise localization inside the human body for body area networks also benefit from TOA-based localization to increase its precision.

Measurement of RSS is very simple and it is often performed by using a small software patch in a smart device. Measurement of the TOA involves waveform design and synchronization between the transmitter and the receiver, which is far more complex than measurement of the RSS. Performance of the RF localization techniques is affected by the received

91

signal-to-noise ratio (SNR) and the level of multipath in the environment. In open areas, where the multipath effects are negligible, SNR governs the precision, and the traditional estimation theory helps us to analyze the system performance. In indoor and urban areas, multipath effects dominate the performance and that has been a new field for study in the recent years. A large part of the indoor geolocation science and technology, which is the focus theme of this book, has evolved for the analysis of extensive multipath in indoor and urban areas on the performance of RF localization. Ranging using RSS is based on measurement of the average received signal strength, and averaging eliminates the effects of fast multipath fading, leaving shadow fading as the only source for distance measurement error. TOA ranging is based on measurement of the time of flight of a waveform between a transmitter and a receiver. Analysis of the effects of extensive multipath in indoor and urban areas on the measurement of TOA involves analysis of the waveform transmission in multipath and involves complex signal processing algorithms.

We begin our discussion on TOA-based systems in this chapter by introducing the principle of TOA ranging. We introduce narrowband and wideband methods for measurement of the TOA, we use CRLB to analyze the performance of TOA measurement systems using different waveforms, and we relate the performance of DOA ranging to the TOA ranging. In Chapter 5, we introduce opportunistic methods for measurement of the TOA using existing spread spectrum, orthogonal frequency division multiplex (OFDM), MIMO and mmWave wireless communication signals. In Chapter 6, we describe the effects of multipath and introduce methods for the analysis of the effects of multipath on TOA ranging.

4.2 Measurement of TOA in Practice

Measurement of the RSS of an RF signal is very simple, it does not need any synchronization between the transmitter and the receiver, and it is independent of the shape of the transmitted waveform. We can simply measure the RSS level of any received RF waveform by passing the received analog signal through a simple low-pass filter, only a resistor and capacitor, or after analog-to-digital conversion, we may average the square of the digitized samples of the received signal amplitude using a few lines of codes. Estimation of distance from RSS is through the model for the behavior of the RSS that involves observation of a logarithmic function, $g(r) = P_0 - 10\alpha \log r$ of the parameter, r, in shadow fading, $X(\sigma)$. Figure 4.1 shows the general

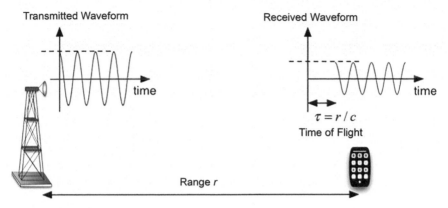

Figure 4.1 General concept of range estimation using the TOA or time of flight.

concept behind measurement of the distance using TOA or time of flight of a waveform. Measurement of the TOA involves with the shape of the transmitted waveform and needs synchronization between the timing of the transmitter and the receiver to measure the time of flight, τ. The distance in TOA ranging is related to the TOA by $r = c \times \tau$, where c is the speed of RF propagation in the medium. Since in free space the speed of RF propagation is the same as speed of light, $c = 3 \times 10^8 \mathrm{m/sec}$, to achieve 1 m precision in measurement of distance, we need to measure the time with an accuracy of $\tau = 3n\,\sec$, which requires atomic clock with precision of $10^{-9}\,\sec = 1\,n\,\sec$, for synchronization between the transmitter and the receiver. When synchronization is established between the transmitter and the receiver, we can measure the time of flight with different waveforms, and precision of this measurement is related to the shape of the waveform and the signal-to-noise ratio of the received signal.

4.2.1 NB and WB Measurement of TOA

In wireless communication and consequently wireless localization, we classify the signals in narrow band (NB) and wideband (WB) signals [Pah13]. In an ideal form, a sinusoid $x(t) = \cos \omega t$ represents a NB signal because its Fourier transform is an impulse, which has a zero bandwidth. An ideal WB signal is an impulse in time domain $x(t) = \delta(t)$, whose Fourier transform is a flat signal with infinite bandwidth. The ideal cosine continues to infinity in time and an ideal impulse continues to infinity in frequency. In practice, as shown in Figure 4.2, we have restrictions in time and frequency. A sinusoid

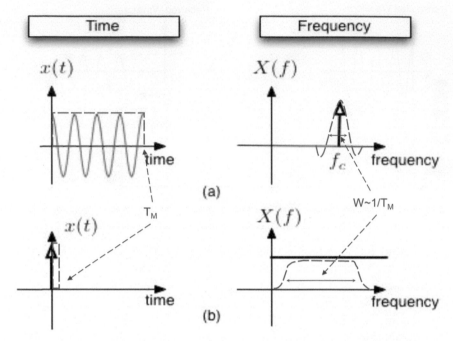

Figure 4.2 Signals used for TOA measurements, in time and in frequency domain, (a) Narrowband signal, (b) Wideband signal.

has a measurement duration of, T_M, and the spectrum of a time bounded sinusoid is a sinc pulse with a bandwidth of $W = 1/T_M$, which is not zero. The WB impulse, in practice, becomes a high-power (amplitude) very narrow rectangular pulse with the duration of T_M and a bandwidth of $W = 1/T_M$. As an example, if the duration of the carrier signal is 10 s, the bandwidth of the pulse is 0.1 Hz, which is an ultra-narrowband signal, and if the duration of the practical pulse is 1 ns, the bandwidth of the signal is 1 GHz, and we have an ultra-wideband (UWB) signal.

We can measure the TOA either with practical NB or WB signals and the method we use for these measurements are different. Figure 4.3 shows the general concept of TOA-based ranging for NB (Figure 4.3(a)) and WB (Figure 4.3(b)) signals. An RF radiating antenna transmits a waveform $x(t)$, and a device receives a delayed version of the signal $y(t) = x(t - \tau)$, the amplitude of the received signal is normalized in the figure. The distance between the transmitter and the receiver is r, which results

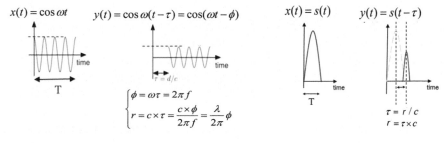

Resembles a tone: With longer T we have a better chance to estimate phase (NB)

Resembles an impulse: With short T we have sharper peak for detection (WB)

(a)

(b)

Figure 4.3 NB versus WB TOA range measurement, (a) NB measurement of the distance using phase of a sinusoid signal, (b) WB measurement of the distance using time of flight of sharp peak of the received waveform.

in a TOA of $\tau = r/c$, which we want to measure. In NB measurement of TOA (Figure 4.3(a)), the transmitted and received signals are given by:

$$\begin{cases} x(t) = \cos \omega t \\ y(t) = \cos \omega(t - \tau) = \cos(\omega t - \phi) \end{cases}, \tag{4.1a}$$

where ω is the angular frequency of the transmitted signal and ϕ is the phase of the received signal. The relation between these parameters is given by:

$$\begin{cases} \phi = \omega \tau = 2\pi f \\ r = c \times \tau = \frac{c \times \phi}{2\pi f} = \frac{\lambda}{2\pi} \phi \end{cases}, \tag{4.1b}$$

where $\lambda = c/f$ is the wavelength of the carrier frequency. If we have the reference of the transmitted signal at the receiver, we can calculate the phase by cross-correlating the transmitted and received signals:

$$\begin{aligned} R_{x,y}(\phi) &= \frac{1}{T_M} \int_0^{T_M} \cos \omega t \times \cos(\omega t - \phi) \, dt \\ &= \frac{1}{2T_M} \int_0^{T_M} [\cos(2\omega t - \phi) + \cos \phi] \, dt = \frac{\cos \phi}{2}. \end{aligned} \tag{4.1c}$$

Here, we assume that the measurement time T_M is an integer multiple of a cycle time of the sinusoid. This assumption forces the first term in the (4.1c) integral to become zero and the second term gives the result. Then, we can calculate the phase from the following equation:

$$\phi = \cos^{-1}(2R_{x,y}). \tag{4.1d}$$

In practice, if we have a duplicate of the transmitted signal at the receiver, calculation of cross-correlation involves a multiplication and an integration (low-pass filtering). This way after calculation of the cross-correlation, we can calculate the phase and from that the delay and consequently the range or distance. Duplication of the transmitted signal at the receiver is either implemented directly, when we have synchronization between the transmitter and the receiver, or by using a transponder at one of the devices. The transponder redirects the received signal back to the transmitter enabling the transmitter access to both transmitted and received signals for cross-correlation. The measured distance in this case involves the two-way propagation and processing time at the receiver.

In WB TOA-based ranging, shown in Figure 4.3(b), the transmitted and received signals are given by:

$$\begin{cases} x(t) = s(t) \\ y(t) = s(t - \tau) \end{cases},$$ (4.2)

where $s(t)$ is a narrow pulse resembling an impulse. At the receiver, we measure the time difference between the peak of the received signal, $y(t) = s(t-\tau)$, and the transmitted signal, τ. The distance is then given by $r = c \times \tau$. This requires synchronization between the timing of the transmitter and the receiver that can be attained either by sharing a universal atomic clock in both devices or by using a transponder. Using atomic clocks is popular in long-distance TOA-based ranging applications such as GPS, and transponders method is used for TOA ranging of sensor networks using ZigBee or UWB technologies. A third method for synchronization, used in Bluetooth technology, is the use of sliding correlator that we discuss in Section 5.5.1.

4.2.2 Measurement Time, Measurement Noise, and SNR

Wireless communication and localization systems are always integrated. GPS is built for positioning, and the signal is also used for wireless communications at low data rates as well. Wi-Fi or cellular networks are deployed for wireless communications, and we use their signal for positioning. For wireless access and localization using radio frequencies, a transmitted waveform with certain bandwidth is modulated over a carrier, a sinusoid signal. For TOA-based ranging, we can use the phase of the carrier signal for NB ranging and delay of arrival of the waveform for WB measurement of the distance between the transmitter and the receiver.

In wireless communications application, we design the waveform to carry a digital communication symbols in a fixed transmission period, T_s, to support a symbol transmission rate of, $R = 1/T_s \, Sps$. To use the same waveform for RF localization, we need to extract features of the symbols pertinent to localization (RSS, TOA, or DOA) as a measure of the relative location of the transmitter and the receiver with respect to reference transmitter antennas. A communication receiver should process the received waveform symbol by symbol to detect the transmitted digitized information. The detection process should be completed in a time less than the period of the signal, so that we can detect the stream of transmitted information in real time. For example, if we transmit at *1Msps*, the processing time for each symbol is 1μsec. Localization process is indeed much slower, and we can wait for seconds or, if needed, for minutes, to obtain a location fix with an acceptable precision. When we start our cars, we do not mind if the first GPS location fix arrives in a few second or even in minutes. This allows the localization receiver of a TOA system, for example GPS, to average the time of flight over a large number of symbols, one million for 1 Mbps. As a result in digital communication literature, we have a single time duration parameter, T_s, which refers to the time that we process a transmitted symbol. In TOA ranging, we have three time duration parameters, the width of the pulse used for measurement of the time of flight, T, period of symbol transmission, T_s, and the TOA measurement time, T_M.

Figure 4.4 shows a periodic triangular pulse and its three associated time duration parameters used for TOA ranging. If we transmit a sequence of these narrow and sharp signals for localization, the transmitted and the received symbols are given by:

$$\begin{cases} x(t) = \sum_{i=0}^{N} s(t - iT_s) \\ y(t) = \sum_{i=0}^{N} s(t - iT_s - \tau) \end{cases} \tag{4.3}$$

To estimate the TOA, we may use as many symbols as we desire. Pulses are transmitted each T_s second, and we use N symbols for a position fix, our measurement time is $T_M = N \times T_s$. If we increase N, we have a better estimate of the TOA because we have multiple observations of TOA in noise. This measurement time for WB signaling is the same as the measurement time in NB signaling, T_M, in Figure 4.1(a). For the NB measurement, we are also transmitting a periodic sinusoid signal and at the receiver, we use many cycles of this sinusoid to have a better estimate of the phase. The relations between the bandwidth and measure time in NB and WB measurements are

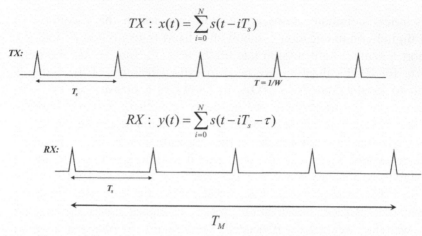

Figure 4.4 Timing parameters involved in TOA measurement of the distance between a transmitter and a receiver.

different. In narrowband measurements, $W \sim 1/T_M$, that is, a very small value, and in WB measurement, $W = 1/T$, where $0 < T \leq T_s$, which has a very large value. In addition, in TOA ranging, we have an independent measurement time that allows us to average over long periods to achieve better precision in estimating the TOA. To further clarify this discussion, it is beneficial to discuss the concept of noise and the received signal-to-noise ratio (SNR) in pulse transmission for TOA ranging.

Figure 4.5 shows the basic relationship among pulse transmission, SNR, and background noise. Figure 4.5(a) shows the common formulation for waveform transmission in additive Gaussian noise:

$$\begin{cases} x(t) = s(t) \\ y(t) = s(t - \tau) + \eta(t) \end{cases}, \tag{4.4a}$$

where $\eta(t)$ is the additive Gaussian thermal noise. The formulation of this problem in terminology of the classical estimation theory is that we have a continuous time observation of function of a parameter in additive thermal noise. In RSS-based localization, we had observation of a discrete value of function of a parameter in shadow fading noise. Standard deviation of shadow fading was something between 5 and 10 dB. The variance of additive thermal noise is something between -80 and -160 dBm for different TOA localization systems. Therefore, we expect TOA to provide extremely high precision ranging estimates.

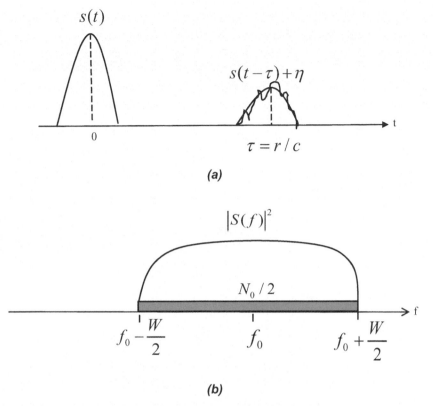

Figure 4.5 Background noise and waveform transmission, (a) transmitted pulse and the received pulse in Gaussian noise, (b) power spectral density of the received signal and the Gaussian noise.

Figure 4.5(b) shows the power spectral density of the received signal as well as the flat spectrum of background thermal noise. The normalized received energy per symbol for a system with bandwidth of W is given by:

$$E_s = \int_0^{T_s} |s(t)|^2 \, dt = \int_0^{+\infty} |S(f)|^2 \, df, \qquad (4.4b)$$

where $S(f) = FT\,[s(t)]$ is the one-sided Fourier transform of the transmitted pulse shape. The received power used for measurement of the TOA is

$$P_s = \frac{E_s}{T_M} = N \times \frac{E_s}{T_s}. \qquad (4.4c)$$

The background thermal noise is a white Gaussian noise with flat spectrum and the two sided spectral height of $N_0/2 = 2KT_K$, where T_K is the temperature in Kelvin (290 Kelvin at room temperature) and $K = 1.381 \times 10^{-23} W/Hz/$Kelvin degrees is the Boltzmann constant. Therefore, a system with a bandwidth of W suffers from a thermal noise with a variance (power) of:

$$\sigma^2 = N_0 W = 4KTW. \tag{4.5a}$$

Calculated in dBm, we can derive the following simple equation for the background noise:

$$10 \log N_0 W = 30 + 10 \log(4KTW) = -168 + 10 \log(W). \tag{4.5b}$$

The above equation for calculation of the thermal noise in dBm is very useful for practical application to calculate the background noise level of different TOA-based localization systems.

Example 4.1: (Noise in Wi-Fi, ZigBee, and GPS)

Direct Sequence Spread Spectrum (DSSS) signals are the most popular signals used for TOA measurement. Systems with different bandwidth using this technology will have different levels of background thermal noise. The legacy IEEE 802.11 standard was supporting DSSS option with the bandwidth of $W = 26MHz$. Using Equation (4.5b), we can calculate the background noise level in this bandwidth to be

$$-168 + 10 \log(W) = -168 + 10 \log(26 \times 10^6) = -94dBm.$$

The IEEE 802.15.4, the ZigBee, uses the DSSS with a bandwidth of 5MHz, and the background noise for this system is -98 dBm. GPS also uses DSSS with a bandwidth of 1 MHz, and it has a background noise of -108 dBm. Obviously, background noise reduces with bandwidth.

The actual signal-to-noise ratio is defined as the ratio of the received signal power to the background noise, given by:

$$SNR = \frac{P_s}{\sigma^2}. \tag{4.6a}$$

For the performance evaluation of wireless access and localization system, we often define signal-to-noise ratio per symbol as:

$$\rho^2 = \frac{2E_s}{N_0} = \frac{2P_s \times T_M}{\sigma^2/W} = 2 \times SNR \times W \times T_M. \qquad (4.6b)$$

We will use this definition of the normalized signal-to-noise ratio in the performance evaluation of the TOA-based ranging in Section 4.4. We can use Equation (4.6b) to calculate the measurement time for a localization system.

Example 4.2: (Measurement time for the GPS)

The received signal power from a GPS satellite on the earth is -156 dBm, the bandwidth of the system is 1 MHz, and the acceptable normalized SNR for operation of the system is $\rho^2 = 4\,(6dB)$. From Example 4.1, the noise level for 1 MHz bandwidth is -108 dBm; therefore, the 10logSNR = $-156 + 108 = -48$ dB. Substituting these values in Equation (4.6b), we have

$$4 = 2 \times 10^{-4.8} \times 10^6 \times T_M \quad \Rightarrow \quad T_M = 2 \times 10^{-1.2} = 0.126\,\text{sec} = 126ms$$

The minimum measurement time of 126ms is needed for a location fix. With a symbol transmission rate of 1 Mbps, we use 126,000 transmitted symbols for a GPS position fix.

4.2.3 Ambiguity in NB and WB TOA Measurements

To measure the TOA, we use periodic signals, periodic pulses, or sinusoids. Any periodic signal used for measurement of the time has an ambiguity interval, which results in ambiguity in measurement of the distance. For WB measurements of the TOA, we send periodic narrow pulses and we measure the delay of arrival or time of flight by measuring the time difference between the peak of the received and transmitted pulses (Figure 4.4). If the TOA is larger than the period of the signal, $\tau > T_s$, receiver measures the delay between the peak of the received signal and a wrong pulse, and the measured range becomes ambiguous:

$$\hat{r} = c \times (\tau - nT_s) = r - c \times nT_s = r - nr_s; \quad 0 < \hat{r} < r_s, \qquad (4.7)$$

where n is any integer and $r_s = c \times T_s$ is the maximum distance, which can be measured without ambiguity. In other words, measurement of the TOA using periodic WB pulses has a distance ambiguity of $r_s = c \times T_s$. Designers of the TOA ranging systems select a minimum value of T_s that can measure the maximum distance for an application without ambiguity.

Example 4.3: (Ambiguity in WB measurements)

For a WB TOA measurement system that sends pulses every millisecond, from Equation (4.7), the distance ambiguity is $r_s = 3 \times 10^8 \times 10^{-3} = 3 \times 10^5 m = 300$ km.

Similar to WB measurement of TOA, the sinusoid used for NB measurement of phase is also a periodic signal and causes ambiguity in measurement. In NB systems, we measure the distance from the measurement of phase of the received signal using Equation (4.1b), if the actual distance is more than λ, the associated phase becomes more than 2π, causing ambiguity in measurement. The equation for finding the measured range using the phase is:

$$\hat{r} = \frac{\lambda}{2\pi}(\phi - 2\pi n) = \frac{\lambda}{2\pi}\phi - n\lambda; \quad 0 < \hat{r} < \lambda. \tag{4.8}$$

The range for non-ambiguous measurement of distance is $r_s = \lambda$. Therefore, in TOA-based ranging using an NB sinusoid, we can only measure the distances without ambiguity if the distance is less than the magnitude of the wavelength. Larger distance alias to a distance smaller than its original value.

Since the frequency of operation regulated by government agencies such as FCC, the NB TOA ranging using the carrier signal remains restricted to short distances.

Example 4.4: (Distance ambiguity for NB measurement)

The carrier signal for wireless communications inside the human body using WVCE is 450 MHz, and the associated wavelength becomes 67 cm. If we use the carrier for TOA measurement, we can measure up to 67 cm without ambiguity. This distance is more than the distance of the WVCE, inside the GI tract, from the body mounted sensors. Therefore, TOA ranging using the carrier frequency of the transmitted signal is practical. For Wi-Fi operating at 2.4 GHz, the wavelength is 12.5 cm, which makes it impractical to use the phase of carrier for indoor geolocation. The GPS for commercial applications uses a carrier frequency of 1.5 GHz with a wavelength of 20 cm, which is not practical. In the latter two cases, one can design a system so that the WB envelop of the signal provides a rough estimate of the distance with no ambiguity. Then, use the NB estimate of the distance to refine the WB estimate of the distance.

4.2.4 Using Two Sinusoids to Control NB Ambiguity

TOA localization using a sinusoid is simple, but it has a short range for distance measurement without ambiguity. A practical approach for using simple sinusoids to estimate the TOA, while controlling the ambiguity, is to design a system with two sinusoids. Assume instead of one sinusoid, we transmit two sinusoids at frequencies f_1 and f_2 and further assume that:

$$f_1 - f_2 < \frac{c}{r}, \tag{4.9a}$$

where r is the desired distance that we want to measure. Then, the two measured phases associated with the two frequencies are:

$$\begin{cases} \phi_1 = \frac{2\pi}{\lambda_1} r - 2\pi n - \frac{2\pi f_1}{c} r - 2\pi n \\ \phi_2 = \frac{2\pi}{\lambda_2} r - 2\pi n = \frac{2\pi f_2}{c} r - 2\pi n \end{cases}. \tag{4.9b}$$

If we subtract the two phases, the ambiguity will vanish and we have:

$$\phi_1 - \phi_2 = \frac{2\pi}{c} (f_1 - f_2) r \tag{4.9c}$$

and the distance is estimated from:

$$\hat{r} = \frac{c}{2\pi} \cdot \frac{\phi_1 - \phi_2}{f_1 - f_2} = \frac{c}{2\pi} \cdot \frac{\Delta\phi}{\Delta f}. \tag{4.9d}$$

Example 4.5: (Distance ambiguity for NB measurement)

The bandwidth of the IEEE 802.11g using OFDM technology is 20 MHz. OFDM is really a multi-carrier system and 802.11g uses 64 carriers for communications each with an approximated bandwidth of $20\,\text{MHz}/64 = 312.5$ KHz. If we use the first and the last carrier phase for TOA measurement, the difference of frequency between these two carriers is approximately $\Delta f = 20$ MHz, and using (4.9a), we can measure up to $3 \times 10^8/20 \times 10^6 = 15$ m. If we use two adjacent carriers $\Delta f = 312.5$ KHz, we can measure up to $3 \times 10^8/312.5 \times 10^3 = 960$ m.

Figure 4.6 shows the power spectrum of the two-tone signal. This system is using tones but it is occupying a bandwidth of $W = f_1 - f_2$; therefore, it is really a WB system, similar to Figure 4.5(b). The one-side power spectral density of the received signal for this WB system is:

$$|S(f)|^2 = \frac{P_s}{4} [\delta (f - f_1) + \delta (f - f_2)] \tag{4.10}$$

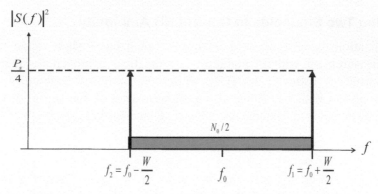

Figure 4.6   Power spectral density of an ideal two tone signal.

with a power of P_s, a noise power of $\sigma^2 = N_0 W$, and a SNR of $SNR = P_s/N_0 W$. The importance of the system is that it is simply implemented by two tones. Then, regardless of the FCC regulated carrier frequency of a transmitted waveform, we can design a waveform tailored to application by controlling the ambiguity using two modulated tones over the carrier frequency.

4.3 CRLB for TOA Ranging

In Section 4.2, we presented NB and WB methods for measurement of the time of flight for TOA ranging and we showed how we can calculate the received SNR. SNR is an important factor for performance evaluation of wireless access and localization systems; however, the real performance evaluation criterion for ranging is the variance of the DME that is determined by the CRLB. In this section, we derive the CRLB for TOA ranging and relate that to the NB and WB signals and the SNR of the received signal. This exercise enables us to intuitively capture the relation among the bandwidth, power, and precision of TOA-based localization systems.

Formulation of the problem for derivation of the CRLB for TOA ranging is very different from the formulation used for derivation of CRLB for RSS ranging. Figure 4.7 shows the differences among the analytical modeling of the RSS- and TOA-based ranging in classical estimation theory point of view to explain why calculations of the CRLB in the two systems are completely different. In calculation of the CRLB for RSS-based ranging (Figure 4.7(a)), our observation is a discrete sample of the average RSS observed in shadow

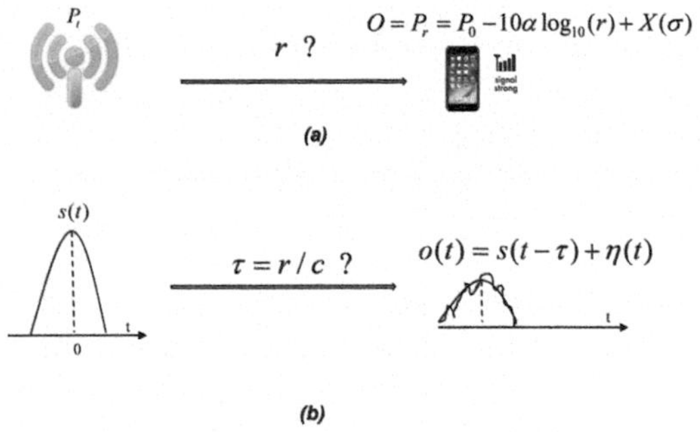

$$O = P_r = P_0 - 10\alpha \log_{10}(r) + X(\sigma)$$

(a)

$$\tau = r/c \ ?$$

$$o(t) = s(t - \tau) + \eta(t)$$

(b)

Figure 4.7 Analytical differences among RSS and TOA estimation approachs, (a) in RSS ranging, observation is a discrete sample of average power in shadow fading, (b) in TOA ranging, observation is a continuous time waveform in continuous thermal noise.

fading modeled as a zero mean white Gaussian noise. Therefore, we model the estimation process as observation of a function of a parameter in shadow fading noise:

$$O = P_r = P_0 - 10\alpha \log_{10}(r) + X(\sigma).$$

The observation used for estimation of the TOA (Figure 4.7(b)) is a continuous function of time distorted by the continuous time thermal noise:

$$o(t) = s(t - \tau) + \eta(t), \tag{4.11a}$$

which is a zero mean Gaussian process with the variance

$$E\left\{|\eta(t)|^2\right\} = \sigma^2 = \frac{N_0}{2}. \tag{4.11b}$$

This observation is a function of time and delay and includes infinite number of samples of a waveform observed in different instances of time in additive noise. As we show in Appendix Section A.4, the continuous time likelihood and log-likelihood functions for such a process are given by:

$$\begin{cases} f(o|\tau) \propto \exp\left\{\frac{1}{N_0} \int\limits_{T_0} [o(t) - s(t - \tau)]^2 dt\right\} \\ \Lambda(\tau) \propto \frac{1}{N_0} \int\limits_{T_0} [o(t) - s(t - \tau)]^2 dt \end{cases}. \tag{4.12a}$$

Then, the FIM for the continuous time observation of function of a parameter becomes:

$$\mathbf{F} = -E\left\{\frac{d^2}{d\tau^2}[\Lambda(\tau)]\right\} = -\frac{1}{N_0}\frac{d^2}{d\tau^2}\int_{T_0}\{E[o^2(t)]dt - 2E[o(t)]s(t-\tau) + s^2(t-\tau)\}dt$$

The first and third terms in the bracket are constant values because the first one is the second moment of the observation and the third term after the integral is the energy of the signal, when we apply the derivative operator only the middle term survives. Since from Equation (4.11a) we have $E[o(t)] = s(t - \tau)$, the FIM becomes:

$$\mathbf{F} = \frac{2}{N_0}\int_{T_0}\frac{d^2}{d\tau^2}s^2(t - \tau)dt. \tag{4.12b}$$

Using Parseval's theorem, we can calculate this integral in the frequency domain. Since derivative in time is equivalent to multiplication with $j\omega$ in frequency domain, we have:

$$\mathbf{F} = \frac{2}{N_0}\int_{T_0}\frac{d^2}{d\tau^2}s^2(t - \tau)dt = \frac{1}{\pi N_0}\int_{-\infty}^{+\infty}|j\omega|^2|S(\omega)|^2\,d\omega, \tag{4.13a}$$

where $S(\omega)$ is the Fourier transform of $s(t)$. The CRLB is the variance of the TOA estimation, σ_τ^2, and it is the inverse of the FIM:

$$\sigma_\tau^2 = CRLB \geq \mathbf{F}^{-1} = \frac{\pi N_0}{\int_{-\infty}^{+\infty}\omega^2|S(\omega)|^2\,d\omega}. \tag{4.13b}$$

Using Equation (4.6b), we define the normalized SNR as:

$$\rho^2 = \frac{2E_s}{N_0}.$$

Since energy per symbol is defined as:

$$E_s = \int_{-\infty}^{+\infty}s^2(t)dt = \frac{1}{2\pi}\int_{-\infty}^{+\infty}|S(\omega)|^2\,d\omega, \tag{4.14a}$$

if we define the normalized bandwidth as:

$$\beta^2 = \frac{\int\limits_{-\infty}^{+\infty} \omega^2 |S(\omega)|^2 \, d\omega}{\int\limits_{-\infty}^{+\infty} |S(\omega)|^2 \, d\omega},$$ (4.14b)

the CRLB in Equation (4.13b) turns to a simple and intuitive formulation:

$$\begin{cases} CRLB = \sigma_\tau^2 \geq \frac{1}{\rho^2 \beta^2} \\ \sigma_\tau \geq \frac{1}{\rho\beta} \end{cases}.$$ (4.15a)

This strong result indicates that the standard deviation of the estimation of TOA is inversely proportional to the normalized bandwidth and normalized SNR. The DME statistics is obtained by translating these TOA errors to the distance by multiplying it with the speed of light. Therefore, the standard deviation of the DME using TOA ranging is:

$$\sigma_r \geq \frac{c}{\rho \times \beta}.$$ (4.15b)

Since the normalized bandwidth and normalized signal-to-noise ratio are functions of the shape of the waveform, the CRLB and variance of the DME for TOA ranging are different for different waveforms. We can explain that by calculating the CRLB for a few popular waveforms that we have presented for practical measurement of the TOA in Section 4.2.

4.3.1 Comparison of TOA- and RSS-Based Ranging

Comparing the performance of TOA ranging versus RSS ranging, we use Equation (4.15b) for standard deviation of the TOA ranging obtained from CRLB, and Equation (2.13c) from Chapter 2 for standard deviation of RSS ranging using CRLB:

$$\sigma_r \geq \frac{\ln 10}{10} \frac{\sigma}{\alpha} r.$$

In RSS-based ranging, the standard deviation of the DME is directly proportional to the standard deviation of shadow fading, σ, and the distance between the transmitter and the receiver, r, and it is inversely proportional to the distance power gradient, α. All of these parameters are out of control of the designer. Shadow fading and distance power gradient are characteristics of the environment and distance is mandated by the application. The RSS is an unreliable metric for estimating the distance and it provides uncontrollable errors on the order of the actual distance, which we want to measure. The only reason that RSS-based localization systems have become so popular is that the measurement of RSS is very simple and can be implemented in any programmable device with a small software patch. As a result, we can use any available RF infrastructure opportunistically to come up with an inexpensive positioning system. Popularity of RSS-based localization benefits from popularity of Wi-Fi and, more recently, widespread IoT devices for the implementation of the smart world.

The accuracy of TOA-based ranging, Equation (4.15b), is inversely proportional to the bandwidth and SNR and a designer can control the accuracy to fit any application by adjusting the power level and the bandwidth of the system. Since variance of estimate does not relate directly to the distance, we can achieve desirable positioning accuracies in application with long distances between the reference points and the target object such as GPS or cell tower localization in open and sub-urban areas.

Example 4.6: (Accuracy of GPS ranging)

In GPS, satellites are approximately at 20,000 km from the earth. With a bandwidth of 1 MHz and 10 dB received signal-to-noise ratio, they can achieve an accuracy of:

$$\sigma_r = c \times \sigma_\tau \geq \frac{c}{\rho \times \beta} = \frac{3 \times 10^8}{10 \times 10^6} = 30\,m.$$

Using TOA-based ranging, GPS provides accuracies within a few tens of meters on the earth using satellites that are tens of thousands of kilometers away. The main challenge for TOA systems is the need for synchronization and a fast clock for digital implementation. For example, GPS maintains atomic clock with an accuracy of 10^{-9} s (1 ns). Each ns error in TOA measurement causes 30 cm DME ($d = c\,\tau$). Synchronization needs complex hardware and it becomes economically feasible when we integrate it with the wireless communication hardware.

UWB positioning systems also use TOA-based ranging to achieve high-precision localization in short-range applications in open areas (in-room) such as manufacturing floors.

Example 4.7: (Accuracy of UWB ranging)

The range of coverage of UWB indoor systems operating at 3.4–10.6 GHz is approximately 15 m. With a bandwidth of 1 GHz and 10 dB received SNR, they can achieve an accuracy of:

$$\sigma_r = c \times \sigma_\tau \geq \frac{c}{\rho \times \beta} = \frac{3 \times 10^8}{10 \times 10^9} = 0.03\,m\,(3\,cm).$$

However, as we discuss in Chapter 6, TOA-based localization is very sensitive to extensive multipath in indoor and dense urban areas, where most of the smart world applications are emerging. RSS-based Wi-Fi localization emerged as an alternative for TOA-based localization in these areas.

Example 4.8: (Accuracy of Wi-Fi indoor ranging)

The average distance between Wi-Fi access points deployed in the offices is approximately 30 m. In a typical indoor area with that size, we have an OLOS environment with a distance power gradient of 3.5 and the standard deviation of the shadow fading is 8–10 dB (according to IEEE 802.11 model). With these parameters, we can achieve an accuracy of:

$$\sigma_r \geq \frac{\ln 10}{10} \frac{\sigma}{\alpha} r = \frac{\ln 10}{10} \frac{8}{3.5} 30 = 16m$$

This accuracy is comparable to that of the GPS and justified using Wi-Fi localization. In reality, GPS signal cannot penetrate deep in indoor areas and its performance in dense urban areas degrades because of multipath and blockage by tree leaf. As a result, Wi-Fi localization has emerged as the most popular localization technique for commercial indoor geolocation in RTLS and WPS systems for smart devices.

Example 4.9: (Accuracy of iBeacon for in-room ranging)

The average distance between IoT sensors inside a room is approximately 5 m. In a typical in-room, we have an LOS environment with a distance power gradient of 2, and the standard deviation of the shadow fading is 5 dB (according to IEEE 802.11 model). With these parameters, we can achieve an accuracy of:

$$\sigma_r \geq \frac{\ln 10}{10} \frac{\sigma}{\alpha} r = \frac{\ln 10}{10} \frac{5}{2} 5 = 2.9m$$

This is a reasonable value for indoor ranging using IoT devices such as iBeacon. As a result, iBeacon product developers often offer in-room localization software as well [V-Est15]. For in-room applications, we have plenty

of LOS conditions with strong direct path that invites precise TOA positioning using UWB signals for applications such as electronic gaming [Zhe16]
to support centimeter precision positioning. TOA based UWB positioning
with centimeter range precision is also useful for shelf-by-shelf positioning
in department stores or in warehouses to direct people to specific shelves
carrying a specific item for sale.

After comparing RSS and TOA based localization using CRLB, we return
to derivation of CRLB for TOA-based ranging for different waveforms,
because unlike RSS based positioning TOA based positioning precision is
dependent on the waveform.

4.3.2 CRLB for TOA Ranging Using the Carrier Signal

If we have a sinusoid with the received power of P_s, then the energy of the
symbol for a measurement time of T_M (Figure 4.2(a)) is $E_s = P_s T_M$. Then,
from Equation (4.6b), the normalized received SNR is:

$$\rho^2 = \frac{2E_s}{N_0} = \frac{2P_s T_M}{N_0}.$$ (4.16a)

The power spectrum of the cosine with power of P_s is:

$$|S(f)|^2 = \frac{P_s}{2}\left[\delta(f - f_0) + \delta(f + f_0)\right].$$ (4.16b)

Using Equation (4.14b), we can calculate the normalized bandwidth of this
power spectrum:

$$\beta^2 = \frac{\int\limits_{-\infty}^{+\infty} (2\pi f)^2 \frac{P_s}{2}\left[\delta(f - f_0) + \delta(f + f_0)\right] df}{\int\limits_{-\infty}^{+\infty} \frac{P_s}{2}\left[\delta(f - f_0) + \delta(f + f_0)\right] df} = 4\pi^2 f_0^2$$

Then, using Equation (4.15a), we can calculate the CRLB and the standard
deviation of TOA estimation:

$$\begin{cases} CRLB \geq \frac{1}{4\pi^2 f_0^2} \times \frac{N_0}{2P_s T_M} \\ \sigma_\tau \geq \frac{1}{4\pi f_0} \sqrt{\frac{N_0}{2P_s T_M}} \end{cases}.$$ (4.16c)

The standard deviation of the TOA measurement is inversely proportional to
the frequency of operation, square root of received power, and square root of

measurement time. The standard deviation of the DME for TOA ranging with a single tone is then given by:

$$\sigma_r \geq \frac{c}{4\pi f_0} \sqrt{\frac{N_0}{2P_sT_M}}. \tag{4.16d}$$

The above calculation is for ideal condition when we can filter the sinusoid and its noise with a zero bandwidth. In practice, localization system has a bandwidth of, W, and from the above equation, it becomes:

$$\sigma_r \geq \frac{c}{4\pi f_0} \sqrt{\frac{N_0W}{2P_sT_M}} \tag{4.16e}$$

Example 4.10: (DME of the GPS using its carrier frequency)

A GPS receiver has the received signal strength of -138 dBm from a satellite and it uses the center frequency of 1.5 GHz with a bandwidth of 1 MHz. Using Equation (4.5b), the background noise is $10 \log N_0W = -168 + 10 \log(10^6) = -108\,dBm$. For hundred millisecond position fix time, $T_M = 100ms = 10^{-1}s$, using Equation (4.16e), the standard deviation of the DME is:

$$\sigma_r \geq \frac{c}{4\pi f_0} \sqrt{\frac{N_0}{2P_sT_0}} = \frac{3 \times 10^8}{4\pi \times 1.5 \times 10^9} \sqrt{\frac{10^{-13.8} \times 10^6}{2 \times 10^{-15.8} \times 10^{-1}}} \approx 40\,cm.$$

Note that with the frequency of 1.5 GHz, the distance measurement ambiguity is 20 cm, and the 40 cm accuracy holds if we have another mean to estimate the distance with 20 cm accuracy to avoid the ambiguity.

4.3.3 CRLB for TOA Ranging Using Two Tones

Figure 4.6 shows the power spectrum of the received signal and noise for a two tone system used for TOA ranging. For the received power of P_s, the normalized SNR is given by Equation (4.6b).

$$\rho^2 = \frac{2E_s}{N_0} = \frac{2P_sT_M}{\sigma^2/W} = 2 \times SNR \times W \times T_M \tag{4.17a}$$

The power spectral density, shown in Figure 4.6, is expressed by:

$$|S(f)|^2 = \frac{P_s}{4} [\delta(f - f_1) + \delta(f - f_2)]$$
$$= \frac{P_s}{4} [\delta(f - f_0 + W/2) + \delta(f - f_0 - W/2)].$$

Using this spectral density, we can calculate the normalized bandwidth using Equation (4.14b):

$$\beta^2 = \frac{\int_{f_0-\frac{W}{2}}^{f_0+\frac{W}{2}} (2\pi f)^2 \frac{P_s}{4}[\delta(f-f_0+W/2)+\delta(f-f_0-W/2)]2\pi df}{\int_{f_0-\frac{W}{2}}^{f_0+\frac{W}{2}} \frac{P_s}{4}[\delta(f-f_0+W/2)+\delta(f-f_0-W/2)]2\pi df} \qquad (4.17b)$$

$$= 2\pi^2 \left[(f_0 - W/2)^2 + (f_0 + W/2)^2\right] = 4\pi^2 \left(f_0^2 + \frac{W^2}{4}\right)$$

Substituting Equations (4.17a) and (4.17b) into Equation (4.15a) for calculation of CRLB of a continuous time signal, we have:

$$CRLB \geq \frac{1}{8\pi^2 \times SNR \times WT_0 \times \left(f_0^2 + \frac{W^2}{4}\right)}. \qquad (4.18a)$$

The bound for the DME for TOA ranging is found from:

$$\sigma_r = c \times \sigma_\tau = c \times \sqrt{CRLB} \geq \frac{c}{2\pi\sqrt{2 \times SNR \times WT_M \times \left(f_0^2 + \frac{W^2}{4}\right)}}. \qquad (4.18b)$$

4.3.4 CRLB for TOA Ranging Using Multi-Carrier Transmission

Figure 4.8 shows the power spectral density of the received signal and noise in a multi-carrier transmission system. The received power, P_s, is equally distributed among N carrier frequencies separated from each other by Δf in frequency domain. The normalized signal-to-noise ratio is the same as Equation (4.17a). As shown in Figure 4.8, the power spectral density given by:

$$|S(f)|^2 = \frac{P_s}{2N} \sum_{i=-\frac{N-1}{2}}^{\frac{N+1}{2}} \delta\left(f - f_0 - i\Delta f\right); \quad \Delta f = \frac{W}{N-1}. \qquad (4.19a)$$

Using this spectral density, we can calculate the normalized bandwidth using Equation (4.14b):

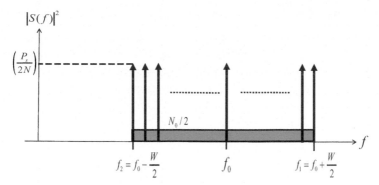

Figure 4.8 Power spectrum of a Multi-Carrier transmission system. The transmitted power is divided equaly among N carriers separated in frequency by Δf.

$$
\beta^2 = \frac{\int_{f_0-\frac{W}{2}}^{f_0+\frac{W}{2}} (2\pi f)^2 \frac{P_s}{2N} \sum_{i=-\frac{N+1}{2}}^{\frac{N+1}{2}} \delta(f-f_0-i\Delta f)\, 2\pi df}{\int_{f_0-\frac{W}{2}}^{f_0+\frac{W}{2}} \frac{P_s}{2N} \sum_{i=-\frac{N+1}{2}}^{\frac{N+1}{2}} \delta(f-f_0-i\Delta f)\, 2\pi df}
$$

$$
= \frac{4\pi^2}{N} \sum_{i=-\frac{N-1}{2}}^{\frac{N+1}{2}} (f_0 + i\Delta f)^2
$$

$$
= 4\pi^2 \left(N f_0^2 + \Delta f^2 \sum_{i=-\frac{N+1}{2}}^{\frac{N+1}{2}} i^2 \right)
$$

$$
= 4\pi^2 \left(f_0^2 + \frac{N(N+1)}{12(N-1)} W^2 \right)
$$

(4.19b)

Substituting Equations (4.17a) and (4.19b) into Equation (4.15a) for calculation of CRLB of a continuous time signal, we have:

$$
\sigma_\tau^2 = CRLB \geq \frac{1}{8\pi^2 \times SNR \times W \times T_M \times \left[f_0^2 + \frac{N(N+1)}{12(N-1)} W^2 \right]}
$$

(4.20a)

and the bound on estimation of the distance using measurement of TOA or time of flight is:

$$
\sigma_r = c \times \sigma_\tau \geq \frac{c}{2\sqrt{2\pi \times SNR \times W \times T_M \times \left[f_0^2 + \frac{N(N+1)}{12(N-1)} W^2 \right]}}.
$$

(4.20b)

The bound for the variance of the DME reduces as we increase the number of elements and it reaches zero as the number of elements approaches infinity.

4.3.5 CRLB for TOA Ranging Using a Pulse with Flat Spectrum

Figure 4.9 shows the flat power spectrum associated with periotic transmission of sinc pulses. The received power, P_s, is uniformly distributed over W, the bandwidth of the system. The normalized SNR is the same as Equation (4.17a), which we used in last derivations of the CRLB for multicarrier systems. However, as shown in the Figure 4.8, the power spectral density is different and it is given by:

$$|S(f)|^2 = \frac{P_s}{2W}; f_0 - \frac{W}{2} < f < f_0 + \frac{W}{2}, \tag{4.21a}$$

where P_s is the transmitted power and W is the bandwidth of the system. Using this spectral density, we can calculate the normalized bandwidth using Equation (4.14b):

$$\beta^2 = \frac{\int_{f_0-\frac{W}{2}}^{f_0+\frac{W}{2}} (2\pi f)^2 \frac{P_s}{2W} 2\pi df}{\int_{f_0-\frac{W}{2}}^{f_0+\frac{W}{2}} \frac{P_S}{2W} 2\pi df} \tag{4.21b}$$

$$= \frac{4\pi^2}{3} \frac{\left(f_0+\frac{W}{2}\right)^3 - \left(f_0-\frac{W}{2}\right)^3}{W} = 4\pi^2\left(f_0^2 + \frac{W^2}{12}\right)$$

Substituting Equations (4.17a) and (4.21a) into Equation (4.15a) for calculation of CRLB of a continuous time signal, we have:

$$CRLB = \frac{1}{\rho^2\beta^2} = \frac{1}{8\pi^2 \times SNR \times W \times T_M \times \left(f_0^2 + \frac{W^2}{12}\right)}, \tag{4.22a}$$

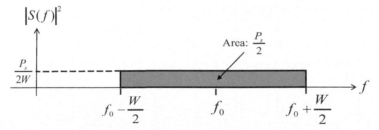

Figure 4.9   Flat spectrum used for calculation of CRLB for the TOA based systems using sinc pulses in time domain. Power is uniformly distributed across the transmission bandwidth of the system.

and the bound on estimation of the distance using measurement of TOA or time of flight is:

$$\sigma_r = c \times \sigma_\tau \geq \frac{c}{2\pi\sqrt{2 \times SNR \times W \times T_M \times \left(f_0^2 + \frac{W^2}{12}\right)}}. \tag{4.22b}$$

Comparing Equations (4.22b) for flat spectrum pulses and (4.18b) for two-tone TOA measurement, we expect similar performances, in particular if we note that in all traditional systems $f_0 \gg W$.

Example 4.11: (Performance of TOA Ranging in 2.4 GHz ISM Bands)

The pioneering TOA-based indoor geolocation system of the PinPoint used the entire 84 MHz bandwidth of the ISM bands at 2.4 GHz. Assuming that the received SNR is 10 dB, measurement time is the same as symbol duration, $W \times T_M = 1$, and considering that, $f_0 >> W$, resulting in $f_0^2 + W^2/12 \approx f_0^2$, we can use Equation (4.22b) and calculate the achievable precision:

$$\sigma_r \geq \frac{3 \times 10^8}{2\pi \times \sqrt{2 \times 10} \times 2.4 \times 10^9} = 4.5 \times 10^{-3}m \approx 0.5cm$$

Comparing this with the values of a few meters error for RSS localization in a typical indoor area, we notice that with the TOA systems, achieving centimeter precision is possible. However, we shall consider that every nanosecond error in measuring the peak of the signal would result in a 30 cm error in accuracy, leaving the burden of maintaining low precision on the implementation of the signal processing parts of the technology. In addition, as we will show in Chapter 6, excessive indoor multipath creates unexpectedly high value of error leaving TOA-based localization for open indoor areas, where the effects of multipath is reasonable. Interactive gaming using large television (TV) screens, localization of WVCE, and localization in operation rooms or manufacturing floors are areas in which we can benefit from high precision ranging offered by TOA based ranging.

4.3.6 CRLB for TOA Positioning

Similar to Section 3.2.1, we can extend the TOA ranging to TOA positioning with multiple reference point (RP)s using classical estimation theory for multiple observations of function of multiple parameters. With N reference points and a two-dimensional positioning problem, we have N-observation of function of two parameters, with the following formulation:

$$\begin{cases} O_i = \tau_i = \frac{r_i}{c} + \eta_{\tau_i}; & i = 1, ..., N \\ r_i = \sqrt{(x - x_i)^2 + (y - y_i)^2} \end{cases} \quad . \tag{4.23a}$$

In this formulation, variance of the observation noise is the variance of the ranging error, given by Equation (4.15a), which depends on the bandwidth of the transmitted signal, the received SNR, frequency of operation, measurement time, and the shape of the waveform used for localization.

Formulating the problem in vector notations, we have *N*-TOA observations

$$\mathbf{O} = \begin{bmatrix} \tau_1 & \tau_2 & . & \tau_N \end{bmatrix}^T$$

in zero mean Gaussian noise:

$$\boldsymbol{\eta} = \begin{bmatrix} \eta_1 & \eta_2 & . & \eta_N \end{bmatrix}^T,$$

relating with:

$$\mathbf{O} = \mathbf{G}(x.y) + \boldsymbol{\eta}, \tag{4.23b}$$

where:

$$\mathbf{G}(x, y) = \frac{1}{c} \begin{bmatrix} \sqrt{(x - x_1)^2 + (y - y_1)^2} \\ \sqrt{(x - x_2)^2 + (y - y_2)^2} \\ . \\ \sqrt{(x - x_N)^2 + (y - y_N)^2} \end{bmatrix} \tag{4.23c}$$

Then, following the derivation in Section 3.2.2, the differential variations of TOA based on location of the reference points are given by:

$$d\tau_i(x, y) = \frac{1}{c} \left(\frac{x - x_i}{r_i} dx + \frac{y - y_i}{r_i} dy \right); \quad i = 1, ...N,$$

and in a vector format, we have:

$$\begin{cases} d\boldsymbol{\tau} = \boldsymbol{H} \times d\boldsymbol{r} \\ \boldsymbol{H} = \nabla_{x,y} \left[\mathbf{G}(x, y) \right] \end{cases}, \tag{4.24a}$$

where:

$$d\boldsymbol{\tau} = \begin{bmatrix} d\tau_1 \\ d\tau_2 \\ . \\ d\tau_N \end{bmatrix}; \quad d\boldsymbol{r} = \begin{bmatrix} dx \\ dy \end{bmatrix}; \quad \boldsymbol{H} = \nabla_{x,y}\boldsymbol{\tau} = \frac{1}{c} \begin{bmatrix} \frac{x-x_1}{r_1} & \frac{y-y_1}{r_1} \\ . & . \\ . & . \\ \frac{x-x_N}{r_N} & \frac{y-y_N}{r_N} \end{bmatrix}$$

Assuming SNRs from all arriving signals are the same, we have:

$$cov\left(d\tau_i,\ d\tau_j\right) = \begin{cases} \sigma_\tau^2, & i = j \\ 0, & i \neq j \end{cases} \ ; \quad i, j = 1, 2, ...N \qquad (4.24b)$$

The FIM matrix for this setting is the same as Equation (3.4e):

$$\mathbf{F} = E\left\{|d\mathbf{r}|^2\right\}^{-1} = \frac{\mathbf{H}^T\mathbf{H}}{\sigma_\tau^2} \qquad (4.24c)$$

and the CRLB is given by:

$$CRLB \geq \mathrm{Tr}(\mathbf{F}^{-1}) = \sigma_r^2 = \sigma_x^2 + \sigma_y^2. \qquad (4.24d)$$

Since measurement noise for TOA ranging is caused by thermal noise, it is very small and location estimate becomes very accurate. In RSS-based positioning measurement, noise is caused by shadow fading, which has a very large variance compared with the variance of the thermal noise.

4.4 Performance of DOA-Based Localization

Design and performance analysis of direction of arrival (DOA) technique is closely related to TOA systems, but each has its own identity. TOA systems measure the distance between the transmitter and the receiver and they need at least three known reference points to position. The DOA systems can locate a device by using two or even one antenna. TOA-based systems also need to establish a time reference between the transmitter and the receiver to synchronize. The DOA system uses antenna arrays to measure the angle of arrival of the signal using the difference of the TOA of the signal from different elements of the antenna array, which does not need for synchronization between the transmitter and the receiver.

Figure 4.10 shows the basic concept behind the measurement of DOA and localization using one or two antennas. In localization using RF devices, as shown in Figure 4.10(a), if we have a directional antenna with a beamwidth, and another metric such as TOA to measure the distance, we can locate an object with an accuracy proportional to $r \tan \theta_s$, where $2\theta_s$ is the error in measurement of the DOA. As shown in Figure 4.10(b), with two directional antennas, we can achieve the same level of accuracy as well. Therefore, the DOA metric can be used with a simple single antenna and another ranging

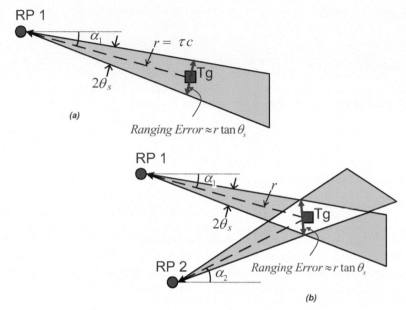

Figure 4.10   Basic concept behind using DOA for location estimation with, (a) one and, (b) two antennas.

method to measure the distance or with two reference points to find the location. We can clearly observe that given the accuracy of the DOA measurement, the accuracy of the position estimation depends on the transmitter position with respect to the receivers. When the transmitter lies between the two receivers, DOA measurements will not be able to provide a position fix. As a result, more than two receivers are normally needed to improve the location accuracy. In macro-cellular environments where the primary scatters are located around the transmitter and far away from the receivers, the DOA method can provide acceptable location accuracy. But dramatically large location errors will occur if the LOS signal path is blocked and the DOA of a reflected or a scattered signal component is used for location estimation. In indoor environments, the surrounding objects or walls inside a building usually block the LOS signal path. Thus, the DOA method will not be useable as the only metric for indoor geolocation. While this is a feasible option in next-generation cellular systems, where smart antennas are expected to be widely deployed to increase capacity in the outdoor areas, it is in general not a good solution for low-cost applications, in particular in indoor areas.

4.4.1 CRLB for DOA Estimation with Two Antennas

Calculation of the CRLB for the DOA is closely related to the calculation of the CRLB for the TOA-based ranging. If we have two receiver antennas deployed close to each other, in a far distance from an RF transmitting antenna (Figure 4.11), we can estimate the angle of arrival using the time difference of arrival (TDOA) of the signal in the two neighboring antennas. As shown in the Figure 4.11, since the distance between the two receiver antennas, l, is small and the transmitter antenna is far away, the paths between the transmitter and the two antennas become parallel lines. If we represent the angle of arrival of these paths by α, the geometrical difference between the length of the two arriving paths is $\Delta r = l \times \sin \alpha$. If the TDOA between the two paths measured at the receiver is $\Delta \tau$, the electronically measured difference between the lengths of the two paths is $\Delta r = c \times \Delta \tau$; therefore:

$$\Delta \tau = \frac{l \times \sin \alpha}{c}. \qquad (4.25a)$$

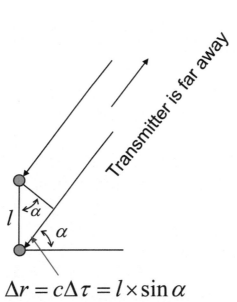

$$\Delta r = c \Delta \tau = l \times \sin \alpha$$

Figure 4.11 DOA measurement using TDOA of two receiving antennas from a transmitter antenna in a far distance, r.

Since measurement of the TDOA, $\Delta\tau$, involves noise, we can formulate our problem in terms of classical estimation theory as:

$$O = \Delta\tau + \eta = \frac{l}{c}\sin\alpha + \eta. \tag{4.25b}$$

This is the classic observation of function of a parameter in noise, described in Appendix Section A.1.2, where

$$g(\alpha) = \Delta\tau = \frac{l}{c}\sin\alpha. \tag{4.25c}$$

The variance of the estimation noise, η, is the variance of TDOA estimation, which is twice the variance of estimation of the TOA, given by Equation (4.15a). Therefore:

$$E\left[|\eta|^2\right] = 2\sigma_\tau^2 = \frac{2}{\rho^2\beta^2}, \tag{4.25d}$$

where (ρ, β) are the normalized SNR and bandwidth defined by Equations (4.6b) and (4.14b), respectively.

As shown by Equation (A.10b) in Appendix A.3.2, the CRLB for observation of function of a parameter in noise is the variance of the measurement divided by magnitude square of the derivative of the function; therefore:

$$\sigma_\alpha^2 = CRLB \geq \frac{2\sigma_\tau^2}{|g'(\alpha)|^2}. \tag{4.26a}$$

Since the derivative of a sin is a cos, we have:

$$\sigma_\alpha^2 \geq \frac{2\sigma_\tau^2}{l^2\cos^2\alpha} = \frac{2}{l^2\cos^2\alpha \times \rho^2\beta^2}. \tag{4.26b}$$

As an example, if we were using a sinc pulse with flat spectrum, using Equation (4.22a), the CRLB for DOA becomes:

$$\sigma_\alpha^2 \geq \frac{2}{l^2\cos^2(\alpha)} \times \frac{c^2}{8\pi^2 \times SNR \times W \times T_M \times \left(f_0^2 + \frac{W^2}{12}\right)} \tag{4.26c}$$

Here, the CRLB is a function of angle of arrival of the waveforms. For $\alpha = 0$, all the paths arrive perpendicular to the plane of the arrays and we have the best estimate for the DOA. For $\alpha = \pi/2$, the transmitter is in line with the plane of the array and the CRLB goes to infinity and we cannot measure the angle.

4.4.2 CRLB for DOA Estimation Using an Antenna Array

Figure 4.12 is the extension of Figure 4.11 to multiple antennas to form an antenna array. If we use an antenna array, the performance of the DOA estimate will improve. In classical estimation theory terminology, we have multiple observations of functions of a parameter in Gaussian noise. The N-observations of N-functions of a parameter in noise are formulated as:

$$O_i = g_i(x) + \eta_i; \quad i = 1, 2, \ldots, N. \tag{4.27a}$$

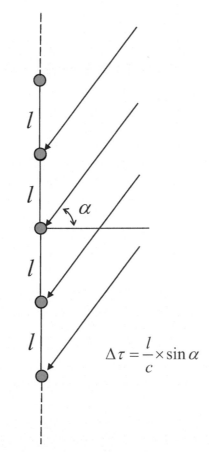

Figure 4.12 DOA estimation using antennas arrays.

In a vector notation, we have:

$$\mathbf{O} = \mathbf{G}(x) + \boldsymbol{\eta}, \tag{4.27b}$$

where:

$$\begin{cases} \mathbf{O} = \begin{bmatrix} O_1 & O_2 & . & . & O_N \end{bmatrix}^T \\ \mathbf{G}(\alpha) = \begin{bmatrix} g_1(\alpha) & g_2(\alpha) & . & . & g_N(\alpha) \end{bmatrix}^T . \\ \boldsymbol{\eta} = \begin{bmatrix} \eta_1 & \eta_2 & . & . & \eta_N \end{bmatrix}^T \end{cases} \tag{4.27c}$$

and:

$$\mathrm{E}\left[\boldsymbol{\eta}^T\boldsymbol{\eta}\right] = \sigma_\tau^2 \mathbf{I}. \tag{4.27d}$$

As shown in Equation (A.18b) in Appendix Section A.5.1, the CRLB for this formulation is given by:

$$CRLB \geq \mathbf{F}^{-1} = \frac{\sigma_\tau^2}{|\mathbf{G}'(\alpha)|^2} \tag{4.27e}$$

For an antenna array with N-elements, if we consider the center element as the reference, we have $N-1$ observation of the TDOA:

$$\mathbf{O} = \mathbf{G}(\alpha) + \boldsymbol{\eta} = \frac{l}{c}\sin\alpha \begin{bmatrix} \frac{N-1}{2} \\ . \\ 2 \\ 1 \\ -1 \\ -2 \\ . \\ -\frac{N-1}{2} \end{bmatrix} + \begin{bmatrix} \eta_{\frac{N-1}{2}} \\ . \\ \eta_2 \\ \eta_1 \\ -\eta_{-1} \\ -\eta_{-2} \\ . \\ -\eta_{-\frac{N-1}{2}} \end{bmatrix} \tag{4.28a}$$

To proceed with our calculations, we begin with calculation of the denominator of Equation (4.25e):

$$\left|\mathbf{G}'(\alpha)\right|^2 = \mathbf{G}'^{T}(\alpha)\mathbf{G}'(\alpha) = \frac{l^2}{c^2}\cos^2\alpha\sum_{-\frac{N-1}{2}}^{\frac{N-1}{2}} i^2.$$

Since

$$\sum_{-\frac{N-1}{2}}^{\frac{N-1}{2}} i^2 = \frac{N(N^2-1)}{12},$$

we have

$$\left| \mathbf{G}'(\alpha) \right|^2 = \mathbf{G}'^T(\alpha)\mathbf{G}'(\alpha) = \frac{l^2}{c^2}\cos^2\alpha \sum_{-\frac{N-1}{2}}^{\frac{N-1}{2}} i^2 = \frac{l^2}{c^2}\cos^2\alpha \frac{N(N^2-1)}{12}.$$

$$(4.28b)$$

Then, the CRLB is given by:

$$\sigma_\alpha^2 = CRLB \geq \frac{\sigma_\tau^2}{\frac{l^2}{c^2}\cos^2\alpha \frac{N(N^2-1)}{12}} = \frac{12c^2}{N(N^2-1)l^2\cos^2\alpha}\sigma_\tau^2. \quad (4.28c)$$

As we increase the number of elements, we have $N = 3, 5, 7, \ldots$ and the combinatorial term in the denominator becomes $N(N^2-1)/12 = 2, 10, 28, \ldots$ increasing the precision of the estimate proportionally.

Assignments for Chapter Four

Questions

1. What are the differences among NB and WB measurements of TOA in terms of bandwidth requirements and measurement ambiguity?
2. What are the advantages and disadvantages of TOA- and RSS-based positioning in terms of precision, complexity of the system, and bandwidth requirements?
3. Why can GPS use satellites that are tens of thousands of meters away from a mobile device and still have accuracies close to existing indoor geolocation systems on smart phones?
4. Why are AOA techniques not popular in indoor geolocation applications?
5. What is the error relationship between ranging and position location?

Problems

Problem 1:
Extend the 2D CRLB for TOA-based localization to 3D scenarios. Name an application where the 3D bounds help.

Problem 2:
Using the parameters of the antenna array defined in Section 4.4.2, show that the CRLB for DOA estimation is given by:

$$\sigma_D^2 \geq \frac{12c^2}{N(N^2 - 1)l\cos(\alpha)}\sigma_\tau$$

where N is the number of elements in the array, l is the distance between the antenna array elements, and α is the angle between the elements line the DOA of the signal.

Problem 3:
A TOA ranging system operates at the center frequency of 2.4 GHz and uses two tones to estimate distances up to 100 m. Determine the accuracy of the system. Assume that the signal-to-noise ratio is 20 dB.

Problem 4:
  a) Plot the CRLB of the TOA-based ranging using a waveform with flat spectrum as a function of bandwidth for the bandwidths of 1, 10, 100, and 1000 MHz. Assume an SNR of 10 dB and a carrier frequency of 3.2 GHz.
  b) Repeat (a) if we use two tones for measurement of the phase.
  c) Calculate and plot the CRLB for four tones and compare it with (a) and (b).

Problem 5:
  a) Using the CRLB of the TOA for flat spectrum, give the variance of the distance measurement error, σ_r, of the ranging error using TOA techniques.
  b) Using CRLB derivations for TOA and AOA, give the ratio of standard deviation of AOA and TOA distance measurement errors in terms of the actual angle of arrival, α, distance between antenna element of, l, and the number of antenna element N.

c) Assuming that we operate at 2.4 GHz center frequency and we have a four-element antenna array with distance separation of $l = \lambda/4$, plot the ratio of variance of distance measurement errors using AOA and TOA as a function of angle of arrival α.

d) In which angle of arrivals α, the AOA technique provides the best and the worst results with respect to the TOA measurements. Give this max and min values for the ratio of the distance measurement errors using AOA and TOA.

Problem 6:

Derive the CRLB for a positioning system that uses TOA for distance and an antenna array for DOA to position a target using only one reference antenna.

5

Opportunistic TOA Positioning

5.1 Introduction

Popularity of TOA positioning systems for military and civil applications began with the GPS using the satellite infrastructure originally deployed for military applications. As a result, the pioneering efforts for indoor geolocation for commercial applications, with companies such as Pin Point, Concord, MA, began in late 1990's by designing a TOA-based system that extends the GPS to indoor environment by using the same direct sequence spread spectrum (DSSS) technology with wider bandwidth of 84 MHz in Instrument System and Medical (ISM) bands to take care of multipath arrivals, and a new infrastructure that can cover indoor areas, where GPS signals cannot penetrate. The idea was novel, but these companies did not succeeds because market had not developed, system was very expensive, and performance was not consistent in all indoor areas. As a result, a company that had raised approximately US $70 million from venture capital was sold for US $700 K, a penny for a dollar. As we described in Chapter 4, later on the less complex and cost-efficient RSS-based Wi-Fi positioning systems using fingerprinting emerged as the first successful commercial solutions for positioning in indoors. In military applications, however, positioning is used for emergency responders operating in unknown environments, where fingerprinting is not applicable. As a result, the TOA-based localization research continued for indoor geolocation military applications and other indoor commercial applications demanding higher precisions with smaller infrastructure, such as interactive electronic gaming, localization of small items inside a retail shopping shelf, and localization in open manufacturing areas and warehouses.

In Chapters 2 and 3, we introduced RSS-based localization systems using signals from RFID tags, iBeacon (LEB), Wi-Fi, and cellular networks. RSS measurement is simple and does not need the knowledge of the details of the communication waveforms. In our presentation of RSS-based localization, we first described models for the behavior and performance evaluation of RSS ranging, then we introduced popular systems using this technology. TOA-based positioning needs more detailed description of the signal used for wireless communications and methods for time synchronization between the transmitter and the receiver to measure the time of flight. In Chapter 4, we described fundamentals of TOA-based positioning by presenting methods for measurement and precision analysis of TOA positioning. In this chapter, we first describe utilization of signals used for direct pulse transition, multi-carrier modulation, and spread spectrum communications, for opportunistic positioning. Then, we explain fundamental techniques for time synchronization between the transmitter and the receiver for measurement of the time of flight. Finally, we address challenges in TOA measurements in non-homogeneous environments.

TOA positioning systems use signals of opportunity for positioning to maintain a low cost for design and deployment. By signals of opportunity, we mean signals that are designed for wireless communication applications in indoor and urban areas and we use them for positioning a device in that area.

5.1.1 Opportunistic Signals for TOA Positioning

Figure 5.1 shows an overview of the wireless communication networking standards and the signals of opportunity that they provide for TOA-based localization in urban and indoor areas. The cellular networking industry emerged from an analog mobile phone designed for cars to a digital service focused on voice applications, in digital format. Part of the 2G (QUALCOMM cdmaOne) and 3G wireless digital communication systems adopted DSSS transmission technology to increase the capacity and integrate the voice and data. The 4G networks were focused on higher data rates for mobile users in the urban and suburban metropolitan areas and highways using orthogonal frequency division multiplexing (OFDM) and multiple input multiple-output (MIMO) antenna systems. The emerging 5G intends to increase the capacity of 4G for orders of magnitude using direct pulse transmission at 60 GHz mmWave. The WLAN industry evolved to support home networking and hot-spot coverage to complement the cellular for high-speed indoor wireless communications. The first-generation legacy wireless local area networks (WLANs) used DSSS, frequency hopping spread spectrum (FHSS) and diffused infrared (DFIR). IEEE 802.11 a/g began using OFDM

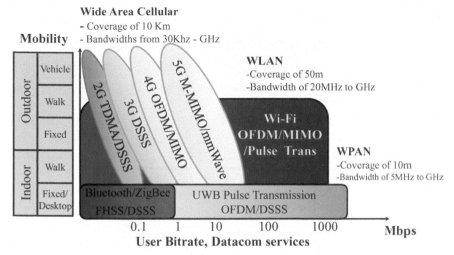

Figure 5.1 Overview of the wireless communication signals of opportunity for TOA based positioning.

signals and 802.11 n/ac extended that to MIMO technology capable of multiple streaming. The IEEE 802.11ad uses direct pulse transmission at 60 GHz mmWave spectrum. Two other wireless technologies emerged to complement the popular and ever growing WLAN industry into wireless personal area networks (WPAN). First were the low-power wireless personal area networks (WPANs), Bluetooth, and ZigBee, which were designed for ad hoc sensor networking. The second class of WPANs focused at ultra-high-speed Gbps transmission using UWB signals

Wireless communication systems using UWB and mmWave signals are direct pulse transmission systems well suited for TOA ranging. The DSSS signal used in wireless communications is a pseudo pulse transmission system, which is the traditional signal for TOA positioning adopted for the GPS. The OFDM signals for wireless communications are multicarrier transmission systems, and the FHSS is a pseudo-multi-carrier system, both suitable for TOA ranging. The MIMO antenna systems are antenna arrays for measurement of DOA. These features of wireless communications signals offer tools for TOA-based positioning using direct pulse transmission, multi-carrier modulation, and array antennas. As we explained in Sections 4.3 and 4.4, regardless of the type of the signal, the CRLB of the TOA-based ranging is inversely proportional to the transmission bandwidth

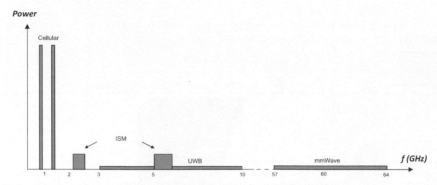

Figure 5.2   Relative power limits and the bandwidths available for cellular, Wi-Fi, UWB and mmWave wireless communication networks.

[see Equation (4.15b)]. To compare the bandwidth of these systems, the original cdmaOne used 1.25 MHz of bandwidth, the 3G system increased that a few times, the 4G systems use 5–20 MHz of bandwidth, and 4G may use mmWaves with bandwidths on the order of GHz. The bandwidth of Wi-Fi is approximately 20 MHz and the bandwidth of the ZigBee is 5 MHz, the UWB systems occupy bandwidths on the order of GHz. Therefore, we can use all of these signals for opportunistic positioning, but UWB and mmWave signals can support much more precise positioning.

Figure 5.2 shows the available bands and relative transmission power of several wireless communication systems. For the cellular networks, we have smaller bands with a maximum of 25 MHz for lower frequencies (800 MHz to few GHz) with the ability to transmit higher power to support more comprehensive coverage. We have unlicensed ISM bands for Wi-Fi standards at 2.4 (84 MHz of bandwidth) and 5.2 GHz (originally 125 MHz), the UWB and mmWave bands have an allocated bandwidth of approximately 7 GHz around 7 GHz and 60 GHz, respectively.

5.2 Pulse Transmission for TOA Positioning

In Section 4.3.4, we derived the CRLB for sinc pulse transmission with a flat spectrum and in Equation (4.22b), we showed that the standard deviation of ranging error is inversely proportional to the center frequency, f_0, and bandwidth, W, of the transmitted pulse by a factor of:

$$\left(f_0^2 + \frac{W^2}{12} \right). \tag{5.1}$$

Therefore, as we increase the bandwidth and center frequency of signal, precision of ranging increases. Hypothetically, with infinite bandwidth, we can achieve infinite precision. In Example 4.7, we showed that precision of 5 cm is achievable in 2.4 GHz ISM bands. As we shift to higher frequencies for UWB and mmWave, we can achieve even higher precision. However, to avoid the cost of design and deployment of TOA-based systems, we need to resort to the signals of opportunity.

Radio frequency (RF) pulse transmission has been used for military radar applications since the 1950s [Tay01]. A new wave of interest in using RF pulses for military and commercial applications began in the late 1990s. This technology evolved with two hypes. The first hype was in the early 2000s, and it began with the emergence of UWB technology for short-range wireless communications in the 3.1–10.6 GHz unlicensed bands. The second hype began in the early 2010s with the availability of millimeter wave (mmWave) for local and wide-area wireless communications at 57–64 GHz unlicensed bands. These pulse transmission techniques are low-power small-size radio technologies needed to support broadband multi-rate wireless communications with potential for accurate opportunistic positioning in indoor and urban areas.

The traditional method and the most popular techniques for TOA ranging is the direct sequence spread spectrum (DSSS) originally invented for secure military communications in World War II. The DSSS technologies played an important role in the emergence of local and wide-area networks in the 1990s, and it is the technology of choice for the GPS. As we will show in Section 5.2.3, the DSSS is indeed a pseudo-pulse transmission technique with its own characteristics. The popular DSSS wireless networking technologies provide a fertile ground for opportunistic TOA positioning using a virtual pulse transmission method. In the remainder of this section, we provide more details on technologies used for UWB transmission at 3.4–10.6 GHZ UWB bands and 57–64 GHZ mmWave bands, as well as DSSS technology used in different wireless communications bands.

5.2.1 UWB Pulses for Opportunistic Positioning at 3.1–10.6 GHz

In 1997, the IEEE 802.15 Wireless Personal Area Networking (WPAN) standardization committee began to define specifications for BodyLANs and eventually adopted Bluetooth as its standard specification, 802.15.1. The same standardization group considered UWB solutions within the IEEE 802.15.3a sub-committee, which focused on defining standard specifications for WPANs operating in the 3.1–10.6 GHz unlicensed bands released by the FCC in 2002. Figure 5.3 shows the *spectral mask* mandated by the FCC for unlicensed UWB communications. This figure also identifies the frequency spectrum considered by the IEEE 802.15.3a proposal. The original petitions for the UWB band covered the entire spectrum and were intended to be used by impulse radio entirely [Win00]. However, after a number of controversial debates concerning interference with other low-power devices, in particular GPS receivers, the FCC decided to specify the mask shown in Figure 5.3, which reduces the radiation in the spectrum between 0.96 and 3.1 GHz, so as to facilitate virtually harmless co-existence with popular existing systems.

As far as standardization activities are concerned, in the IEEE 802.15.3a group, several options were evaluated for UWB communications, among which are the historical impulse radio technology, as well as the Direct Sequence (DS-UWB) and Multi-Band OFDM (MB-OFDM) systems. This standard lost its momentum later on and was dissolved in January 2006. However, the technical work produced by the committee has educational and

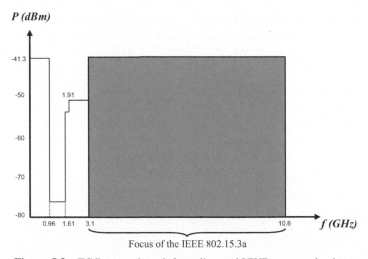

Figure 5.3   FCC spectral mask for unlicensed UWB communications.

pedagogical value in its consideration of alternative technologies for implementation of precise indoor geolocation capabilities. Because of its ultra-wide bandwidth, the UWB signal isolates (resolves) multipath components resulting in a stable received signal power with reduced multipath fading effects. Using bandwidths on the order of GHz for UWB systems allows achieving gigabit wireless transmission and precise indoor ranging in the order of centimeter. Both of these features were perceived to be essential in support of emerging video gaming and home entertainment industries. More details on UWB communications are available in [Gha04][Opp04][Pah05] and other references cited in these references.

Implementation of the UWB signal transmitter does not involve modulation, and if carefully designed, the transmitter can be much simpler than those in traditional narrowband or wideband spread spectrum systems used for TOA ranging.

Example 5.1: (UWB Pulse Shape in Impulse Radio)

Time Domain Corporation (TDC) is one of the pioneering companies engaged in the design of impulse radio UWB systems [TDCweb]. The pulse shape used in their pioneering system was a monocycle mathematically described by:

$$x(t) = 6A\sqrt{\frac{e\pi}{3}}\frac{t}{\tau}e^{-6\pi\left(\frac{t}{\tau}\right)^2}, \tag{5.1a}$$

where A represents the peak amplitude of the pulse, τ is a constant determining the width of the pulse, and t is the time variable. The spectrum of the monocycle pulse in frequency domain is given by:

$$X(f) = -j\frac{2f}{3f_c^2}\sqrt{\frac{e\pi}{2}}e^{-\frac{\pi}{6}\left(\frac{f}{f_c}\right)^2}, \tag{5.1b}$$

where $f_c = 1/\tau$ is the center frequency of the pulse. Figure 5.4(a) shows a typical graph of the pulse for $\tau = 0.5$ ns associated with a center frequency of $f_c = 2$ GHz. The half-power (3dB) bandwidth of the pulse occupies about 2 GHz (Figure 5.4(b)). The pulses are transmitted periodically, and the information is encoded in the location or position of the pulses, implementing a pulse position modulation system. Figures 5.4(c) and (d) illustrate the pulse repetition and its effects on the spectrum of the signal. The typical pulse width used for this implementation is between 0.2 and 1.5 ns, and the time interval between consecutive pulses ranges from 25 to 100 ns.

One of the challenges in the design of an UWB system is the choice of antennas. This is in particular more challenging for the original impulse

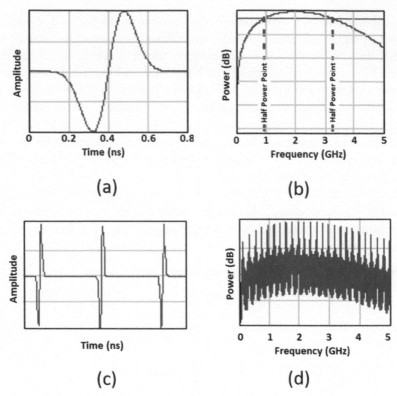

Figure 5.4   (a) The original UWB pulse used in pioneering impulse radio system designed by Time Domain Corporation, (b) the spectrum of the pulse, (c) repetition of the pulses, (d) effects of repetition on the spectrum of the signal [TDCweb].

radio systems that covered a spectrum ranging from very low frequencies up to a few GHz. The antennas designed for these bands need to have a flat response across a wide spectrum and they also have to be compact. Since these conditions are hard to maintain, the antennas used for impulse radio transmission actually change the shape of the transmitted pulse.

Example 5.2: (Pulse Shape and Antennas for Impulse Radio)

The pulse shape introduced in Example 5.1 is in fact a received pulse shape. Figure 5.5(a) represents the actual transmitted pulse. When this pulse is applied to a bow-tie antenna, shown in Figure 5.5(b), the pulse propagates with distortion and the received signal will appear as shown in Figure 5.5(c). The antennas behave as filters, and even in free space, a differentiation of the pulse occurs as the wave radiates [Win98].

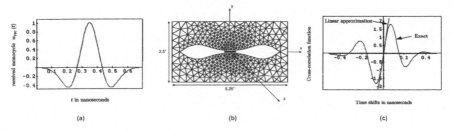

Figure 5.5 (a) Transmitted impulse, (b) a bow-tie UWB antenna, (c) the received signal after antenna [Win98] ©IEEE 1998.

The legacy impulse radio system, described in Examples 5.1 and 5.2, used bow-tie patch antennas with a size of 2.5 × 5.25 inches (Figure 5.5(c)), and its spectrum does not fit in the FCC UWB template (Figure 5.3). Smaller antennas with different shapes are commercially available. Figure 5.6 shows a small cone antenna and a smaller path UWB antenna with dimensions of around an inch. The two leading proposed UWB technologies for the IEEE 802.15.3a, DS-UWB and MB-OFDM, use narrower spectrums than Impulse Radio, and both standards chose more than one channel in the spectrum. The DS-UWB used two bands with approximately 2 and 4 GHz bandwidth (Figure 5.7(a)) and MB-OFDM used 15 bands, divided into five groups (Figure 5.8) to provide flexibility in supportable data rates for communication [Pah13]. Figure 5.7(b) shows the two wavelet pulses for DS-UWB in time domain. The shorter pulses belong to the wider bandwidth and both pulses look like a modulated waveform with a carrier because these are time responses of a band-pass filter. We can use all of these narrow UWB

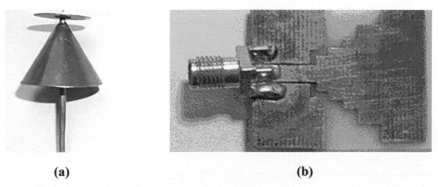

(a) **(b)**

Figure 5.6 Small UWB antennas, (a) a cone antenna with height of around an inch, (b) a small path antenna with a length of less than two inches.

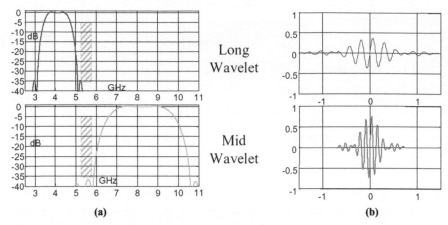

Figure 5.7   (a) Frequency and (b) time response of the two basic pulses used in DS-UWB system proposal fitting with the FCC UWB template used by IEEE 802.15a standard [Koh04].

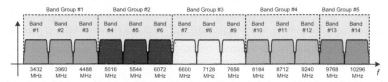

Figure 5.8   Overview of the bands and their grouping in MB-OFDM proposal for IEEE 802.25.3a at 3.1–10.6 GHz band specified by FCC.

pulses for direct calculation of the TOA using the time displacement of the transmitted and the received signals and achieve high precision localization. The challenge with these UWB technologies is their limitation in coverage. On the average with the FCC-regulated maximum transmission power, these systems will cover around 15 m in indoor areas, which restricts indoor geolocation applications intending to use externally mounted antennas using fire engines or other emergency response applications.

5.2.2 mmWave UWB Pulses for Positioning at 57–64 BHz

In 2011, FCC released unlicensed bands in the 57–66 GHz, not all of which is available all over the world, but parts are available in most countries. This spectrum is called the 60 GHz mmWave bands because the 30–300 GHz are referred to as mmWave frequencies. The mmWave has attracted considerable attention as a solution for high-capacity next-generation wireless networks. The 60 GHz band is used for the IEEE 802ad, aj, ay, az for Gbps wireless because more bandwidth is available, and at these frequencies, it is easier to design small directional antennas for multiple streaming. The available bandwidth at 60 GHz is literally ultra-wide and approximately the same as available bands at 3.4–10.6 UWB spectrum, resulting in waveforms capable of Gbps transmission and centimeter precision. Comparing with UWB systems in 3.4–10.6 GHZ, at 60 GHz, antenna is much smaller, allowing higher number of elements enabling of more focused beam forming to support more streams for communications and more precise estimation of DOA for localization. Figure 5.9 shows characteristics of a typical 60 GHz antenna. Figure 5.9(b) shows a typical 8×8 patch antenna array with 64 elements. At 60 GHz, the wavelength is $\lambda = 5$ mm, and with antenna element separation of $\lambda/4 = 1.25$ mm, we can design this huge array with dimension of around 1 cm, the size of a small coin. Figure 5.9(a) shows the directional pattern of the antenna, and Figure 5.9(c) visualizes the potential beam formations for multiple streaming in 3D. Comparing with the size of UWB antennas shown in Figure 5.7, we can observe this benefit of 60 GHz mmWave over the UWB band operating in 3.4–10.1 GHz.

The benefit of antenna arrays is that it focuses the direction of energy but extremely focused energy can result in harmful effects when they are exposed to the body. As a result, federal government agencies in charge of spectrum allocations regulate the maximum Equivalent Isotropic Radio Propagation (EIRP) for the design of antennas in these bands. The EIRP is summation of the maximum transmitted power with the maximum allowable antenna gain. Table 5.1 provides Tx power, antenna gain, and EIRP regulations for 60 GHz operation in different regions.

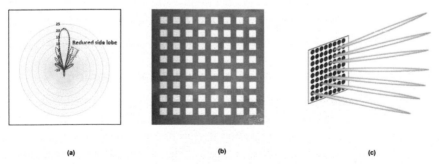

(a)   (b)   (c)

Figure 5.9 (a) Beam pattern of two different 8×8 antenna arrays with different side lobes at 60 GHz, (b) a 64-element patch antenna array with approximate dimensions of 1 cm, (c) 3D visualization of multiple streams radiating from the antenna.

Table 5.1   Transmit power, antenna gain, and EIRP regulations for 60 GHz operation in different regions of the world

Region of Operation	Tx Power dBm	Antenna Gain dBi	EIRP
US/Canada	27	15	43
Japan/Korea	10	48	58
EU	13	44	57

Example 5.3: (EIRP in ISM and mmWave bands)

Comparing with Wi-Fi devices operating at 2.4 GHz ISM bands with 20 dBm Tx power and Omni-directional antennas with antenna gain of 1.6(2dBi), the EIRP for those operation were 22 dB. At 60 GHz (Table 5.1), we have 21–36 dB increase in EIRP. However, we have more path-loss in the first meter at 2.4 GHZ, which is inversely proportional to the frequency of operation. The path loss in the first meter for the 2.4 GHz operation is approximately 40 dB, while at 60 GHz, we have a first meter path loss of 52 dB. Including the 12 dB difference in the comparison, 60 GHz still has an edge of 9–24 dB in path loss for coverage.

Designers of the next generation of wireless networks expect using the wide bandwidths at mmWave with massive MIMO antennas, small cell size, and multiple streaming to respond to the exponential growth of user demands for higher speeds. The use of massive antenna arrays, as we described in Equation (4.28c) for CRLB of DOA of an antenna arrays, has the potential to bring down positioning estimates to mm ranges and make it available to the cellular telephone network users. The main challenge for this approach for precise localization in multipath rich indoor and urban area environment is the effects of multipath, which we discuss in Chapter 6. Figure 5.10 shows the 60 GHz mmWave band allocations in different countries and segmentation of the bands by standardization organization into four bands to make it available for all countries.

5.2.3 Pseudo DSSS Pulses for Opportunistic Positioning

Spread spectrum technology was invented during the Second World War, and it has dominated military communication and positioning applications, where it is attractive because of its resistance to interference and interception, as well as its amenability to high-resolution ranging. The main difference between spread spectrum transmission and traditional radio transmission technologies is that the transmitted signal in spread spectrum systems occupies a

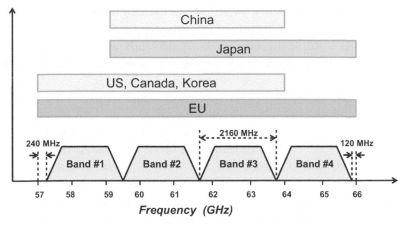

Figure 5.10 60 GHz mmWave band allocations in different countries and segmentation of the bands by standardization organization into four bands to make it available for all countries.

much larger bandwidth than the traditional radio modems. Compared to baseband UWB pulse transmission techniques, the occupied bandwidth by spread spectrum is still restricted enough so that the spread spectrum radio can share the medium with other spread spectrum and traditional radios in a frequency division multiplexed format. There are two basic methods for spread spectrum transmission: direct sequence spread spectrum (DSSS) and frequency hopping spread spectrum (FHSS). We discuss DSSS in this section, FHSS is discussed in Section 5.3.2. GPS uses DSSS technology to support satellite positioning. In the early part of the 1980s, commercial applications of DSSS technology were investigated for wireless indoor networks [Pah85] and it emerged as the transmission technique for legacy Wi-Fi, 3G cellular networks, ZigBee, some options for UWB systems, and a number of proprietary cordless telephones.

The 3G cellular industry adopted DSSS to support CDMA networks as an alternative to TDMA/FDMA networks in order to increase system capacity, provide a more reliable service, and achieve soft handoff for cellular connections. In the legacy Wi-Fi, spread spectrum technology was adopted primarily because the first unlicensed frequency bands suitable for high-speed wireless communication were ISM bands, which were initially released by the FCC under the condition that the transmission technology must use spread spectrum to control the interference [Mar85]. In ZigBee and UWB, DSSS is adopted because of simplicity of implementation and low power consumption, which is necessary for ad-hoc and sensor networks implementations. Availability of these signals of opportunity for localization and the fact that DSSS is used in the GPS systems demands an understanding of the fundamentals of this technology for anyone interest in learning about positioning systems.

Figure 5.11 shows the basic principles of a DSSS transmission system, the transmitted information bits with duration, T_b (Figure 5.11(a)) are multiplied by N pseudo noise (PN) sequence (Figure 5.11(b)) with duration of $T_c = T_b/N$, before they are modulated on a carrier signal. At the receiver, the received signal is first demodulated to recover the baseband envelope of the signal and remove the carrier. The envelope of the received signal is passed through a correlator to calculate its autocorrelation function (ACF). The autocorrelation of a PN sequence $\{b_i\}$ of length N follows the following

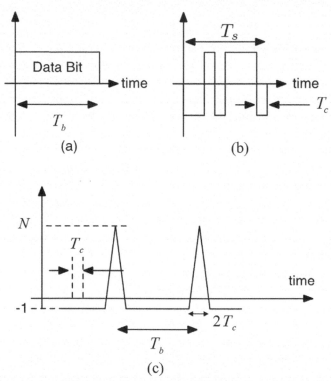

Figure 5.11   Principles of DSSS systems, (a) a transmitted data bit (b) the N-chip spreading PN-sequence code, (c) circular autocorrelation function of the code formed at the receiver.

characteristics (Figure 5.11(c)):

$$R(k) = \sum_{i=0}^{N-1} b_i b_{i-k} = \left\{ \begin{array}{ll} N; & k = mN \\ -1; & \text{otherwise} \end{array} \right. . \tag{5.2}$$

For any transmitted bit, the output of the correlator function at the receiver produces a narrow pulses with a height of N and a base that is twice the chip duration, $2T_c$. Therefore, a DSSS system can be thought of as a *pseudo-pulse transmission technique*, where we receive a sharp narrow pulse designating each transmitted bit. Comparing with traditional pulse transmission techniques, in impulse radio, the transmitted energy is impulsive and of low energy, which restrict the maximum coverage to 15 m. Higher level of coverage needs very high energy pulse transmission. In DSSS, the pulse energy is distributed over a huge number of chips with low amplitudes, and the received pulse is formed or reconstructed by autocorrelation process at the receiver. This allows high power transmission designed for long distances used for satellite communication and navigation. The peak of the received pulses is used for time of flight estimation for TOA-based ranging.

Figure 5.12 illustrates the difference between a DSSS (narrower in bandwidth) and an impulse radio signal in time and in the frequency domain. In the DSSS, the transmitted bit with duration of T_b is divided into chips with a

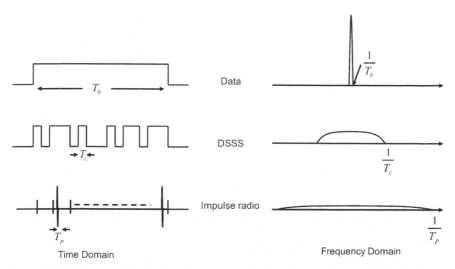

Figure 5.12 Comparison between the time and frequency characteristics of narrowband data, DSSS and impulse radio UWB signals.

duration of $T_c = T_b/N$ with considerably smaller duration. Therefore, with the same transmitted power and data rate, we will have a lower power spectral density and a bandwidth expansion equal to the processing gain of the DSSS system, N. In the case of an impulse radio system, the transmitted pulses, T_p, occupy a fraction of a chip, causing another expansion in the bandwidth, a consequent reduction in the power spectral density, and increase of the bandwidth to UWB.

5.3 Multi-Carrier Transmission for TOA Positioning

In Section 4.3.2 and in Equation (4.20), we showed that the variance of ranging error using multi-carrier signals is inversely proportional with

$$\frac{N(N+1)}{12(N-1)}.$$

Therefore, multi-carrier transmission with a large number of carriers can enhance the performance of the TOA-based ranging significantly. Hypothetically, with infinite number of carriers, even with limited bandwidth, W, we can achieve ranging with zero precision.

As shown in Figure 5.13, a multi-carrier transmission system with a large number of carriers, N, at center frequencies:

$$f = f_0 + k\Delta f; \quad k = 0, 1, \ldots \ldots N - 1$$

can be viewed as a discrete frequency transmission system. The normalized received signal (Figure 5.13(a)) is:

$$
\begin{cases}
Y(f) = \sum_{k=0}^{N-1} Y(k)\delta(f - f_0 - k\Delta f) \\
Y(k) = e^{j\phi_k} = Y(f)|_{f=k\Delta f}; \quad k = 0, 1, 2, \ldots, N-1
\end{cases}, \quad (5.3a)
$$

where ϕ_k is the measured phase by each carrier and Δf is the separation of the frequencies used in the multi-carrier transmission. This is equivalent to transmitting a complex valued, the magnitude and phase of the carriers, signal as a discrete sequence. We can take its inverse discrete frequency Fourier transform to find the periodic time response of the channel:

$$y(\tau) = \sum_{k=0}^{N-1} Y(k)e^{j2\pi\tau(f_0 + k\Delta f)}. \quad (5.3b)$$

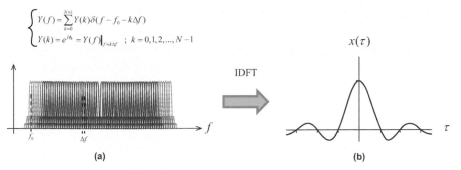

$$\begin{cases} Y(f) = \sum_{k=0}^{N-1} Y(k)\delta(f - f_0 - k\Delta f) \\ Y(k) = e^{j\theta_k} = Y(f)\big|_{f=k\Delta f} \; ; \; k = 0,1,2,...,N-1 \end{cases}$$

IDFT

$x(\tau)$

f_0 Δf $\longrightarrow f$

τ

(a)   (b)

Figure 5.13 Relation between OFDM and pulse transmission, (a) the transmitted carriers in frequency domain, (b) the IDFT of the complex valued samples of the frequency response, representing a since pulse in time domain.

As we show in Section 10.3.3, this signal is actually a periodic sinc pulse, shown in Figure 5.13(b). Therefore, there is a duality between TOA ranging in time and frequency domain measurements using multiple carriers. Frequency domain multi-carrier transmission is also a virtual pulse transmission method measuring time of flight with a different technique. The fundamental concept in multi-carrier measurement of the TOA is that we send a carrier and we measure its phase at the receiver if the carrier is transmitted with different frequencies, we observe multiple samples of the phase. We can calculate the distance from each of these phases and average them over all carriers or we can take the IDFT of the phases and measure the time of flight from the sinc pulse, which was obtained from Equation (5.3b).

To avoid implementation costs, we need to find signals of opportunity for design and deployment of multi-carrier TOA positioning systems. OFDM signals are used in wireless communications and OFDM is a multi-carrier system implemented using the fast Fourier transform. Therefore, OFDM wireless communication signals are suitable for TOA-based positioning, and we describe them in Section 5.3.1. Another signal of opportunity for multi-carrier measurement of the TOA is the FHSS signal used for wireless communications. FHSS is a pseudo-multi-carrier transmission system and we describe it in Section 5.3.2.

5.3.1 OFDM Signals for Opportunistic Positioning

OFDM is the most popular transmission technologies used in existing wireless communication systems. Availability of OFDM signals for Wi-Fi in indoor areas and for LTE in urban outdoor area is very valuable for local-ization using signals of opportunity. Figure 5.14 shows the basic principles of OFDM used for wireless communications. The incoming bits of information are mapped to complex valued symbols in the so-called signal constellation. The signal constellation is a grid of points in a complex plane each associated with a unique string of bits and represented with a complex valued point in the constellation. The constellation in Figure 5.14(a) has 64 symbols each representing a unique six bit of incoming data. The inverse Fast Fourier Transform of a group of these complex symbols forms the transmitted signal in time domain. This is equivalent to modulating the transmitted symbols over a multi-carrier system, shown in Figure 5.14(b). Therefore, OFDM technology is indeed a simple implementation of multi-carrier transmission using the Fast Fourier Transform (FFT). This multi-carrier system provides for opportunistic TOA ranging using the OFDM signal.

OFDM was first introduced for voice-band data communications in the early 1980s [Pah88]. In the late 1990s, it found its way in DSL modems, cable modems, Wi-Fi, LTE, and UWB technologies. As shown in Figure 5.14, the basic concept for multi-carrier transmission is very simple, instead of transmitting a single stream at a rate of R_s symbols/s, we use N streams over N-carriers spaced by about R_s/N Hz, each carrying a stream at the rate R_s/N symbols/s. The primary advantage of multicarrier transmission and its implementation using OFDM technology is its ability to cope with severe channel conditions such as high attenuations at higher frequencies in long copper lines or the effects of frequency-selective fading in wireless channels. To measure the TOA using this multi-carrier system, we can transmit one of

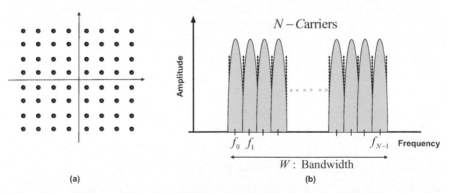

Figure 5.14 Basic principles of OFDM, (a) signal constellation with 64 information symbols, (b) multiple carriers carrying symbols in parallel.

the symbols repeatedly, then the received signal is a set of carriers with the same magnitude by a phase reflecting the frequency of sub-carrier carrying the signal, which can be used for TOA measurement.

5.3.2 FHSS for Opportunistic Positioning

The original spread spectrum technology, invented during the Second World War, by the movie star Hedy Lamarr, was indeed an FHSS transmission. As shown in Figure 5.15, an FHSS NB transmitter (Figure 5.15) shifts the center frequency of the transmitted signal with every packet transmission. The shifts in frequency or frequency hops occur according to a random pattern that is only known to the transmitter and the receiver. If we move the center frequency randomly among one hundred different frequencies, then the required transmission bandwidth is hundred times more than the original transmission bandwidth. We call this new technique **a spread spectrum technique** because the spectrum is spread over a band that is hundred times larger than original traditional radio. FHSS can be applied to both analog and digital communications, but it has been applied primarily for digital transmissions. For multi-carrier TOA ranging, phase of the FHSS received carriers at different hop frequencies provides the same information that OFDM signals provides. Therefore, FHSS spread spectrum can provide for a pseudo-multi-carrier TOA range measurement system.

Application of FHSS in modern wireless communications systems began by adoption of the FHSS as one of the three options of the legacy

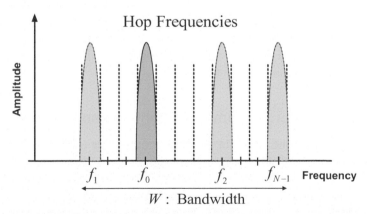

Figure 5.15 Basic principles for FHSS, a NB transmission system sends a packet at a carrier frequency and then hops randomly to another carrier to transmit the next packet.

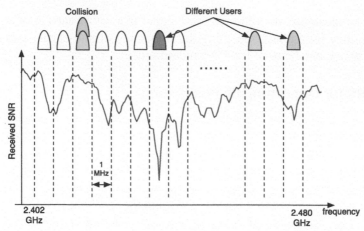

Figure 5.16 Multi carrier reception of FHSS in legacy IEEE 802.11 and Bluetooth systems.

IEEE 802.11. The most popular application of the FHSS is in the IEEE 802.15.1a, the Bluetooth. The FHSS signals are available for opportunistic TOA-based positioning. IEEE 802.11 FHSS and Bluetooth both use the same bandwidth of 1 MHz and the same 78 channels in the 2.4 GHz ISM bands, as shown in Figure 5.16. Members of each network coordinate their transmissions and the hopping patterns in their own network, while there is no coordination among the users in the two different network and they simply co-exist. The multi-user interference occurs when two different users transmit on the same hop frequency. If the hopping codes are random and independent from one another, the "hits" will occur with some calculable probability. If the codes are synchronized and the hopping patterns are selected so that two users never hop to the same frequency at the same time, multiple-user interference is eliminated. When we have multiple users with lower data rates, this approach allows them to share the medium.

5.4 Synchronization for Time-of-Flight Measurements

Measurement of the time of flight or TOA requires an established timing between the transmitted and the received signal. Since 1 ns error in time caused by poor synchronization causes 30 cm error in measurement of the distance, we need atomic clock synchronization for accuracies on the orders of meter. GPS signals are synchronized by atomic clock and cellular networks use them for synchronization. The most popular method for synchronization in spread spectrum positioning systems is to use a sliding correlator. Sliding correlators calculated the cross correlation between the received signal and a local duplicate of the transmitted signal to measure a time reference. Another approach to avoid synchronization is to use the time difference of arrival. Two receivers measuring the absolute value of arrival time of a signal from a device use the time difference of arrivals as a metric for localization. In short-distance TOA measurement for sensor networks, the receiver may use a transponder and automatically relay the signal back to the transmitter as a reference for measurement of the round trip time of flight.

5.4.1 Sliding Correlator for Time Synchronization

In DSSS systems, measurement of time of flight is performed by cross correlating the received PN-sequence with a locally generated PN-sequence. The peak of the cross correlation is used to establish a time reference for measurement of time of flight. Efficient implementation of the cross-correlator is one of the important design issues in DSSS systems. Digital implementation of this correlation function requires a very high sampling rate to accommodate the wide transmission bandwidth and capture the waveform narrow pulses in time. Correlation function is a parametric convolution integral, direct calculation of this integral involves numerical integration of the integral for different delay values or use of Fourier transform techniques that are both very computationally extensive at high sampling rates. A relatively simple analog implementation of a cross correlator that is used extensively in spread spectrum systems is the so-called *sliding correlator.* Sliding correlators are used to implement the correlation function for DSSS systems. GPS receiver devices read the DSSS signal from different satellites and using sliding correlators measure the TOA to calculate the distance from each of the satellites. Sliding correlators are also used for time synchronization of FHSS devices such as Bluetooth and other DSSS systems.

Figure 5.17 shows the basic concept of a sliding correlator. The cross correlation function between the received signal, $x(t)$, and the reference signals, $x(t - \tau_s)$, is given by:

$$R_{x,y}(\tau) = \frac{1}{T_p} \int_0^{T_p} x(t)x(t - \tau_s)dt. \tag{5.4a}$$

The delay of the reference signal is designed to be much smaller than the period of the signal, $\tau_s \ll T_p$, which implies that the sliding correlators run the transmitter and receiver PN-sequences with slightly different clocks

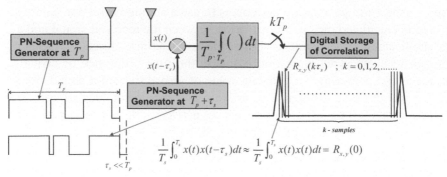

Figure 5.17   Basic concept of a sliding correlator for calculation of the correlation function of DSSS.

so that at the end of each transmitted PN-sequence period the receiver PN-sequence is a small value, τ_s, more than the transmitter. This way, if we start the transmitter and the receiver at the same time, in the first period, the reference is almost the same as the original signal, $x(t - \tau_s) \approx x(t)$. When we multiply the transmitted and the received signal over the first period, the integral we have:

$$\frac{1}{T_p} \int_0^{T_p} x(t)x(t)dt = R(0). \tag{5.4b}$$

In the second period of integral, the reference signal has shifted for, τ_s, over the transmitter signal and it is $x(t - \tau_s)$, when we integrate, we have:

$$\frac{1}{T_p} \int_0^{T_p} x(t)x(t - \tau_s)dt = R_{x,y}(\tau_s). \tag{5.4c}$$

In the next period, the output of the integrator is $R(2\tau_s)$, and each T_p second, we calculate a sample of the correlation function at a sampling rate of τ_s in virtual time. When $k\tau_s = T_p$ we have calculated k samples of the correlation function, we have virtually one full period computed, and we can detect the peak for location estimate or any other synchronization time references stored in the digital form. Since in real time it took us T_p seconds to compute one digitally stored sample, the total computation time is $T_0 = kT_p$ seconds. Detailed example of implementation of sliding correlator for multipath channel measurement using DSSS is available at [Pah05].

We can use a sliding correlator to calculate sample of cross correlation function of any periodic signal. We can use sliding correlator to calculate samples of cross correlation of a sinusoid or sample of cross correlation function of FHSS signal. If an FHSS receiver runs the clock for the hopping of frequencies slightly faster, then at the start of the process, all hops overlap and the integral gives a large number. As time passes, hops shift their locations and there is no overlap with a low value at the output of the integrator. This situation continues until the delay between the original and reference signals becomes a multiple of transmission period. Bluetooth uses this concept to discover users identified with a specific hopping pattern.

5.4.2 U-TDOA for Positioning with Relative Timing

Many cellular telephone network service providers use Uplink Time Difference of Arrival (U-TDOA) techniques to enable E-911 positioning services. Figure 5.18 shows an overall block diagram of a U-TDOA system. Cell tower antennas receive the signal from a device, time stamp it, and send it to the mobile switching center. At the center, the differences between the TOA from different antennas, $\tau_i - \tau_j$, are used for positioning the device. Cellular service providers using GPS atomic clock for synchronization can calculate the time differences within nanosecond precision. The CDMA-based 3G networks use DSSS and calculate the cross-correlation of the signal sent from the device with the reference signal at the cell tower antenna for wireless communication. The TOA of the peak of this signal can be time-stamped for the localization purpose taking place in the central switch. The U-TDOA is used for many other applications. For example, for wireless electronic fencing application, one may install three reference receivers in three corners of a

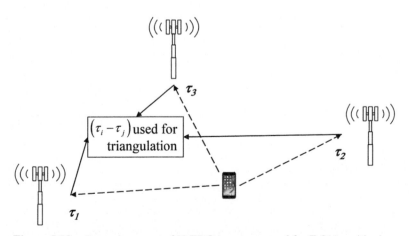

Figure 5.18   General concept of U-TDOA systems used for E-911 positioning.

house and tag a pet with a device radiating DSSS signal. The reference points use a sliding correlator to calculate the TOA and send it to a central device, such as a smart phone, a pad, or a laptop to calculate and display the location of the pet.

5.4.3 Using Transponders to Avoid Synchronization

Less complicated sensor networks operating with low energy in smaller areas use active or passive transponders to report the arriving time of the received signal to the transmitter. The transmitter then uses the arrival time of the returned signal for calculation of the round trip and consequently the distance between the transmitter and the receiver. Figure 5.19 provides an overview of measurement of TOA for ranging using the round trip. An RFID reader or a ZigBee devices transmits a periodic PN sequence and the RFID tag or another ZigBee device returns the signal to the transmitter. The transmitted and the received signals are cross correlated to determine round trip TOA (RTTOA) using the peak of the cross-correlation. To calculate the actual TOA, the Receiver Processing Time (RPT), the time taken by the receiver to retransmit the signal is subtracted from the RTTOA and then divided by 2:

$$\tau = \frac{RTTOA - RPT}{2}. \tag{5.5}$$

The distance between the transmitter and the receiver is then calculated from $r = c \times \tau$.

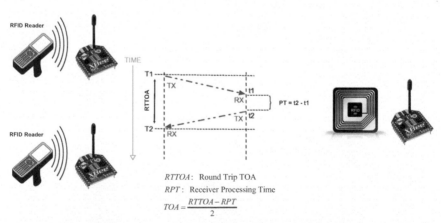

Figure 5.19 General concept of a round trip TOA ranging.

5.5 TOA Positioning in Non-Homogeneous Environment

In our analysis of TOA ranging, we have calculated the distance from the TOA using $r = \tau \times c$, where τ is the time of flight of the direct path between the transmitter and the receiver and c is the speed of light. In reality, radio propagation velocity is expressed as a function of the relative permittivity:

$$\begin{cases} v(\omega) = \dfrac{c}{\sqrt{\varepsilon_r(\omega)}} \\ r = \dfrac{c}{\sqrt{\varepsilon_r(\omega)}} \times \tau \end{cases}, \tag{5.6}$$

where velocity, v, is a function of permittivity, ε_r, and permittivity is a function of the (radial) frequency of operation, ω. Our assumption in our analysis was that the medium for transmission is the homogeneous free space with permittivity of one for all frequencies; consequently. the speed of radio wave propagation in our analysis was the same as speed of the light. This assumption is actually correct for most urban and indoor radio areas, in which the direct path is either in line of sight or obstructed with a few narrow obstacles such as walls or furniture in which conductivity is close to 1. If we have homogeneous environments with different conductivity, for example, in under water localization, the same equation is applied with the adjustment on the conductivity of the medium. Measurement of time of flight in non-homogeneous environments is more complex, and we have a few important positioning cases in non-homogeneous media.

In GPS positioning, signal from the satellites passes through ionosphere and troposphere layer scintillations before reaching surface of the ground. The speed of propagation in these media is lower than the speed of light causing additional delays for arrival [Coc91]. In UWB precise short-range positioning, when the signal passes through a wall or a piece of furniture, the speed of propagation inside these objects is less than light causing additional delay resulting in loss of expected precision [Dar08]. In RFID positioning, when the tag is attached to a specific product inside a large container or, when a temperature sensor is inside a food container [Nad12], we need to use TOA to attain reasonable precision. When the container has fruits or other material with high conductivity, we have a non-homogenous radio propagation environment that affect TOA positioning. The propagation speed inside the liquid-like fruits and free space gaps between the solid materials are quite different. Inside the human body is a non-homogeneous liquid-like environment, where the speeds of radio propagation in different organs are different [Pah12]. In this section, we explain two examples for non-homogeneous positioning applications. The first example is related to the effects of atmospheric layer delays on GPS positioning and the use of the differential GPS to control these effects. The second example is modeling of the effects of non-homogenous environment inside the human body on the TOA-based positioning.

5.5.1 Non-Homogeneous Medium and Differential GPS

In non-homogeneous environment, each subject between the transmitter and receiver has different characteristics of conductivity and relative permittivity, if we do not know the exact velocity of propagation and the thickness of each media, it causes error in time-of-flight estimations. Assuming that the direct path between a transmitter and a receiver passes through different medium with their own depth, r_i, the total or the actual distance is given by:

$$r = \sum_{i=1}^{N} r_i = \sum_{i=1}^{N} \frac{c}{\varepsilon_i(\omega)}, \tag{5.7}$$

where r_1 to r_N are the distances travelled in each media between the transmitter and the receiver.

The 24 GPS satellites on the orbit each transmitting their own PN sequence are used for positioning. Satellites are synchronized in time with atomic clock, and we know their locations, (x_i, y_i, z_i). As shown in Figure 5.20, GPS signal passes through ionosphere and troposphere layers surrounding the earth. The speed of radio wave propagation in these layers is lower than the speed of propagation in free space. These non-homogeneities in propagation medium cause additional unknown delays in the TOA of the received signal. Therefore, in addition to the location parameters, (x, y, z), a GPS receiver on the earth needs to measure a time correction factor, T_c, to compensate for the atmospheric delays. Assuming delays from all satellites have the same value, T_c, we can solve this problem by measuring our distance from at least four satellites:

$$\begin{aligned} c(\tau_i - T_c) &= r_i - cT_c \\ &= \sqrt{(x - x_i)^2 + (y - y_i)^2 + (z - z_i)^2}; \quad i = 1, 2, 3, 4 \end{aligned} \tag{5.8}$$

where τ_i is the TOA of the signal from the *i-th* satellite. The clock correction compensates for the additional delay error caused by lower velocity of propagation in ionosphere and troposphere and eliminates the need for the receiver to know the timing of each of the satellites. In this formulation, the effects of velocity change in non-homogeneity of the medium are averaged over all

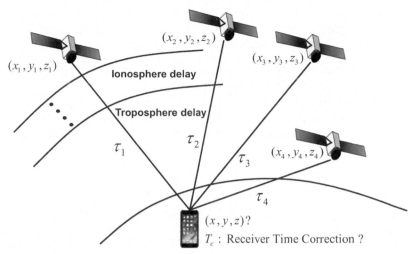

Figure 5.20 General scenario for positioning in GPS and the role of ionosphere and troposphere delays.

satellites and provide accuracies of approximately 15 m for commercial GPS applications.

One way to improve the performance is to correct the ranging error due to non-homogeneity of the medium by estimating the velocity for individual satellite links using a ground station at a known location. Figure 5.21 shows the basic idea behind this concept, a ground station measures its distance from satellites estimated by normal calculations and compares it with its actual distance from that satellite. The ground station could be installed for this purpose or can be an existing infrastructure such as a cell tower. Differential GPS systems install the station for measurement and broadcast their error information for local users of the GPS to enable them to bring the accuracy to as low as tens of centimeters [Mis06]. Cell tower-assisted GPS systems use a similar basic concept to collected data in a fixed location to extend the coverage of GPS in urban and indoor areas [Zan09].

The fundamental RF positioning idea behind differential measurements is to collect vital data in known locations and use that to improve precision of a localization system and it can be used for any positioning system. In RSS-based systems using fingerprinting, we have a software implementation of the same concept because we go to know locations and measure the RSS to improve the performance against the shadow fading. In TOA-based localization, we use the concept of measurement in a known location to adjust for uncertainties on the velocity.

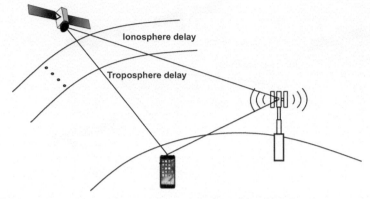

Figure 5.21   Using a ground station to measure the velocity of radio wave propagation in a non-homogeneous medium.

5.5.2 Time of Flight Inside the Non-Homogeneous Human Body

In Section 3.2.2, we showed that using RSS-based localization one may achieve accuracies around a few centimeters inside the human body. To achieve more precision in localization, we need to consider the TOA-based localization. However, human body is a non-homogeneous medium with a variety of permittivity for different organs. Inside the human body, we do not know the length of the portion of the direct path between transmitters that passes through different organs. Therefore, we need to use an average speed on the measured TOA to estimate the distance. To calculate the average speed, one may use the *average* permittivity to estimate the average propagation velocity:

$$\begin{cases} \bar{\varepsilon}_r = \sum\limits_{i=1}^{N} p_i \varepsilon_i \\ \bar{v} = \frac{c}{\sqrt{\bar{\varepsilon}_r}} \end{cases}, \tag{5.9a}$$

where p_i is the probability of having a specific organ in between the WVCE and body mounted sensors. We can assign that probability, for example, based on the average volume size of the organs inside the abdomen.

The estimated distance in this case is expressed as:

$$\begin{aligned} \hat{r} = \tau\bar{v} &= (\hat{\tau}_1 + \hat{\tau}_2 + \cdots + \hat{\tau}_n)\frac{c}{\sqrt{\bar{\varepsilon}_r}} \\ &= \sum_{i=1}^{n} \frac{r_i}{v_i} \frac{c}{\sqrt{\bar{\varepsilon}_r}} = \sum_{i=1}^{n} \frac{r_i}{c/\sqrt{\varepsilon_i}} \frac{c}{\sqrt{\bar{\varepsilon}_r}} = \frac{1}{\sqrt{\bar{\varepsilon}_r}} \sum_{i=1}^{n} \sqrt{\varepsilon_i} r_i \end{aligned} \tag{5.9b}$$

Table 5.2   Permittivity and volume of organs in mm^3 used for simulation of the effects of non-homogeneity of the human body

Permittivity and Volume of Organs in mm^3		
Intestine (50.7, 3936.3)	Stomach (67.8, 357)	Gall bladder (52.3, 12.4)
Lung (23.77, 4320)	Heart (65.97, 625.4)	Kidney (68, 325.1)
Spleen (63.1, 160.2)	Liver (51.15, 1357)	Muscle (47.8, 32403.4)

The difference between d, in Equation (5.7), and $\hat{d}$, in Equation (5.9b), is the ranging error that is caused by the non-homogeneous propagation environment inside the human body:

$$DME_{N-H} = \hat{r} - r = \frac{1}{\sqrt{\bar{\varepsilon}_r}} \sum_{i=1}^{n} \sqrt{\varepsilon_i} r_i - \sum_{i=1}^{N} \frac{c}{\varepsilon_i(\omega)}. \qquad (5.10)$$

This error between the actual distance and the distance measured by TOA and average velocity of the propagation is caused by using a single velocity rather than actual multiple velocities.

To show the expected amount of ranging error in TOA-based localization due to the effects of non-homogeneous medium inside the human body, in [Ye12], a 3D simulation inside the human torso was conducted. The human torso includes eight major organs of the human body with different volume and conductivity (Table 5.2). Figure 5.22 shows the scenario for collecting the data. Approximately 500 locations on the torso are selected randomly. The direct path between pairs of locations in different sides of the body was established, and each path was segmented into paths through different organs. Then, using Equations (5.7) and (5.9b), the estimated and the actual distances were calculated, and using Equation (5.10), the ranging error caused by non-homogenous medium inside the body was determined. The average permittivity used for time-of-flight estimation is calculated by weighting the permittivity of each organ according to their volume, and the average permittivity was 46.35 in the torso environment. The permittivity and volume of different organs used for this simulation are shown in Table 5.2.

Figure 5.23 presents the results of simulation of TOA-based ranging error caused by non-homogeneous characteristics of the human body and the best fit Gaussian distribution to the results of simulations. The standard deviation of the ranging error is 24.3 mm, while the mean value is −3.92 mm. The mean value of ranging error is a negative value because the largest organ in the torso

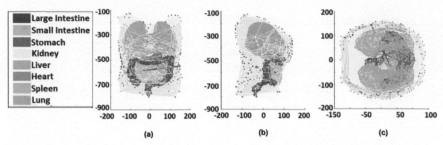

(a)   (b)   (c)

Figure 5.22   The human torso simulation scenario for 3D measurements of the ranging error due to non-homogeneity of the human body for radio propagation (a) front view, (b) side view, (c) top view [Ye12].

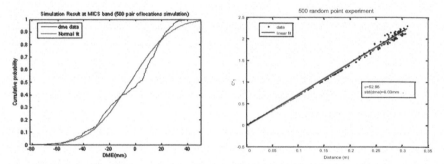

Figure 5.23   The CDF of ranging error for TOA-based localization caused by non-homogeneous characteristics of the human body and best fit Gaussian distribution to the data.

cavity is the lungs, which have a much smaller permittivity value than the average permittivity of human tissues. Hence, the signal propagates faster in the lungs than the average speed of signal propagation inside the human body. When we use the average propagation to calculate the estimated distance, the value is smaller than the real distance, because we underestimated the distance signal went through the lungs. In practice, the value of bias in the estimation does not play an important role because it can be removed easily. The standard deviation of 2.43 mm is caused by variations in the velocity of the wave propagation across the entire torso. In applications such as movement of an endoscopy capsule inside the small intestine, this value can be reduced because variations in the medium are expected to by much smaller than variations in the entire torso.

Assignments for Chapter Five

Questions

1. Name opportunistic signal designed for wireless communication, which can be used for TOA-based positioning.
2. What are the advantages and disadvantages of pulse transmission opportunistic TOA-based positioning in UWB spectrum and in mmWave spectrum?
3. Why DSSS is referred to as a pseudo pulse transmission technique for positioning?
4. Why FHSS is referred to as pseudo multi-carrier modulation for TOA-based positioning?
5. Name two approaches for time synchronization in TOA-based ranging and compare them in terms of complexity of implementation and applicability for large distance ranging.
6. What a sliding correlator work and what is the benefit of this technique?
7. What are the challenges in TOA measurement in non-homogeneous environments such as inside the human body?
8. What is a transponder and how it helps in TOA-based positioning?
9. What is U-TDOA and which industry uses that for positioning? Explain the reason why it is used.
10. Compare precision of wireless positioning in different bands for wireless communications shown in Figure 5.2, if we design a TOA-based positioning system using these signals.

Problems

Problem 1:

The chip rate of the DSSS of the cdmaOne is 1.24 Mcps, and the length of the PN-sequence is 127.

a) Sketch and label the periodic autocorrelation function of this waveform.
b) If we use the waveform of full-period of the PN-sequence for distance measurements, what would be the length ambiguity of the distance measurement?
c) If the system is operating at 900 MHz what would be the length ambiguity if we use the carrier phase for measuring the distance? Compare with that of (b) and comment on how we can use both.

d) Calculate the background noise of the system in dBm, and using that, calculate the receiver sensitivity if the minimum required SNR is 6 dB. Include the processing gain in your calculation.

Problem 2:

a) What is the distance ambiguity of a phase measurement using a single-tone localization system operating at center frequency of 2.4 GHz

b) If we want to use two tones to estimate distances up to 100 m, what is the bandwidth of the system. Use the frequency difference between the two tones as the bandwidth.

c) Calculate the background noise in dBm as a function of ambiguity range of TOA measurements using two tones. Plot the ambiguity range in meters as a function of background noise in dBm.

d) If we use the first and last tones of the IEEE 802.11g OFDM signal, which has 64 tone in 20 MHz of bandwidth, for measuring TOA, what is the distance ambiguity and how long would it take to measure one TOA? The symbol transmission rate for the 802.11 g is 250 KSps.

Problem 3:

An UWB indoor system uses triangular pulses with a width of 1 ns and a period of 500 ns with a transmitted power of 0 dBm

a) What are the bandwidth and the distance ambiguity of the system

b) What is the background noise of the system in dBm.

c) If the minimum required signal-to-noise ratio is 10dB, what is the coverage of the system in an open area with a distance power gradient of 2.

6

TOA Positioning in Multipath

6.1 Introduction

Using CRLB in Section 4.3, we calculated the precision of TOA ranging in additive white Gaussian thermal noise. Equation (4.15b) in this section showed that the standard deviation of TOA-based ranging error caused by thermal noise is inversely proportional to the bandwidth and SNR. This feature of TOA-based ranging enables us to achieve high-precision ranging in long distances by controlling the bandwidth and power, the SNR. This fundamental simple relationship among precision, bandwidth, and power changes substantially in multipath, and Equation (4.15b) cannot explain the behavior of TOA-based systems in multipath. As a result, analysis, measurement, and modeling of the effects of multipath on the behavior of TOA ranging in different multipath scenarios has been a focal point of research for performance analysis of the TOA positioning systems in the past couple of decades. Analysis of the effects of multipath becomes extremely important in design of TOA-based positioning systems for multipath rich indoor and urban areas, which are the focal point of this book, because extensive multipath has drastic impacts on the performance. This chapter is devoted to the analysis of the effects of multipath on the TOA-based ranging. We discuss multipath conditions in different positioning systems, we analyze effects of multipath on range estimation methods, and we introduce methods for measurement and modeling of the effects of multipath on TOA ranging in urban, indoor, and inside the human body. In Chapter 9, we provide algorithms used for mitigation of the multipath effects on TOA-based ranging.

6.2 Multipath and Positioning Systems

Multipath in open areas with direct line-of-sight (LOS) communication between a transmitter and a receiver shifts the location of the peak of the pulse as a reference for TOA measurement, causing degradation in precision of TOA-ranging. In obstructed LOS (OLOS) multipath conditions, the received power of the direct path is reduced and the peak of the received pulse is shifted further, causing more drastic reduction in precision of TOA-based ranging. In particular, if the obstruction is too strong, the received signal power from the direct path goes below the detection threshold of the receiver misleading the receiver to detecting first reflected path as the direct path causing a huge unexpected ranging errors. The variance of these errors is not represented by CRLB of Equation (4.15b), and these errors are unexpectedly higher than the expected error predicted by this equation. As a result, understanding of the nature of multipath in indoor and urban areas is essential for understanding of the behavior of TOA-based ranging in these multipath-rich environments. We begin our analysis by explaining the causes of multipath before we discuss how it effects precision of TOA ranging.

6.2.1 Causes of Multipath Propagation

Electromagnetic waves at frequencies used in wireless networks can be treated as rays, each representing a propagation path between a transmitter and a receiver [Pah13]. In free space, LOS is the only propagation path connecting the transmitter and the receiver. In indoor and urban areas, multiple propagation paths carry the signal in different directions. Figure 6.1 shows the main four mechanisms causing multipath in indoor and urban areas, demonstrated in the layout of an indoor area. These mechanisms are transmission, reflection, diffraction, and scattering.

Reflection and transmission: Upon reflection or transmission, a ray attenuates by factors that depend on the frequency, the angle of incidence, material used in construction, and thickness of the walls [Pah95]. Reflection and transmission mechanisms often dominate radio propagation in indoor applications. In outdoor urban area applications, the transmission mechanism often loses its importance because it involves transmissions through buildings with multiple walls that reduce the strength of the signal to negligible values.

Diffraction: Rays that are incident upon the edges of buildings, walls, and other large objects can be viewed as exciting the edges to act as a secondary line source. Diffracted fields are generated by this secondary wave source and propagate away from the diffracting edge as cylindrical waves. Since diffractions are secondary waves propagation, their strengths are negligible as they are compared with transmission and reflection. Consequently, diffraction

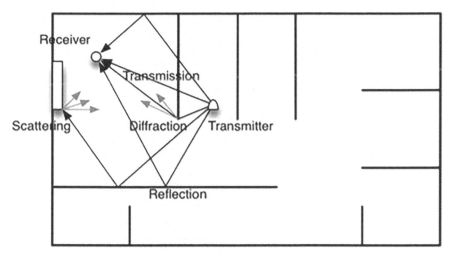

Figure 6.1 Propagation paths and mechanisms causing them.

is an important phenomenon for outdoor application in high-rise urban areas where signals transmission through buildings is virtually impossible. But it is less consequential indoors where a diffracted signal is extremely weak compared to a reflected signal or a signal that is transmitted through a relatively thin wall.

Scattering: Irregular objects such as wall roughness and furniture in indoor areas and vehicles and foliage in outdoor areas scatter rays in all directions in the form of spherical waves. This occurs especially when objects are of dimensions that are on the order of a wavelength or less of the electromagnetic wave. Propagation in many directions results in reduced power levels especially far from the scatterer. As a result, this phenomenon is not that significant unless the receiver or transmitter is located in a highly cluttered environment. For example, this mechanism dominates diffused infrared or any other light propagation when the wavelength of the signal is so high that the roughness of the wall results in extensive scattering. In satellite and mobile radio applications, tree leaves often cause scattering.

6.2.2 Effects of Multipath on TOA Ranging

Figure 6.2 shows the fluctuations of the RSS with distance due to short-term multipath fading and formation of multipath in an indoor area as a mobile terminal moves away from an RF propagating antenna. When the

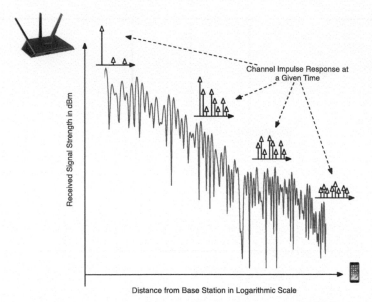

Figure 6.2   Relation between power in the multipath components and RSS fluctuations with distance in wireless channels.

receiver terminal is close to the transmitter, the power of the LOS path is the dominating one and the power in all other paths are negligible. As the terminal moves away from the transmitting antenna, the power in the direct path becomes comparable to those of the other paths and we see several multipath components with similar signal strength. When some objects such as walls, human body, or furniture obstruct the LOS (OLOS) between the transmitter and receiver, the direct path receives significant attenuation and its power may even become less than that of some of the other multipath components. As the distance is increased further, we finally reach a point that direct path is not detectable anymore, but still we can detect other paths arriving at the receiver through windows and reflections from certain walls. This last situation is very alarming for TOA ranging, because we measure the length of an erroneous path and we assume it as the length of the direct path, which is much shorter if it was detected. In indoor areas when a large metallic object such as an elevator blocks the direct path, we may observe a DME in the order of several meters, while the actual distance is only a few meters. Similarly in outdoor urban areas, when a buildings blocks the LOS path, large DME occurs.

6.2.3 Multipath and Positioning System Technologies

At the time of this writing, the most popular RF positioning systems are GPS, using TOA, cellular positioning systems (CPS), using Uplink Time Difference of Arrival (U-TDOA) or RSS, Wi-Fi localization using RSS- and TOA-based ad-hoc localization systems using ZigBee or UWB. Figure 6.3 provides an overview of these positioning systems, their technologies, and their coverage. GPS uses TOA ranging and it has a global coverage in LOS outdoor areas. It uses TOA ranging, because the ratio of coverage to precision of positioning is very high and a simple RSS-based localization is not practical. GPS provides accuracy of approximately 15 m in urban areas, while satellites are approximately 20,000 km away.

The traditional CPS, designed to support E-911 services with an accuracy of around 100 m, have comprehensive coverage of residential areas of the world using the U-TDOA technology. Coverage of cell towers used as the reference point for CPS localization is around 10 km. Although the coverage of CPS is much less than the satellites and it uses the same TOA-based technology, it has much less precision because it is designed for urban and indoor areas, where the multipath is extensive and direct LOS path gets blocked frequently. We can use the CRLB of Equation (4.15b) to calculate

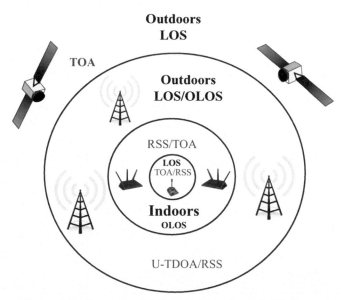

Figure 6.3   Coverage of different positioning technologies.

precision of the GPS because it operates in LOS open areas with negligible multipath effects. CPS is operating in mixed LOS/OLOS environments, and the extensive multipath in these environments is the dominant source of ranging error and these effects are not reflected in Equation (4.15b). As we explained in Section 3.4.5, RSS-based CPS are also used in smart phones and these systems can achieve a precision of a few hundred meters (see Figures 3.22 and 3.23). The reason that the RSS-based CPS provides a performance comparable to the TOA-based CPS is that the urban and indoor areas suffer from extensive multipath. If we use CRLB in thermal noise to compare the performance of these two systems, we will find that these similarities in performance are very unexpected.

We divide indoor positioning systems into systems with coverage of a building and the systems designed for localization inside a room. For localization inside a building, an accuracy of a few meters is adequate to distinguish the position of objects in different rooms. In-building positioning is dominated by OLOS situations posing a challenge to more precise TOA-based positioning, and RSS-based Wi-Fi localization using fingerprinting is by far the most popular technology. Wi-Fi access points are deployed approximately every 30 m inside the buildings, and achieving an accuracy of a few meters is achievable if we go through fingerprinting. For in-room localization, accuracies of around a meter satisfy many positioning applications, and we may need accuracies of up to a few centimeters for applications such as electronic gaming [Zhe15]. However, in-room positioning is dominated by LOS conditions, using TOA-based positioning with ZigBee, or UWB technologies, these precisions are achievable. As we discuss in Section 3.3.3, RSS-based localization using RFID and iBeacon technologies are also used for in-room localization, when accuracy of a few meter is adequate for the application.

6.2.4 Characteristics of Indoor Multipath and TOA Positioning

In wireless networks, we transmit a signaling waveform or electronic *symbol* from a transmitter antenna, and at the receiver, we analyze the received waveform to either extract certain information about the channel and transmitted symbol or to estimate the distance between the transmitter and the receiver. The multipath nature of the channel results in several waveforms arriving at the receiver along different paths, each having a different amplitude, phase, and time of arrival. Therefore, the received waveform changes according to the multipath structure of the environment. In TOA-based systems, the information needed for estimation of the distance between the transmitter and receiver is obtained from the time of arrival of the received signal. Therefore, wireless localization is highly affected by the multipath nature of the channel caused by the different mechanisms introduced in Section 6.2.1. We need to understand these effects to design algorithms for precise TOA-based positioning in these environments.

Walls reflect RF wave propagation like a dark mirror, and in indoor areas, walls are built in parallel. Parallel mirrors produce infinite images each leading to a transmission path between the transmitter and the receiver. Figure 6.4 illustrates the parallel mirroring effects causing a cluster of paths. Figure 6.4(a) shows the physical trace of direct path as well as a second- and third-order reflected path between a transmitter and a receiver inside two parallel walls. Figure 6.4(b) shows associated cluster of arriving paths at the receiver. Infinite number of images produce infinite number of paths with reduced intensities. Parallel walls exist in urban and indoor areas causing clusters of arriving paths. In indoors, these situations occur more often because radio signal penetrates the walls and we have many intertwined parallel walls. Since indoor walls are only a few meters apart the inter-arrival delay between the paths are only a few nanoseconds demanding ultra-wideband transmission to resolve the paths.

In Section 3.3.3, we divided indoor positioning systems into those designed for operation in a room and those which have a wider coverage of an entire building. Inside a room, we have parallel walls and many multipath

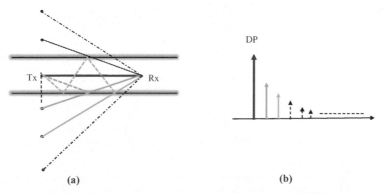

(a)   (b)

Figure 6.4 Parallel mirroring effects causes an infinite cluster of multipath receptions in indoor areas, (a) physical trace of the paths, (b) associated cluster of arriving paths.

arrivals, but RF communication links are LOS leaving the first path as the strongest path. RFID tags, iBeacon, ZigBee, and UWB technologies are often used for in-room short-range localization. Figure 6.5 shows a typical application of an RFID system for short-range proximity broadcasting in a museum. In this application, the user carries an RFID reader to collect information from RFID tags attached to different items in the museum. Because RFID tags operate in very short distances with LOS conditions, multipath effects are minimal and using TOA ranging is possible. If we use multiple antennas at the reader, we can focus the beam and measure the DOA as well. Measuring the distance from TOA and the direction from DOA, we can position the objects in a few meters range of coverage of the RFID systems. iBeacon and Zigbee technologies can provide similar positioning services within a larger coverage area by installing an infrastructure of fixed access points. Coverage of the UWB devices are larger than the RFID and in the range of iBeacon and ZigBee devices, however, UWB system bandwidths are much wider allowing more precise positioning in shorter distances and the capability to operate in LOS/OLOS situations.

To design a positioning system for in-building or indoor geolocation, we should consider that in this environment, we mostly face OLOS situations. Since RF signal can pass through the walls in OLOS situations, sometimes we can detect the direct path and sometime we cannot. When we have an undetected direct path (UDP) condition, we expect

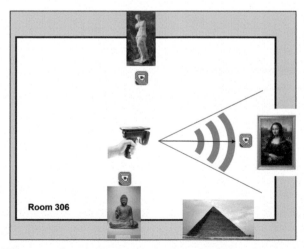

Figure 6.5   RFID tags and readers operating in close proximity of each other with minimal multipath.

large and unexpected ranging errors. The most popular in-building positioning system is RSS-based Wi-Fi positioning, which operates reasonably well at a very low cost in excessive LOS/OLOS and UDP situations inside the buildings. These systems rely on fingerprinting that is not suitable for emergency first responder applications. UWB technology is the better choice because it is a TOA-based system that can resolve multipath components and operate in LOS/OLOS in-building operations [Pah13]. In short-range in-room applications, UWB supports applications such as interactive electronic gaming demanding centimeter accuracies [Zhe15].

6.2.5 Characteristics of Outdoor Multipath and TOA Positioning

In outdoor urban areas, shown in Figure 6.6, the antenna heights are high and buildings obstruct the LOS. Since a building blocking the direct path does not allow the signal to pass through, we have two different multipath conditions: LOS and OLOS. In LOS, the direct path is dominant and the strongest among the paths, the OLOS situation always creates an UDP condition. This can be a blessing in the analysis, as we compare with the indoor, because in outdoors we can associate UDP conditions to physical layout of the buildings. In indoor areas, signals can penetrate through the wall creating a more complex procedure to predict the UDP conditions.

Figure 6.7 shows a typical positioning scenario for a CPS in a campus area. A device in coverage of three cell towers has established a connection with all three of them; however, the signal arriving to two of the towers is arriving from a reflected path. The reflected path shifts the location of the

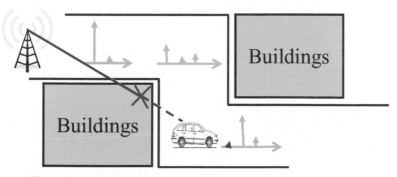

Figure 6.6   Multipath in LOS and OLOS conditions in urban areas.

Figure 6.7   Example of multipath formation in cell tower localization.

peak of the arriving pulse used for TOA-based localization and that shift causes error in estimation of the distance. The DME caused by the multipath is usually much higher than the estimated DME using the CRLB. Also, the error caused by multipath cannot be reduced by increasing the power because increase in the power of direct path also increases the power of the multipath component.

Satellite positioning also suffers from multipath conditions; Figure 6.8 shows the multipath scenario for two GPS satellites. The satellite on the left provides the direct path as well as a reflected path from a wall. The satellite on the right has its direct path blocked by a high-rise building and a reflected path from a building, and diffracted path from a tree is detected by the receiver. In both scenarios multipath is formed. For the satellite on the left,we have two paths but the direct path is dominating and is stronger than the reflected path. For the satellite on the right, direct path is blocked and we have two erroneous weaker paths with similar strengths causing substantial error in estimating the distance.

Cellular networks have their own hierarchy for deployment and coverage of outdoor antennas. A femto cell covers approximately 50 m to cover a building from outside, a micro-cell covers approximately 500 m to cover

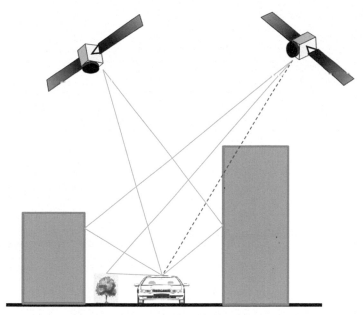

Figure 6.8   Blocked direct path and multipath caused by reflection from the buildings in the GPS.

streets in dense urban canyons, a macro-cell covers a few kilometers for sub-urban areas and highways, and a mega-cell uses satellites for intercontinental coverage. Here, the complexity of multipath conditions decreases by increasing the coverage. A femto cell is exposed to more multipath than a micro-cell and hence the least complex multipath conditions occur for satellite positioning. This situation is the reverse of what we observed in indoors, where the smaller cells produce smaller multipath effects.

6.3 NB and WB Estimation of TOA in Multipath

In Section 4.2, we explained how we can measure the distance based on phase of the received carrier frequency and we referred to that as narrowband (NB) measurement of the TOA. We also explained that the TOA can be measured using periodic narrow pulse transmission, which we referred to as wideband (WB) measurement of TOA. In that section, we also explained that both methods have distance ambiguity while the distance ambiguity of NB measurement is governed by frequency of operation and the distance ambiguity of WB measurement is controlled by the period of transmission of the pulses. In Section 4.3, we derived the CRLB for a variety of NB and WB signals and demonstrated that they produce similar precision, which is in general extremely higher than RSS-based ranging. The analysis presented in Chapter 4 did not include the effects of multipath. In this section, we explain how multipath affects the NB and WB measurement of the TOA and how it affects calculation of the CRLB to analyze the precision of TOA-based ranging in multipath.

To begin this discussion, we need a model for the multipath character-istics of the channel. In the wireless communication literature [Pah95], the multipath characteristics of the channel for a given location of the transmitter and receiver is given by:

$$h(t) = \sum_{i=0}^{L} \beta_i \, \delta[t - \tau_i] \, e^{j\phi_i}, \tag{6.1a}$$

where β_i, Φ_i, and τ_i represent the amplitude of the path, phase of the carrier, and delay associated with the arrival of each path between the transmitter and the receiver, respectively. The phase and delay are related by $\phi_i = \omega_0 \tau_i$ (see Equation (4.1a) and Figure 4.3(a)) with $\omega_0 = 2\pi f_0$ the angular frequency of carrier signal. The actual distance between the transmitter and the receiver is $d = \tau_0 \times c$, and $d_i = \tau_i \times c$; $i = 1, 2, ..L$, where L is the length of multipath components.

Using this formulation for the channel, the average received signal strength used for RSS-based ranging is:

$$P_r = 10 log\,[RSS] = 10 \log \left[\sum_{i=0}^{L} \overline{|\beta_i|^2} \right]. \tag{6.1b}$$

The average over the amplitude square is the time average of observed RSS in a location when either the receiver is moved or objects are moving around the transmitter and the receiver. The measurement of the average RSS is independent of the bandwidth of the measurement device, and therefore, the measured distance using RSS is independent of the bandwidth. In the case of TOA-based ranging, we estimate the time of flight, τ_0, either by measuring the phase of the received carrier, ϕ_0 (NB measurement) or by measuring the time shift of a reference point of a transmitted pulse (WB measurement).

6.3.1 Multipath and NB TOA Ranging

For NB measurement of the phase in practice, we transmit a sinusoid, $x(t) = \cos \omega t$, and we measure the phase of the received signal, $y(t) = \cos(\omega t - \phi)$, then we calculate the distance from $d = \frac{\lambda}{2\pi}\phi$ (Figure 4.3(a)). Figure 6.9 shows the NB measurement process in a multipath channel. As shown in Figure 6.9(a), for a transmitted sinusoid, $x(t) = \cos \omega t$, in a multipath channel represented by Equation (6.1a), the received signal is:

$$y(t) = \sum_{i=0}^{L} \beta_i \cos(\omega t - \phi_i) \atop = \beta \cos(\omega t - \phi)$$ (6.2)

This signal is the summation of *L+1* cosines with the same frequency and different magnitudes and phases. Since the system represented in Figure 6.1(a) is a linear time invariant system and sinusoids are eigenfunctions of these systems, the output is still a cosine with the same frequency and unknown magnitude and phase. Figure 6.9(b) shows the phasor diagram that can be used to calculate these unknown values. As we explained in Section 4.2, we can measure the phase of this received cosine, ϕ, by cross-correlating the transmitted and the received signals. However, the actual phase that we want to measure is the phase of the direct path, ϕ_0, and

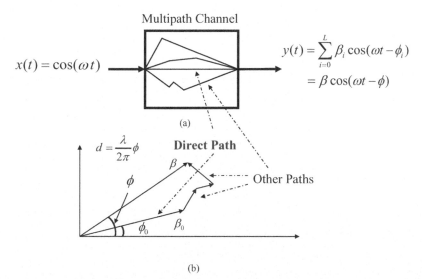

(a)

(b)

Figure 6.9 NB measurement of TOA in multipath, (a) the transmitted sinusoid and the received signal in multipath, (b) the phasor diagram for calculation of magnitude and phase of the received signal.

relation between ϕ and ϕ_0 involves accurate estimate of magnitude and phase of all multipath components. Therefore, NB measurement of the TOA in multipath-rich indoor and densely populated urban areas is not practical. NB measurement of the TOA is practical for GPS and for short-range LOS RFID localization, where direct path is very strong compared to multipath components. Also, NB measurement of the TOA is practical for short-range OLOS localization inside the human body, where the multipath components arrive very close to one another.

6.3.2 Multipath and WB TOA Ranging

Figure 6.10 shows the basic concept of WB TOA ranging in a multipath channel. The transmitted signal is designed to be a sharp narrow pulse $x(t) = s(t)$. In the absence of multipath, we use the time displacement of the peak of the received signal, $y(t) = s(t-\tau)$, with respect to the transmitted signal to determine the time of flight. In multipath arrival, using the channel model given by (6.1), the complex envelope of the received pulse becomes:

$$y(t) = \sum_{i=0}^{L} \beta_i s(t - \tau_i) \, e^{j\phi_i}. \qquad (6.3)$$

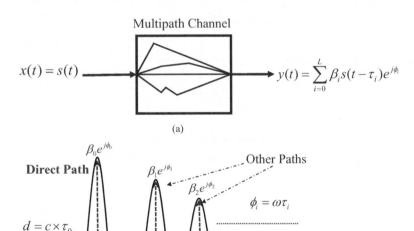

(a)

(b)

Figure 6.10   WB measurement of TOA in multipath, (a) the transmitted narrow pulse and the complex envelop of the received pulses in multipath, (b) multiple arriving pulses.

Figure 6.10(a) shows how this signal is formed, and Figure 6.10(b) shows a sample arrival in multipath. We are interested in measurement of TOA of the direct path, τ_0, which is the first arriving path in Figure 6.10(b). If the duration of the pulse, T, is narrower than the inter-arrival time of the paths, $\tau_1 - \tau_0 > T$, arriving pulses from different paths do not overlap and we can measure the TOA without any error caused by the multipath. The bandwidth of the measurement system is $W = 1/T$; therefore, if $W > 1/(\tau_1 - \tau_0)$, we have non-overlapping pulses. The inter-arrival of the paths is related to the environment and the distance between reflecting objects. In indoor areas, walls are a few meters apart, and in outdoor urban areas, exterior walls of the buildings are tens of meters apart. In open areas, mountains and hills which are hundreds of meters apart cause the reflections. Therefore, bandwidth requirement for operation in multipath changes for different application environments. To have a quantitative idea about the relation between bandwidth and delay of arrival of the paths in an environment, we resort to a simple example.

Example 6.1: (Bandwidth and Path Lengths)

Consider a two-path channel where the received signal consists of the direct path and a path reflected from a wall and assume that we are transmitting a triangular pulse to measure the TOA (Figure 6.11). Figure 6.11(a) shows the multipath scenario, and Figure (6.11(b)) shows the received signal. The TOA or the time of flight is indeed the delay of arrival of the direct path, τ_0. The difference between the length of the two paths is $\Delta d = d_1 - d_0$, and we have $\Delta \tau = \Delta d/c$. Therefore, to avoid overlap between the two pulses, we need a

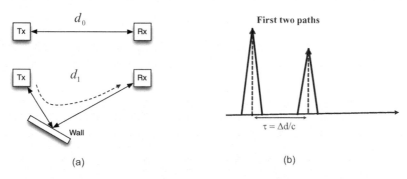

(a)   (b)

Figure 6.11 WB ranging, multipath, and bandwidth requirements, (a) two paths between the transmitter and the receiver, and (b) the received waveform form a transmitted triangular pulse.

bandwidth of:

$$W > \frac{c}{\Delta d} = 3 \times 10^8 = 300 \; MHz/m.$$

This means that if the multipath components are 1 m different in a typical indoor area, we need 300 MHz of bandwidth to resolve the paths, and for an open outdoor area with the second path reflecting from a nearby hill producing 1 km distance between the two paths, we need 300 KHz of bandwidth. For 1 mm difference in distance inside the body, we need 300 GHz of bandwidth!

6.3.3 CRLB of TOA-Based Ranging in Multipath

Our tool for analytical performance evaluation of localization systems is the CRLB, which we borrowed from the classical estimation theory. In Section 4.3, (4.11), we modeled the observed received signal as:

$$o(t) = s(t - \tau_0) + \eta(t),$$

where $\eta(t)$ is the additive Gaussian thermal noise with the variance:

$$E\left\{ |\eta(t)|^2 \right\} = \frac{N_0}{2}.$$

Then, we calculated the CRLB for TOA ranging, and in (4.13b), we show that

$$\sigma_\tau^2 \geq CRLB = \mathrm{F}^{-1} = \frac{\pi N_0}{\int_{-\infty}^{+\infty} \omega^2 |S(\omega)|^2 \, d\omega}.$$

In this section, we present calculation of the CRLB in the multipath environment. Using the channel impulse response model of (6.1a), with a normalized first path amplitude, the observed signal in multipath is given by:

$$o(t) = \sum_{i=0}^{L} \beta_i e^{j\phi_i} s(t - \tau_i) + \eta(t) = s(t - \tau_0) + \sum_{i=1}^{L} \beta_i e^{j\phi_i} s(t - \tau_i) + \eta(t),$$

$$(6.4a)$$

and we want to estimate the TOA of the direct path, τ_0. In (6.4a), in addition to the thermal noise, $\eta(t)$, we also have a multipath arrival noise:

$$\eta_m(t) = \sum_{i=1}^{L} \beta_i e^{j\phi_i} s(t - \tau_i). \qquad (6.4b)$$

In extensive multipath, using central limit theorem, we can approximate the multipath noise as a Gaussian distributed noise. Then, the CRLB is the same as (4.13b), with the thermal noise variance replaced by total variance of thermal and multipath noise:

$$\sigma_\tau^2 \geq CRLB = \mathrm{F}^{-1} = \frac{\pi N_0 + 2\pi E\left[\eta_m^2(t)\right]}{\int_{-\infty}^{+\infty} \omega^2 \left|S(\omega)\right|^2 d\omega} \approx \frac{2\pi E\left[\eta_m^2(t)\right]}{\int_{-\infty}^{+\infty} \omega^2 \left|S(\omega)\right|^2 d\omega}$$

(6.4c)

because in practice, multipath noise dominates thermal noise (i.e. $E\left[\eta_m^2(t)\right] >> N_0$). The thermal noise is a function of power and bandwidth, and we could calculate it numerically based on the specification of the signal used for pulse transmission (see (4.6b)). The multipath noise requires knowledge of the characteristics of the channel impulse response, which is a difficult task, and it depends on the specifics of the channel multipath conditions. More importantly, how can we calculate the CRLB when the direct path is blocked and we have an UDP multipath condition? More details on challenges in calculation of CRLB in multipath and an overview of literature in that area is available in [Gez05].

In general, calculation of CRLB of TOA estimation in multipath becomes very complex and dependent on the multipath profile in a given location as well as the accuracy of the measurement of channel multipath profile. In multipath-rich indoor areas, where we may have tens of paths, computation becomes complex and we need validation of the results using empirical measurements. Because of these reasons, for performance evaluation of TOA-based ranging in multipath, we resort to direct analysis of the DME using empirical measurement and modeling or using computational techniques. These approaches are dependent on multipath condition of the environment, similar to path-loss models, and we need different models for applications in different environments. In empirical measurement approach, we collect a massive empirical database using an UWB measurement system, and we use that database to analyze the behavior of TOA ranging error or to develop models for DME in TOA ranging [Ala06, Als07, Hei09, Gen16]. In computational approach, we emulate the effects of multipath by numerical solution to the multipath radio propagation theories and we validate our results with limited empirical measurements [Ask17, Kha18]. In the next two sections, we address these two techniques.

6.4 Measurement of Multipath Channel Characteristics

Empirical performance evaluation of TOA ranging in multipath relies on the results of wideband measurements of the multipath characteristics of the channel. Wideband measurements provide information on magnitude and time of arrival of the individual paths. Using the TOA of a path, τ_i, we can calculate the phase of the path from $\phi_i = \omega\tau_i$. Wideband measurements of channel characteristics can be performed either in the time domain by transmission of a narrow pulse resembling an impulse or in the frequency domain by measurement of the frequency response of the channel to different carrier frequencies. Using Fourier transform techniques, the time- and frequency-response measurements are exchangeable.

6.4.1 Time-Domain Measurement of Multipath Arrivals

The objective of time domain-measurement systems is to measure the complex impulse response of the channel represented by (6.1a). In practice, if we transmit a narrow RF pulse with envelop, *s(t)*, resembling an impulse, we can measure the complex envelope of the received signal given by (6.3), from which we can extract the amplitude, arrival time, and phase of the individual paths. In this procedure, we assume that the received pulses are non-overlapping (Figure 6.10) and practically:

$$y(t) = \sum_{i=0}^{L} \beta_i \delta(t - \tau_i)\, e^{j\phi_i} \approx \sum_{i=0}^{L} \beta_i s(t - \tau_i)\, e^{j\phi_i}. \tag{6.4}$$

To implement a system to perform this measurement, we can either transmit a wide-band (WB) direct sequence spread-spectrum (DSSS) signal and use a sliding correlator on the received signal (Section 5.5.1) or directly transmit a narrow ultra wideband (UWB) pulse (Section 5.2). As we discussed in Section 5.5.1, the DSSS is a virtual method for implementation of pulse transmission. In both cases, the time resolution of the measurements is inversely proportional to the bandwidth of the measurement system.

In DSSS, we transmit a steady stream of bits, and the ratio of peak to average transmitted power is unity. With the UWB pulse transmission method, an RF pulse is transmitted periodically with a low duty cycle and the ratio of peak to average power is very high. As a result, given amplifiers designed for identical peak power operation, we can achieve greater coverage with the DSSS approach. In practical implementations of the two systems, we will achieve a better coverage with the DSSS and a better resolution and acquisition time with UWB pulse transmission. Consequently, for areas less than a few tens of meters in radius, the UWB pulse sounding technique is more popular, and for larger areas, the DSSS technique is more typically used. For most indoor geolocation applications, path distances of interest are typically up to a few tens of meters, and thus the UWB pulse transmission is applied. For outdoor areas such as cell tower localization, path distances are longer, and the spread-spectrum technique is more typically used. These systems are also used for multipath analysis for wireless communication applications, and more details of their implementation is available in [Pah95].

6.4.2 Frequency-Domain Measurement of Multipath Arrivals

In frequency-domain measurements of radio propagation characteristics, the frequency response of the channel is measured, and using the inverse Fourier transform, the channel impulse response representing the multipath profile is calculated. If we consider (6.1a) as the objective of time-domain measurements, in the frequency domain, we measure the Fourier transform of this function on the delay variable that is given by:

$$H(f) = \int_{-\infty}^{\infty} h(t)e^{-j\omega t}dt = \sum_{i=1}^{L} \beta_i(t)e^{-j\omega\tau_i}e^{-j\phi_i}, \qquad (6.3a)$$

where similar to (6.1a), $\phi_i = \omega_0\tau_i$. This function is the Fourier transform of the complex envelope of the ideal channel impulse response given by (6.1a), which is a time-limited waveform beginning at τ_0 and ending in τ_L. Since the Fourier transform of a time-limited channel has infinite duration, we cannot ideally measure characteristics of the channel, but if we use a bandwidth of $W > c\Delta\tau_{\min}$, as we explained in Example 6.1, we can resolve multipath components and avoid inter-path interference. Anyways, measurement systems are band-limited and what we measure is actually a windowed frequency characteristics of the channel defined as:

$$H(f) = H(f,0) = W(f) \sum_{i=1}^{L} \beta_i e^{-j\omega\tau_i}e^{-j\phi_i}, \qquad (6.3b)$$

where $W(f)$ represents the frequency domain characteristics of the RF filter used in the measurement system.

6.4.3 UWB Multipath Arrivals Measurement Using a VNA

Vector Network Analyzer (VNA) is the most popular instrument used for measurement of multipath characteristics in local areas. The VNA were originally designed for measurement of characteristics of two-port electronic networks such as filters or amplifiers. Later on, it was discovered that it is very instrumental for measurement of multipath characteristics of wireless indoor communication systems [Pah89, How90] and analysis of the effects of multipath on TOA-based indoor geolocation systems [Pah98]. The VNA basically measures the frequency response of a two port device connected to the instrument by sweeping a carrier at discrete frequencies at the input port of the device and measuring the phase and amplitude of the received signal at the output port of the device.

Figure 6.12(a) shows the general setup of a VNA for measurement of multipath channel characteristics. The output of the VNA is connected to an antenna and the input is connected to another antenna located in a different location. This way, the VNA measures the frequency characteristics of the wireless channel between the two antennas. For measurement of the characteristics at a given center frequency and bandwidth, we need to purchase or design antennas for that frequency with the given bandwidth and simply connect the antennas to the VNA ports. Figure 6.12(b) shows three typical

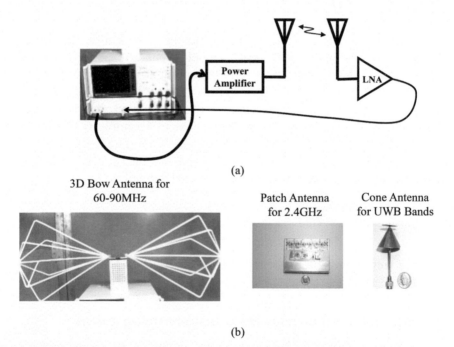

(a)

3D Bow Antenna for
60-90MHz

Patch Antenna
for 2.4GHz

Cone Antenna
for UWB Bands

(b)

Figure 6.12   Measurement of TOA using a vector network analyzer (VNA), (a) system setup, (b) picture of sample antennas at different center frequencies.

antennas at different frequencies, which are used for wideband measurements of multipath characteristics of indoor radio propagations. If the coverage of the system is not adequate, one may add an appropriate power amplifier at the transmitter or a linear noise amplifier (LNA) at the receiver side. By taking the inverse Fourier transform of the frequency response, VNAs also measure the impulse response, which shows the multipath characteristics of the medium between the two antennas. Since the VNA, the antennas, and amplifiers are all off-the-shelf items, design of the system is very simple and practical and we can use that for empirical measurement and modeling of the effects of multipath on TOA indoor geolocation.

The largest bandwidths commonly used for RF signaling for wireless networks are several GHz available at UWB frequencies of 3.4–10.1 GHz and mm Wave frequencies of 57–64 GHz. For measurement of the indoor multipath characteristics, the UWB bands at 3.4–10.1 GHz are more suitable, because at 60 GHz, signal cannot penetrate through the walls and we cannot measure the multipath characteristics in OLOS conditions. At the time of this writing, 60 GHz is a very popular band considered for 5G systems and UWB technology has not gained much of momentum for wireless communication application development. However, UWB measurement of the multipath characteristics of the indoor radio channel is very popular and useful for scientific experiments in multipath modeling for TOA-based localization [Ala06, Als07, Gen16, etc.].

6.5 Empirical Analysis of the TOA-Based DME

Figure 6.13 shows a typical result of an UWB measurement of the radio channel characteristics using a VNA in the frequency band of 3–6 GHz using a VNA. Figure 6.13(a) shows the frequency response in the selected band, which reveals the frequency selectivity of the channel caused by multipath conditions. In a multipath channel, magnitude and phase of each carrier are calculated from a phasor diagram, shown in Figure 6.10(b). For certain carrier frequencies, phase of multipath components align against each other, resulting in a very small amplitude for the received waveform. This phenomenon caused by multipath arrival of the paths is referred to as *frequency selective fading* because it occurs at certain random frequencies [Pah95].

Figure 6.13(b) shows the inverse Fourier transform of the frequency response in the 3–6 GHz band (Figure 6.13(a)), which is the time response of the channel and demonstrates the multipath arrival. Figure 6.13(b) also shows the time associated with occurrence of the peak of each path and we use that

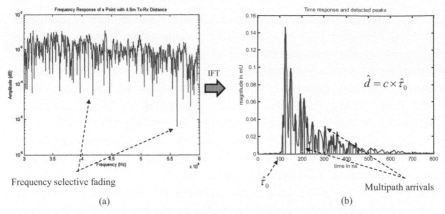

Figure 6.13   UWB measurement of TOA for a frequency band of 3–6 GHz, (a) frequency response with frequency selective fading, (b) multipath profile with the TOA of the direct path.

as the estimate of the arrival time of the paths. The time of occurrence of the first peak, $\hat{\tau}_0$, provides an estimate of the TOA of the direct path and we use it to estimate the distance as $\hat{d} = c \times \hat{\tau}_0$. The expected arrival time of the direct path is $\tau_0 = d/c$, where d is the actual physical distance between the transmitter and the receiver. The TOA-based ranging error or the distance measurement error (DME) is the difference between the estimated distance and the actual distance

$$DME = \varepsilon_{d,w} = \hat{d}_w - d \qquad (6.4)$$

 which depends on the distance between the transmitter and the receiver, the physical layout of the environment, and the bandwidth of the signal.

Figure 6.14 shows the cumulative distribution function (CDF) of the DME of a set of VNA measurements at 3–6 GHz in a room with LOS condition between the transmitter and the receiver. Each point in the CDF represents the DME in a given location; the square of this value is comparable to the CRLB given by (6.4b). The variance of DME is the empirical equivalent of the results of CRLB analysis.

6.5.1 Empirical Analysis of the Effects of Bandwidth on DME

One of the major questions in the design of a TOA-based positioning system is the selection of bandwidth. In Section 4.3, we derived the CRLB for precision of the TOA-based ranging and we showed that it is inversely proportional to the bandwidth. In multipath, variance of the noise is also dependent on the bandwidth, but as shown in (6.4c), this relationship is very

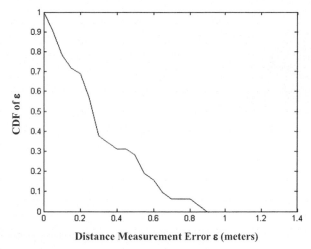

Figure 6.14 CDF of DME at different locations in a LOS indoor area for frequency range of 3–6 GHz.

complex and dependent on the multipath arrivals, which changes from a location to another. Empirical analysis of the effects of bandwidth on TOA ranging error using a VNA is an appropriate approach to address this problem. Empirical measurement of the DME using a VNA does not need to be an UWB measurement. Using an appropriate antenna, we can measure the DME for any frequency band (Figure 6.12(b)) and analyze the effects of bandwidth on TOA ranging in a multipath environment.

Figure 6.15 shows the results of empirical analysis of TOA ranging using VNA measurements in a typical residential building at the Worcester Polytechnic Institute, Worcester, MA [Ben99]. Figure (6.15(a)) shows the outside photograph of the building and floor plan of the first floor of this building. Locations of the transmitter and the receiver are also identified on the floor plan. Figure (6.15(b)) shows four multipath profile and anticipated TOA (read lines) for different bandwidths and center frequencies taken in the specified locations on the floor plan. The 3D bow-tie antenna in Figure 6.12(b) is used for 60 and 90 MHz measurements, and two different dipole antennas are used for measurements at 500 MHz and 1 GHz. The difference between the anticipated TOA (red line) and the peak of the first path (measured TOA of direct path) is the error in measurement of TOA, and if we multiply that with the velocity of light, we have the DME of the measurement in this specific experiment. If we repeat the experiment as people or objects move close to

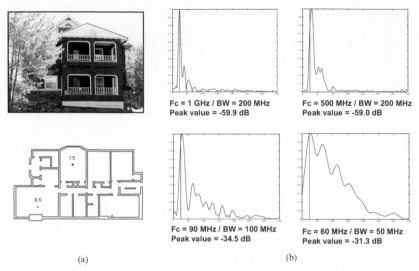

Fc = 1 GHz / BW = 200 MHz
Peak value = -59.9 dB

Fc = 500 MHz / BW = 200 MHz
Peak value = -59.0 dB

Fc = 90 MHz / BW = 100 MHz
Peak value = -34.5 dB

Fc = 60 MHz / BW = 50 MHz
Peak value = -31.3 dB

(a)   (b)

Figure 6.15   Empirical analysis of TOA ranging using VNA measurements, (a) a residential building at WPI campus and its first floor plan with given location of transmitter and the receiver, (b) four multipath profile and anticipated TOA for different bandwidths and center frequencies.

the transmitter or the receiver, the channel multipath profile changes and the associate DME is changed. The variance of these errors caused by changes in multipath condition for fixed locations of the transmitter and the receiver is the empirical value of the CRLB. The larger the bandwidth, the smaller is the mean and variance of the DME, and statistics of quantitative values for these measurements are available in [Ben99].

6.5.2 Empirical Analysis of TOA in LOS and OLOS Conditions

In Section 6.3.1 we used intuitive observations from Figure 6.2 to categorized indoor multipath conditions into LOS and OLOS situations in three classes of dominant direct path, non-dominant direct path, and undetected direct path. We can use VNA measurements for empirical validation of actual occurrence of these conditions. Figure 6.16 shows empirical measurement of occurrence of these three conditions measured by a VNA measurement system [Pah98]. Figure 6.16(a) demonstrates a dominant direct path condition in LOS, where we can easily observe the direct path with minimal errors. Figure 6.16(b) shows a non-dominant direct path condition in an OLOS situation where the

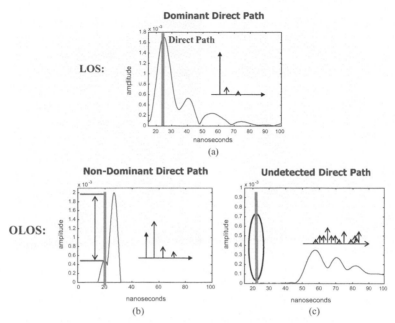

Figure 6.16 Empirical WB measurement of the indoor multipath conditions, (a) LOS measurement with dominant direct path, (b) OLOS condition with non-dominant direct path, (c) OLOS with undetected direct path.

direct path is still the first path but not the strongest, resulting in a noisier estimate of the arrival of the direct path. Figure 6.16(c) shows an undetected direct path OLOS situation with a large DME, for which the direct path has gone under the noise threshold and we erroneously detect the first arriving path as the direct path.

6.5.3 Empirical Analysis of the Effects of Human Body

The most intuitive measurement analysis for the effects of human body on TOA is to use different human bodies with different weights between the same transmitter and receiver. Figure 6.17 shows measurement scenario for the analysis of the effects of human body on TOA measurement. Figure 6.17(a) depicts the scenario when nobody is between the transmitter and the receiver. Figure 6.17(b) shows the scenario when a human body obstructs the direct path between the transmitter and the receiver. The VNA is connected to a pair of 900 MHz dipole antennas, and it measures the

(a)   (b)

Figure 6.17   Measurement scenario for the analysis of the effects of human body on TOA measurement, (a) no body between the transmitter and the receiver, (b) a human body obstructing the direct path.

frequency response between 850 and 950 MHz and calculates its inverse Fourier transform to determine the time response of the channel. Figure 6.18 shows four measurements of the time response of the channel, when there is nobody between the transmitter and the receiver (green line) and when a 110, 130, and 230 pound person is in between. The TOA of the peak and the strength of the first received path is changing as the weight of the person obstructing the direct path increases. In Section 6.7.2, we present an empirical model for the effects of human on TOA ranging, and in Section 2.8.3, we use diffraction and creeping wave theory to analyze the behavior of the received signal in these scenarios.

6.5.4 UWB Limitations for Optimal DME

Analysis of the effects of bandwidth on the performance of TOA-based systems is very complex, and empirical measurements are needed to validate intuitive observations using mathematical analysis. One of the intuitive observations regarding TOA system, which we have examined so far in this book, is that as we increase the bandwidth, the performance of the TOA-based positioning systems improves. We began development of this intuitive

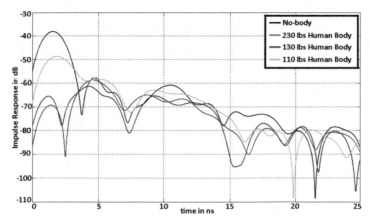

Figure 6.18 Channel time response obtained from measurements with human subjects of different weights between a transmitter and a receiver.

understanding by calculation of the CRLB in open areas in Section 4.4, and using Figure 6.15, we demonstrated that this intuitive observation is correct in multipath conditions as well. In this section, we provide an example to demonstrate that if we increase the bandwidth beyond certain limits, the overall performance of a TOA positioning system in rich multipath will degrade.

Figure 6.19 shows two UWB measurements of multipath characteristics of an indoor area in the same OLOS locations of the transmitter and the receiver for bandwidths of 500 MHz and 3 GHz. In Figure 6.19(a), with 500 MHz of bandwidth, the direct path is above the detection threshold. In Figure 6.19(b), with a bandwidth of 3 GHz, we have resolved more multipath rays and that has reduced the strength of the direct path bringing it under the detection threshold. As a result, the DME with 500 MHz in this location is smaller than the DME with 3 GHz of bandwidth. This counter-intuitive observation only occurs in extensive OLOS indoor multipath, where direct path is not the strongest arriving path.

Figure 6.20 provides the spatial distribution of the DME in a 25 m × 50 m building in a scenario with four outside reference points (RP). Figure 6.20(a) shows the results for a positioning system with 500 MHz of bandwidth, and Figure 6.20(b) shows the results for 3 GHz of bandwidth [Li03]. The empirical positioning DME is calculated for different locations of the building, and they are color-charted with red representing high DME and blue representing

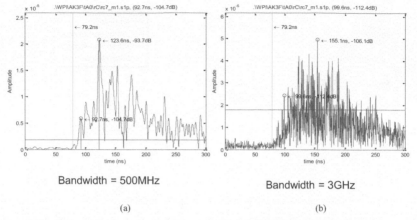

Figure 6.19 Two measurements in the same OLOS locations of the transmitter and the receiver, (a) at 500 MHz direct path is detected, (b) at 3GHz the direct path is under detection threshold.

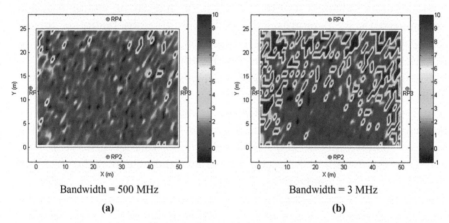

Figure 6.20 Spatial distribution of the DME in a 25 m×50 m building in a scenario with four outside reference points (RP), (a) for 500 MHz bandwidth, (b) for 3 GHz bandwidth.

the low DME. Appearance of more red spots for 3 GHz bandwidth demonstrated the extent of degradation in precision as they are compared with red spots for 500 MHz bandwidth. This empirical experiment demonstrates that increasing bandwidth beyond 500 MHz should be carefully examined in application environment.

6.5.5 An Existing UWB Indoor Measurements Database at NIST

Empirical UWB measurements are time-consuming and expensive, but they are needed for performance evaluation of positioning systems. A database of the results of UWB frequency domain measurements of indoor radio channel characteristics is prepared by the National Institute of Standards and Technology (NIST), Washington DC, and is available at https://www-x.antd.nist.gov/uwb/. The database provides the time- and frequency-domain characteristics of numerous measurement scenarios in different NIST buildings with a user-friendly interface and logical classification of data. This is a very useful publicly available database, in particular for educational purposes. We will use this database in a project at the end of this chapter.

6.6 UWB Modeling of TOA Ranging Error

In Section 6.4, we introduced UWB measurements using a VNA, and in Section 6.5, we used UWB measurements for empirical analysis of TOA-based ranging error. In this section, we provide two example applications of empirical UWB modeling of the behavior of DME for TOA-based ranging. The first example uses UWB measurements in a building to design an empirical model relating the DME to the distance between the transmitter and the receiver and the bandwidth of the system [Ala06]. The second model uses UWB measurements in a room to model the effects of human body on measurement of TOA [Gen13]. These models provide empirical results in multipath channels that are equivalent to calculation of CRLB in additive Gaussian noise channels.

6.6.1 UWB Empirical Modeling of DME in Indoor Areas

In LOS conditions, the first path in the profile is the strongest path and also the representative of the direct path between the transmitter and the receiver, and the timing of the peak of this path is measured to determine the TOA. Under multipath conditions, the peak of this pulse will shift from its expected value because of the effects of other multipath components that are close to the first path. The shift in the peak causes an error in estimating the TOA and consequently the estimated distance between the transmitter and the receiver. This error is a function of the width of the pulse and consequently the bandwidth of the system that is inversely proportional to the width of the pulse.

In OLOS conditions, sometimes the direct path is blocked by objects, such as elevators, metallic furniture and doors, or even human body, and if the strength of this path falls below the detection threshold of the receiver, we have an undetected direct path (UDP) condition that causes a large error in the measurement of the TOA (Figure 6.16(b)). In principle, this error will occur no matter how large the bandwidth of the system is. To understand the behavior of the TOA systems in multipath-rich areas, we need to model the relation between the DME and the multipath conditions in an environment. Since this relationship is very complex, in a manner similar to other statistical models for RF propagation, we need to resort to statistical and empirical modeling. We proceed with the description of one of these models, described in [Ala06], which uses the empirical UWB measurements to model the DME to the bandwidth and the distance in a typical office environment.

The model in [Ala06] defines the DME for a system with a bandwidth of w, with (6.4):

$$\varepsilon_{d,w} = \hat{d}_w - d,$$

where d is the actual distance between the transmitter and the receiver and $\hat{d}_w = c\hat{\tau}_{0,w}$ is the estimate of the distance obtained from measurement of the timing of the first peak of the received channel profile for a given bandwidth. The model divides the DME in a typical indoor office area into two components: one caused by multipath arrivals close to the first detected peak, $\varepsilon_{m,w}$, and the other component the UDP error that is added to the multipath error whenever an object blocks the direct path and a UDP condition occurs, $\varepsilon_{U,w}$:

$$\varepsilon_{d,w} = \varepsilon_{m,w} + \xi(d)\,\varepsilon_{U,w} = \gamma_w \cdot \log(1+d) + \xi(d) \cdot \varepsilon_{U,w}. \qquad (6.5a)$$

The multipath error has two components: a log(1+d) scaling factor that adjusts the amount of error with distance using a logarithmic scale starting with a minimum of zero and logarithmic growth after that, and a Gaussian random variable γ_w with the mean and variance of m_w and σ_w^2 with the probability density function:

$$f_{\gamma_w}(x) = G(m_w, \sigma_w) = \frac{1}{\sigma_w \sqrt{2}} e^{-\frac{(x-m_w)^2}{2\sigma_w^2}}. \qquad (6.5b)$$

The statistics of this random variable adjusts the error to the bandwidth of the system. This approach isolates the effects of distance and bandwidth on the distance measurement error. The UDP error is multiplied by a binary random

Table 6.1 Parameters used for the modeling of the ranging error using (6.5) [Ala06]

W(MHZ)	m_W(m)	σ_W(cm)	$P_{closeU,W}$	$P_{farU,W}$	$m_{U,W}$(m)	$\sigma_{U,W}$(cm)
20	3.66	515	0	0.005	-12.83	0
50	1.57	205	0	0.009	24.48	21.1
100	0.87	115	0	0.091	5.96	358.5
200	0.47	59	0.006	0.164	3.94	289.0
500	0.21	26.9	0.064	0.332	1.62	80.9
1000	0.09	13.6	0.064	0.620	0.96	60.4
2000	0.02	5.2	0.070	0.740	0.76	71.5
3000	0.004	4.5	0.117	0.774	0.88	152.2

variable $\xi(d)$ that reflects the probability of occurrence of a UDP condition as a function of distance d and another Gaussian random variable $\varepsilon_{U,w}$ with the mean and variance of $m_{U,w}$ and $\sigma_{U,w}^2$, and a probability density function:

$$f_{\varepsilon_{U,w}}(z) = G(m_{U,w}, \sigma_{U,w}) = \frac{1}{\sigma_{U,w}\sqrt{2}} e^{-\frac{(z-m_{U,w})^2}{2\sigma_{U,w}^2}}. \tag{6.5c}$$

The binary random variable $\xi(d)$ is 1 if a UDP condition occurs and otherwise it is zero. The probability of occurrence of the UDP is further partitioned into two segments at 10 m break point. As the distance increases, the probability of occurrence of the UDP condition increases. The variance of the error decreases with increase in the bandwidth of the system. Table 6.1 provides the parameters for this model collected through a measurement campaign, using the measurement system shown in Figure 6.12(a) with the UWB cone antenna of Figure 6.12(b). The measurement scenario was at random location in the Atwater Kent Laboratories, Worcester Polytechnic Institute, Worcester, MA, and a total of 405 UWB measurements in one floor of the laboratory were used for calculation of the parameters of the model. The database is partitioned into two sets based on their distances with respect to the breakpoint of 10 m and each partitioned database is further separated into detected direct path and UDP cases to calculate different parameters of the model. Figure 6.21 compares the CDF of the results of empirical measurements versus the values produces by the model for bandwidths of 200 MHz and 1 GHz. As shown in the figure, this model fits well with the empirical DME measurements.

This model can be used for emulation of the DME for a TOA ranging system with a specific bandwidth and a given distance between the transmitter and the receiver reflecting the effects of multipath on TOA ranging for indoor geolocation. Model serves a similar purpose to the CRLB for TOA ranging

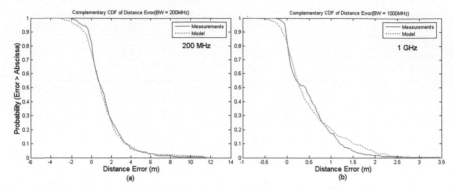

Figure 6.21   CDF of the empirical data versus the model for distance measurement error of the TOA based systems for bandwidths of (a) 200 MHz, (b) 1 GHz.

in the absence of multipath. In the absence of multipath, the performance is a function of bandwidth and power. In the presence of multipath, we have a much more complex situation, which is dominated by multipath and in particular occurrence of the UDP, while the effects of signal power is negligible.

6.6.2 UWB Empirical Modeling of the Effects of Human Body

Body area networks connect the health monitoring body-mounted sensors to the IoT. In these applications, short-range body-mounted sensors communicate with wall-mounted access point to connect to the IoT. This situation provides an opportunity for inexpensive and accurate localization of the patient in a room. Accurate localization can be achieved by TOA-based localization, but even in this short-distance application, human body can block the direct path and cause significant TOA ranging error. Figure 6.22

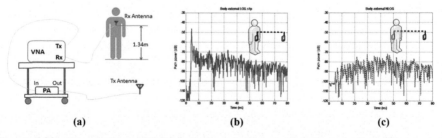

Figure 6.22   Measurement of TOA between a body mounted sensor and a wall mounted access point, (a) the measurement scenario, (b) LOS and strong direct path, (c) OLOS and a very weak direct path.

[Gen13, Ask16] shows the details of multipath profile measurement of the TOA between the body-mounted sensor and the wall-mounted access point in two scenarios. Figure 6.22(a) shows the measurement setup, the transmitter antenna at the wall mounted unit sweeps the channel, and the received signal at the body-mounted sensor is sent to the VNA to determine the frequency response and its inverse Fourier transform, which is the multipath profile shown in parts (b) and (c). In Figure 6.20(b), the body-mounted sensor is facing the access point and we have an LOS condition with strong first path dominating other paths arriving from reflection from surrounding walls. In Figure 6.22(c), human body is blocking the LOS and the strength of the direct path is reduced significantly. Here, the radio link between the transmitter and the receiver is established through the so-called creeping wave creeping around the human body [Ask16]. UWB measurements using a VNA can help to analyze this situation and model the behavior of TOA for different orientations of the human body.

Figure 6.23 shows the measurement scenario for modeling of the effects of the human body on TOA ranging in a room [Gen13]. In the measurement scenario, the transmitter antenna is mounted on a wall and the receiver antenna on the chest of a human subject at a height of 1.34 m, where a body-mounted sensor operates for health monitoring. The distance between the wall-mounted antenna and the body-mounted sensor antenna is kept fixed

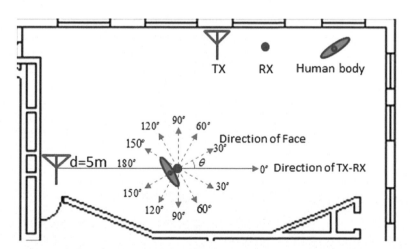

Figure 6.23 Measurement scenario for modeling the effect of human angular orientation on TOA ranging for the radio channel between a body mounted sensor and a wall mounted access point.

Table 6.2   Parameters of the model for DME caused by human body

W (GHz)	$\mu_{U,w}$(m)	$\sigma^2_{U,w}$	a_w	b_w	$SNR_{Th,w}$(dB)
5	0.010	0.005	5.10	5.49	30.4
3	0.009	0.001	3.98	6.69	30.4
1	0.072	0.058	6.21	11.76	29.0
0.5	0.138	0.143	14.69	10.62	27.5

at 5 m, and the human subject rotates its angular position in different angles while keeping the distance of the body-mounted antenna fixed. For each position of the body, 30 degrees apart from each other, 600 measurements are obtained with different bandwidths ranging from 50 MHz to 5 GHz. This database is then used for empirical modeling of the DME and its relation with the angular orientation of the body and bandwidth of the measurement system.

To model the DME, similar to (6.5), errors are divided into errors caused by multipath components close to the direct path and errors caused by blockage of the direct path:

$$\varepsilon_{\theta,w} = \varepsilon_{m,w} + \xi(\theta) \cdot \varepsilon_{U,w} = G(\mu_{m,w}, \sigma_{m,w}) + \xi(\theta) \cdot G(\mu_{U,w}, \sigma_{U,w}), \quad (6.6a)$$

where:

$$\xi(\theta) = \begin{cases} 1, & \theta \in [0°, 90°] \\ 0, & \theta \in [90°, 180°] \end{cases} \quad (6.6b)$$

The mean and variance of the Gaussian random variable representing the undetected condition when the body blocks the direct path is also modeled as a function of signal-to-noise ratio and two new parameters, a_w, b_w:

$$\mu_{U,w} = \frac{a_w}{\Delta SNR} \times \cos^3(\theta)$$
$$\sigma_{U,w} = \frac{b_w}{\Delta SNR} \times \cos^3(\theta),$$

where $\Delta SNR = SNR_{LOS} - SNR_{Th,w}$. Table 6.2 shows parameters of the model [Gen13].

6.7 Ray Tracing for In-Room RF Positioning

In Section 6.3.3, we explained challenges in using CRLB for the analysis of the effects of multipath on precision of TOA-based ranging and the need for direct analysis of the DME using empirical measurement or computational techniques. In Sections 6.4–6.6, we discussed empirical measurement techniques for analysis of the multipath effects on TOA ranging, and we gave example of this approach for modeling the DME in an office building and inside a room when direct path is blocked by human body. In this section, we discuss application of computational techniques for the analysis of the multipath effects. Computational techniques provide a solution to Maxwell's propagations and we can divide them into ray tracing and direct numerical solution of the equations. In ray tracing approach, we use ray optics approximation to determine the arrival of the paths using the geometry of the building. In direct calculation of the Maxwell's equation, we use numerical techniques such as Finite Difference Time Domain (FDTD) to solve the Maxwell's radio propagations equations directly. Ray tracing is practical for the analysis of the TOA positioning systems in open areas, outdoor, and indoor [Hei06]. It is also used for the analysis of radio propagation around the human body [Ask17]. Computational techniques are used for analysis of radio propagation inside and around the human body [Kha18]. In this section, we begin by providing fundamentals of ray tracing and its benefits to analyze the behavior of the TOA positioning systems for indoor geolocation and body area networking. In the following section, we address computational techniques using numerical solutions to Maxwell's equations and its application to localization inside and around the human body.

6.7.1 Fundamentals of Ray Tracing for Multipath Analysis

Ray tracing is one of the oldest methods used for explaining behavior of radio propagation for wireless communications and positioning. Indeed, all of TOA-based discussions that we have provided in this book refer to direct path (ray) and multipath (other rays) and we have already used that approach to explain the multipath phenomenon (e.g. see Figure 6.4). In this section, we explain the basic science behind ray tracing to analyze the effects of multipath on TOA-based indoor geolocation.

Transmission, reflection, diffraction, and scattering cause multipath arrival (Figure 6.1). In indoor areas, transmission through and reflections from the walls are the dominant mechanisms in causing multipath arrival. Diffraction is instrumental in analyzing the effects of human body on short-range TOA positioning. In any case, if we identify a geometric path between a transmitter and a receiver, from any mechanisms, and calculate the length of that path, then we can calculation the magnitude, phase, and TOA of the received signal from that path. This way, using ray tracing, we can reconstruct the channel impulse given by (6.1a):

$$h(t) = \sum_{i=0}^{L} \beta_i \, \delta(t - \tau_i) \, e^{j\phi_i}.$$

Using the distance between the transmitter and the receiver, d_0, calculation of the amplitude, delay, and phase of the direct path, β_0, τ_0 and ϕ_0, in LOS is very simple. The delay and phase of the received signal are inter-related and they are given by:

$$\tau_0 = d_0/c \ \text{ and } \ \phi_0 = \frac{2\pi d_0}{\lambda}. \tag{6.7a}$$

The amplitude of the received signal in LOS follows the Friis equation for free space propagation [Pah13]:

$$\beta_0 = \sqrt{P_r} = \frac{\sqrt{P_0}}{d_0}, \tag{6.7b}$$

where P_r is the RSS and $P_0 = P_t \, (\lambda/4\pi)^2$, with P_t representing the transmitted power. In NB analysis of the effects of multipath on TOA estimation (Figure 6.10), these values are used in the phasor diagram to represent the direct path. In WB measurements, β_0, τ_0 and ϕ_0 represent amplitude, TOA, and phase of the direct path, respectively (Figure 6.24(a)). If we can calculate the length of any other path, with some modifications to (6.7), we can calculate parameters of that path.

The interior and exterior of the buildings are flat walls, and a flat wall reflects the RF signal like a mirror. Using this principle, we can calculate

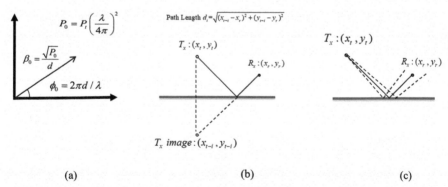

(a)   (b)   (c)

Figure 6.24   (a) Parameters of the direct path, (b) calculation of length of a reflected path using mirroring, (c) finding the reflected path using ray shooting.

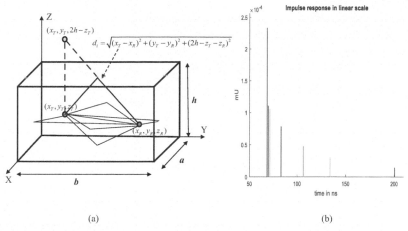

(a)   (b)

Figure 6.25 Using geometric optics for 3D ray tracing inside a room with direct path and six reflected first order paths, (a) calculation of path lengths, (b) a sample calculated channel impulse response.

the length of a reflected path using two methods, mirroring or ray shooting. Figure 6.25(b) shows the mirroring technique in which we find the image of the transmitter assuming wall is a mirror and by connecting the image to the location of the receiver, we identify the location of indecent. The length of the path is the same as the distance between the image of the transmitter and the receiver. In ray shooting, Figure 6.25(c), we transmit two rays with a small angle between them and we trace to see if the receiver is between the reflected paths [Pah95].

When the radio wave meets a wall or surface of an object, some of the signal energy gets reflected, a small part is transmitted through the surface (creeping wave), and some part propagates through. If the wall or object is thin, part of the signal passes through and continues its propagation. Using these simple principles, we can reasonably trace the paths in many useful applications and use that for the analysis of wireless waveform transmission in multipath conditions. If we trace a path between the transmitter and the receiver, we can use the length and number of transmissions and reflection through the walls to determine the path amplitude, TOA, and phase, with some simple adjustments to calculation used for the direct path. Each reflection from a wall adds 180° to the phase and each reflection or transmission multiplies the amplitude with a reflection or transmission coefficient. There are different approaches to calculate the reflection and transmission coefficients and a few of them are outlined in typical wireless communication textbooks [Pah95, Rap02, Mol12].

For indoor applications, effects of diffraction and scattering are commonly considered negligible, and in most applications, the location of antennas and the users are restricted to one floor allowing 2D ray tracing [Hol92]. In urban areas where buildings begin to block direct path between the transmitter and the receiver, diffraction becomes more important and 3D solutions are needed with more complexity of emulation [Yan94]. A number of software emulation tools for analysis of wireless communications using the ray tracing algorithms have been designed by research laboratories and commercial entities and are available in the literature or as products. In wireless communications, we are interested in the multipath spread and its effects on the data rate, which need calculation of all significant paths between the transmitter and the receiver. In TOA-based ranging, we are interested in the direct path and its few neighboring paths, which effect the shape of the received waveform from direct path, in particular its peak used for measurement of the TOA. To analyze the effects of multipath on TOA ranging, we may use our own design of simpler ray tracing scenarios or use the more complex laboratory designed and commercially available ray tracing software developed for wireless communication and tailor them to our TOA ranging application. We proceed first by a simple in-room ray tracing example to show interesting features of this tool for the analysis of the relation between TOA ranging, multipath, and bandwidth.

6.7.2 MATLAB Code for Ray Tracing Inside a Room

If the transmitter and the receiver are in the same room in LOS conditions, the direct path is the strongest path. In this situation, we may neglect the effects of paths with multiple reflections and only count the first-order reflections using seven paths: direct path and six reflected paths from walls, ceiling, and the floor (Figure 6.25(a)). We know the distance travelled by the direct path and we can use the principles of ray optics to determine the distance traveled by the reflected paths. To determine the length of a reflected path using geometric optics, we assume that walls, floor, and ceiling are flat surfaces acting as a dark mirrors. Then, as shown in Figure 6.25(a), the length of the first-order reflected paths are the same as the distance between images of the transmitter from the flat surface to the receiver. In Figure 6.25(a), if the dimensions of the room are *a, b,* and *h,* and location of the transmitter and receiver are (x_T, y_T, z_T) and (x_R, y_R, z_R), respectively, we can calculate the length of the dominant seven paths from:

$$d_0 = \sqrt{(x_t - x_r)^2 + (y_t - y_r)^2 + (z_t - z_r)^2}$$
$$d_1 = \sqrt{(-x_t - x_r)^2 + (y_t - y_r)^2 + (z_t - z_r)^2}$$
$$d_2 = \sqrt{(x_t - x_r)^2 + (-y_t - y_r)^2 + (z_t - z_r)^2}$$
$$d_3 = \sqrt{(x_t - x_r)^2 + (y_t - y_r)^2 + (-z_t - z_r)^2}$$
$$d_4 = \sqrt{(2a - x_t - x_r)^2 + (y_t - y_r)^2 + (z_t - z_r)^2}$$
$$d_5 = \sqrt{(x_t - x_r)^2 + (2b - y_t - y_r)^2 + (z_t - z_r)^2}$$
$$d_6 = \sqrt{(x_t - x_r)^2 + (y_t - y_r)^2 + (2h - z_t - z_r)^2}. \tag{6.8a}$$

From this set of equations, we can calculate the strength, TOA, and phase of the received paths. Equation (6.7) provides for the amplitude, phase, and delay of direct path. The amplitude, delay, and phase of other six first-order reflected paths are calculated similarly using:

$$\beta_i = R(\theta_i) \times \frac{\sqrt{P_0}}{d_i}; \quad \tau_i = d_i/c \text{ and } \phi_i = \frac{2\pi d_i}{\lambda} + \pi; \ i = 1, 2, \ldots 6.$$
$$\tag{6.8b}$$

The difference between (6.8) for first-order reflected paths and (6.7) for direct path is that the amplitude of the reflected paths multiplies by the reflection coefficient, $R(\theta_i)$, and phase has an additional π shift compensating for phase shift in reflection. Reflection coefficient from the walls is in general a function of angle of incidence [Pah95] and takes values between 0.3 and 1. Figure 6.25(b) provides a sample ideal channel impulse response, defined by (6.1), for our simple in-room model for RF propagation using principles of geometric optics.

   Appendix 6.A provides MATLAB code for generation, storage, and plot of the channel impulse response between a transmitter and receiver in a room using geometric optics with inclusion of the direct path and six reflected first-order paths. Distances associated to each path are calculated from (6.8a), and strength, TOA, and phase of each path from (6.7) and (6.8b). The dimensions of the room for this code are 20 m × 50 m × 5 m, and the locations of the transmitter and the receiver are at (10,20,2) and (5,40,2), respectively. Reflection coefficient for all paths is assumed to be 0.7, and the transmitted power is normalized to 1. The program stores the channel impulse response in a file named "sample.raw". Two separate figures produce the channel impulse response in linear and in dB scale. The linear plot of the channel impulse response is the one represented in Figure 6.25(b).

6.7.3 Ray Tracing and Effects of Bandwidth on TOA Ranging

One of the most important issues in the analysis of TOA-based positioning is the understanding of the effects of bandwidth of the system on the precision of ranging estimates. The CRLB is very useful for this analysis when the multipath is negligible. For calculation of CRLB in multipath environments, in Section 6.5.1, we used empirical measurements to demonstrate the effects of bandwidth on TOA ranging. In this section, we use ray tracing to analyze the effects of multipath on precision of TOA ranging with different bandwidths. In ray tracing, we calculate magnitude, TOA, and phase of the paths and represent each of them with an impulse (Figure 6.25(b)). This implies that the bandwidth of the system is infinity. In practice, we send and transmit band-limited waveforms and the received signal, as shown by (6.3):

$$y(t) = \sum_{i=0}^{L} \beta_i s(t - \tau_i) \, e^{j\phi_i}$$

for which we need to define a pulse shape, $s(t)$, and its Fourier transform, $S(f)$, to relate to the transmission bandwidth. Raised cosine pulses are the most popular waveforms used in RF waveform transmission systems for communication and localization applications. Spectrum of the signal has a flat top and two half-raise cosine shapes to smoothly connect the passband of the signal to its stop band (Figure 6.26(a)). The smoothness of the connection is controlled with the parameter β. This spectrum is mathematically represented by:

$$S(f) = F\{s(t)\} = \begin{cases} T, & 0 \le |f| \le \frac{1-\beta}{2T} \\ \frac{T}{2} \left[1 - \sin \frac{\pi T}{\beta} \left(|f| - \frac{1}{2T} \right) \right], & \frac{1-\beta}{2T} \le |f| \le \frac{1+\beta}{2T} \end{cases} \cdot$$

$$(6.9a)$$

For $\beta = 0$, transition takes place abruptly and the signal occupies the minimum band and for $\beta = 1$, we have the smoothest transition with signal occupying the maximum band. The occupied bandwidth is $W = (1 + \beta)/T$, where T describes the time characteristics of the waveform.

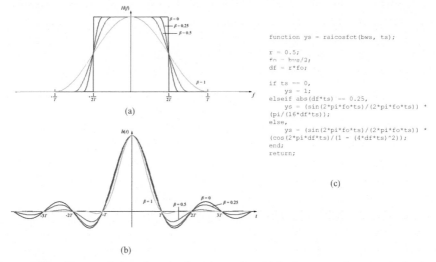

```
function ys = raicosfct(bws, ts);

r = 0.5;
fo = bws/2;
df = r*fo;

if ts == 0,
    ys = 1;
elseif abs(df*ts) == 0.25,
    ys = (sin(2*pi*fo*ts)/(2*pi*fo*ts)) *
(pi/(16*df*ts));
else,
    ys = (sin(2*pi*fo*ts)/(2*pi*fo*ts)) *
(cos(2*pi*df*ts)/(1 - (4*df*ts)^2));
end;
return;
```

(c)

(a)

(b)

Figure 6.26 Characteristics of a raised cosine pulse, (a) in frequency domain, (b) in time domain, (c) MATLAB code for implementation of raised cosine pulses.

The time response of the raised cosine pulses are the inverse Fourier transform of the frequency response, given by:

$$s(t) = F^{-1}\{S(f)\} = \frac{\sin \pi t/T}{\pi t/T} \times \frac{\cos \beta \pi t/T}{1 - 4\beta^2 t^2/T^2}. \quad (6.9b)$$

Figure 6.26(b) shows the time characteristics of raised cosine pulses for different values of β. In general, this pulse has a sharp peak and then damps down with oscillations with a zero crossing at every T. Duration of the pulse for all values of β is $2T$. The sharpest of the pulses is for $\beta = 0$, which is a sinc pulse with ideal box spectrum. Figure 6.26(c) provides a simple MATLAB code function for implementation of the pulses in time domain with bandwidth and sampling rate used as parameters for the function.

Using impulse response computed from the ray tracing and the raised cosine pulses, one can emulate the received signal waveform for any bandwidth and analyze the behavior of the TOA ranging in multipath. All that is needed is to generate the impulse response for a given location of the transmitter and the receiver and use (6.3) to form the received waveform. Then, we can run a peak detection algorithm on the received waveform to detect the TOA of the first path. Comparison of this value with the actual TOA of the ideal impulse response of (6.1a) provides the information for calculation of the DME for the given multipath profile and the system bandwidth.

Appendix 6.B provides the MATLAB code to plot the channel profile for different bandwidths for the ideal path information stored in data file "sample.raw". When we run the program, first it replots the linear channel impulse response (Figure 6.26(c)) from the "sample.raw" and requests the user to enter the bandwidth in MHz

Enter the bandwidth in MHz =>

The user enters a number like 100 and pushes the return; program plots the profile with actual paths (Figure 6.27(a)), profile with the detected peaks (Figure 6.27(b)), and prints the TOA/distance of actual direct path and estimated direct path with the resulting DME (Figure 6.27(c)). Figure 6.28 provides the similar results for a bandwidth of 10 MHz.

Analyzing the results of Figure 6.7, we observe that from the seven actual paths with 100 MHz of bandwidth, we detect five paths while with 10 MHz, we can only detect two paths. The associated DME values for 100 MHz and 10 MHz are 20 cm and 68 cm, respectively. If we repeat for bandwidth of 1 GHz, we can achieve an accuracy close to 1 cm. Comparing ray tracing results (Figure 6.27) with empirical UWB measurements (Figure 6.13(b)), ray tracing provides more insight into the formation of multipath and its

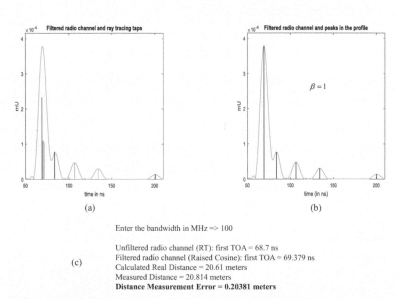

Enter the bandwidth in MHz => 100

(c)

Unfiltered radio channel (RT): first TOA = 68.7 ns
Filtered radio channel (Raised Cosine): first TOA = 69.379 ns
Calculated Real Distance = 20.61 meters
Measured Distance = 20.814 meters
Distance Measurement Error = 0.20381 meters

Figure 6.27   Results of multipath ray tracing with bandwidth of 100 MHz using raised cosine pulses with $\beta = 1$ and a peak detection range of 30 dB, (a) channel profile and actual paths, (b) channel profile with detected paths, and (c) calculation of DME.

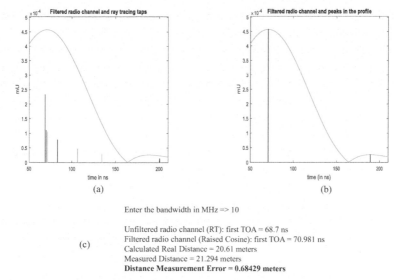

Enter the bandwidth in MHz => 10

(c)

Unfiltered radio channel (RT): first TOA = 68.7 ns
Filtered radio channel (Raised Cosine): first TOA = 70.981 ns
Calculated Real Distance = 20.61 meters
Measured Distance = 21.294 meters
Distance Measurement Error = 0.68429 meters

Figure 6.28 Results of multipath ray tracing with bandwidth of 10 MHz using raised cosine pulses with $\beta = 1$ and a peak detection range of 30 dB, (a) channel profile and actual paths, (b) channel profile with detected paths, and (c) calculation of DME.

relation to the surrounding environment, but it lacks modeling the effects of furniture and human bodies in the area. The empirical measurements captures all the details of the environment as is, but it lacks association of paths to the environment. The best practice for reliable analytical research for the effects of multipath on TOA ranging is to validate the results of ray tracing with empirical measurements.

6.8 Ray Tracing for Indoor Geolocation

In Section 6.7, we introduced a simple 3D ray tracing using seven paths for LOS applications inside a room and used that to analyze the effects of bandwidth on performance of TOA ranging in multipath conditions. The basic assumptions for adequacy of seven paths for modeling radio propagation in a room was that in the LOS, first path is very strong and dominant over other paths, so we only use the first reflected paths and neglect the higher-order reflections usually arriving far from the direct path. This assumption enabled us to use the simple seven path model to emulate the 3D indoor LOS environment of a room. In OLOS indoor areas, the direct path strength is no more dominant all the time and other multipath arrivals become important as well. This situation forces the ray tracing in OLOS conditions to include multiple reflections as well as transmission through the walls into consid- erations. However, effects of reflections from the ceiling and floor become negligible and we can pursue our ray tracing in 2D to reduce complexity of computations. Figure 6.29 shows a sample result of 2D indoor ray tracing with multiple reflections and transmissions in an OLOS conditions. Figure identifies the rays produced from single and multiple reflections and trans- missions; the smaller figure on the lower left corner shows the resulting impulse response. To generate the rays, if we use the same optical imaging technique (Figure 6.24(a)), which we used for inside a room, keeping track of all images and associated rays becomes complex and scales exponentially with the number of walls. Ray shooting method (Figure 6.24(b)) is an alternative computational method with a linear growth in complexity. In ray shooting method, a pincushion of rays is sent from the transmitter, and the progress of each ray is traced through the environment until the rays have either intersected the receiver or lost enough power that their contribution to the received signal is negligible. The time of arrival, intensity, phase, and direction of arrival are recorded for each ray that intersects the receiver.

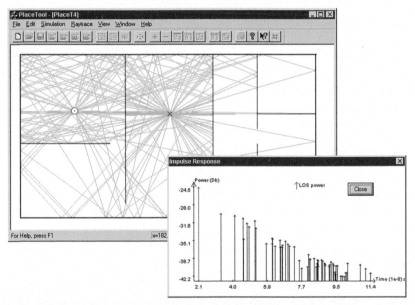

Figure 6.29   Multipath conditions with multiple reflections and transmissions in an OLOS scenario with multiple transmissions and reflections.

Once every ray has been traced to completion, the channel impulse response is formed. In the remainder of this section, we provide examples of using a 2D ray shooting algorithm, described in [Hol92], to analyze the effects of multipath in TOA positioning inside a typical office building.

We begin our discussion on benefits of 2D indoor ray tracing for analysis of RF positioning with demonstration of multipath effects. Figure 6.30 [Ala03] uses the ray shooting software to show the effects of multipath on TOA estimation. Figure 6.30(a) demonstrates how multipath components close to the arrival of the direct path cause shift in the peak of direct path resulting in multipath errors in TOA ranging. Figure 6.30(b) shows an undetectable direct path (UDP) condition, when the direct path has shifted under the receiver threshold of detection causing large DME. This figure demonstrates the same concept that was shown in Figure 6.16 using empirical measurements. However, in empirical measurement (Figure 6.24), we can only trace the expected arrival time of the direct path from physical distance between the transmitter and the receiver. Using ray tracing (Figure 6.30) provides more insight because we can trace arrival time, magnitude, and phase of all arriving paths.

The major challenge in performance evaluation of algorithms used for TOA ranging in indoor areas is analysis of the effects of un-detected direct path (UDP) condition to learn how often they occur, how we can detect them, and how a proposed algorithm to remedy this situation performs in those conditions? To answer these questions, using empirical measurements or ray tracing, we need a physical indoor layout scenario, within which we can anticipate the occurrence of UDP conditions so that we can validate

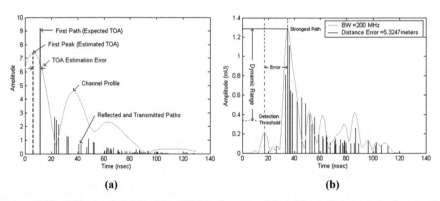

Figure 6.30 Effects of multipath on TOA estimation, (a) multipath components close to the arrival of the direct path, (b) undetectable direct path.

our theories with empirical or emulated results. Therefore, we proceed by defining a sound scenario for analysis of the behavior of TOA ranging in indoor areas.

Large metallic objects such as elevators or Faraday cages block the direct path and we can anticipate occurrence of a UDP when they are located between the transmitter and the receiver. Figure 6.31 shows the layout of the third floor of the Atwater Kent Laboratory at Worcester Polytechnic Institute. On the left side of the picture, we have a Faraday cage in Room 320, which is used to shield the interference for RF measurements (a photograph of the Faraday cage is also shown in the top left of the figure). In the right side of the picture, we have an elevator. Both cage and elevators are large metallic objects capable of blocking direct path for radio propagation. The blue dashed line represents the path and direction of movement in the corridor of the floor for empirical measurements or ray tracing analysis. Three access points (AP) are located opportunistically in the floor to create different propagation conditions. The red lines in the figure demonstrate the top portion of the route, where we anticipate the occurrence of a UDP condition for AP2 and AP3 caused by the two metallic objects. AP2 and AP3 should be able to detect the direct path in the rest of the route. The direct path is not blocked at all for the AP1. Using this scenario, we continue our discussion on the analysis of the effects of multipath on TOA indoor positioning.

6.8.1 Analysis of the Effects of Large Metallic Objects

UDP conditions occur either when a large metallic object blocks the direct path between the transmitter and the receiver or when the signal from direct path is not detectable but still there are other paths with a detectable signal. Characteristics of these two classes of UDP are different, and the methods needed to avoid them are different. In this section, we use ray shooting algorithm in [Hol92] and the scenario of Figure 6.31 to explain these phenomena. Figure 6.32 shows the results of ray shooting to emulate the signal transmitted from AP1 and provides the strength of the received signal, strength of the first detected path, strength of the direct path, and the DME for TOA ranging. Figure 6.32(a) shows the results for upper route in Figure 6.31, where the cage has blocked the direct path for a large portion of the path, resulting in UDP conditions with large DMEs of up to 8 m. When the direct path is dropped with blockage, as shown in the figure, power of the direct path drops abruptly; in addition, the rms delay spread of the received signal increase significantly. Therefore, we can use abrupt changes in strength of the direct path and rms

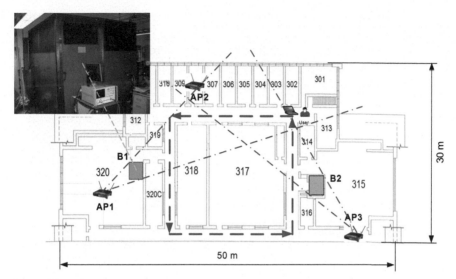

Figure 6.31 Layout of the third floor of the Atwater Kent Laboratory, WPI with an elevator and a Faraday cage.

multipath spread of the channel to detect occurrence of the UDP conditions caused by unexpected blockage [Hei09]. Once we know that a UDP condition is in progress, we can design algorithms to remedy the situation (see Chapter 9).

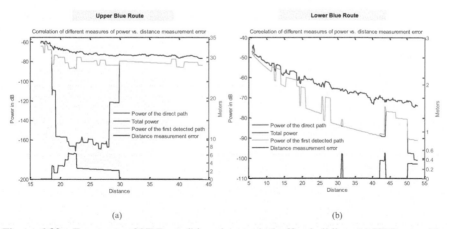

Figure 6.32 Two types of UDP conditions in a typical office building, (a) UDP caused by large metallic objects, (b) UDP caused by natural reduction of power due to distance.

Figure 6.32(b) shows the results of signal processing for the lower route of Figure 6.31, where the path between AP1 and location of the receivers on the route is not blocked by a large metallic object. On this route, we do not expect to observe a UDP condition, but actually, as we see in the figure, we have a few occasions in which direct path cannot be detected. These UDP conditions, not caused by large metallic objects, occur at the fringe of coverage of the signal when the strength of the direct path goes below the detection threshold of the receiver and they have different characteristics. We refer to these UDP conditions as natural UDP conditions. Durations of natural UDP conditions are much shorter and they produce smaller values of error (on the orders of half a meter). These values are still considerable, and we can achieve higher TOA ranging precision if we can find a way to avoid them.

Figure 6.33 provides the rational for occurrence of natural UDP conditions. In OLOS conditions, the received signal power for the direct path decays with a distance power gradient close to 2, similar to the free space propagation. However, the total power in OLOS, according to the 802.11 model, decays with the distance-power gradient. Therefore, coverage of direct path, R_1, is less than coverage of the entire received signal from all paths, R_2. As a result, in the area when distance is between these two coverages, $R_1 < d < R_2$, we cannot detect direct path, but we detect a signal and use its first path erroneously for calculation of distance. This is the exact definition of natural UDP condition. Figure 6.33(b) shows the area for occurrence of this natural UDP condition. This type of UDP condition is difficult to detect because it occurs randomly, and during its occurrence, the received total

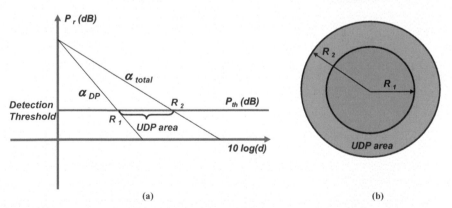

(a)   (b)

Figure 6.33   Occurrence of natural UDP conditions, (a) calculation of coverage for direct path, (b) the area for occurrence of natural UDP.

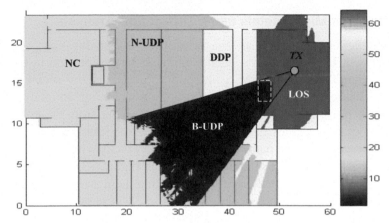

Figure 6.34 Emulation of different multipath conditions for TOA ranging in the third floor of the Atwater Kent Laboratory, WPI.

power and rms delay spread do not change substantially. A simple method to avoid natural UDP conditions is to increase the threshold of signal detection and only use the received signal with stronger power for TOA ranging.

Figure 6.34 [Hei06] shows the results of emulation of different multipath conditions for TOA ranging in a typical office building at the third floor of the Atwater Kent Laboratory, WPI. The transmitter is located in the center of the CWINS laboratory, where the Faraday cage is installed (dashed white line). The magnitude of the direct path in a dense grid of locations on the floor is calculated and compared with a threshold that is adjusted with the peak of the received signal multipath profile to decide on detectability of the direct path. The black area is where the metallic object (cage) has blocked the direct path creating a blocked UDP (B-UDP) condition. In the LOS, direct path is the strongest path and always it is detected. In the yellow, OLOS area, just outside the room, we can still detect the direct path (DDP). In the green area after that, we can detect a signal, but we cannot detect the direct path, which we refer to it as natural UDP (N-UDP). Finally, the blue area shows the area with no coverage (NC). This figure provides an insight into the extent of occurrence of different multipath conditions in a typical building. Existence of a large metallic object creates large areas with B-UDP condition and areas with N-UDP conditions are considerable. This observation underlines the importance of TOA-based algorithms in the absence of multipath (Section 9.5).

6.8.2 Effects of Micro Metals using Diffraction Analysis

Large metallic objects, such as an elevator or a Faraday cage, create large areas with UDP conditions, challenging indoor geolocation in OLOS situations inside the buildings. Micro-metallic objects, such as metal doors or cabinets, cause similar effects in short-range positioning. In analysis of large metallic objects, we neglected the diffracted signal from the edges of the metallic objects because the length of these paths are comparable to the length of other reflected paths making them very weak with respect to the other paths. In short-range positioning, as shown in Figure 6.35, the diffracted paths are much shorter than reflected paths, making them arrive much earlier and having an acceptable power level. Figure 6.35(a) shows the received diffracted signal from a metallic door and the direct path and how DME is created by the difference in the length of the direct path and diffracted path. Figure 6.35(b) shows a path reflected from a wall to compare its length with that of the diffracted path. Therefore, if the direct path is blocked by a micro-metallic object, the direct path will be blocked and the first detected path at the receiver is diffracted from corners of the metallic object.

The UWB empirical measurement system described in Section 6.4.3 can be used to validate our discussion regarding diffraction from micro-metals. Figure 6.36 shows the results of UWB measurement of the effects of diffraction using three scenarios for a door between the transmitter and the receiver (Figure 6.26(a)). The first scenario is associated with free space propagation when the transmitter and the receiver are 1.21 m apart, in the second scenario, a metallic frame of a door is in between, but the metallic door is in open position, and in the third scenario, the door is closed.

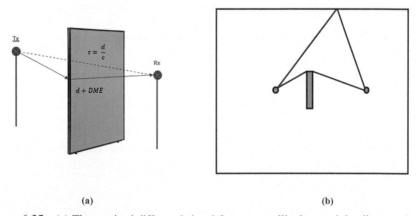

(a)   (b)

Figure 6.35   (a) The received diffracted signal from a metallic door and the direct path, (b) path reflected from a wall and the reflected path.

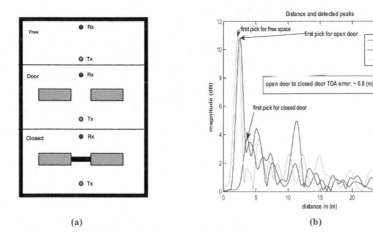

(a)   (b)

Figure 6.36 UWB measurement of the effects of diffraction, (a) three scenarios for a door between a transmitter and the receiver, (b) results of UWB measurements in the three scenarios.

Figure 6.36(b) shows the results of UWB measurements in these three scenarios. In the second scenario (red line), the peak of the direct path is shifted slightly and its strength is reduced slightly. These are due to the effects of multipath arrival from reflections from the frame. In the third scenario, the first path is not the strongest path and it attenuated and shifted significantly because the metallic door has blocked the direct path and only diffracted path is measured at the receiver.

Since diffraction from micro-metals plays an important role in the analysis of accuracy of short-range measurements of TOA, it is useful to develop a theoretical model for its analysis. In [Ask11], the uniform theory of diffraction originally designed for analysis of diffraction in cellular networks [Ber03] is extended to indoor application and blockage by micro-metallic objects. Figure 6.37 shows the abstract 3D view of diffraction scenario from corners of a micro-metals defined with three related angles. The rays that illuminate the edge in turn generate cylindrical waves that propagate behind the object. The Path Gain resulting from diffraction is given by [Ber94, Har99, Siw03]:

$$PG = \frac{P_r}{P_t} = \left(\frac{\lambda}{4\pi}\right)^2 \frac{|D_{G0}(\theta)F(S)|^2}{\cos^2 \psi} \frac{1}{rr_0(r + r_0)}, \tag{6.10a}$$

where θ is the angle of diffracted waves away from the edge (between the image of the incident wave and the diffracted wave), φ represents the angle between the incident wave and the perpendicular axis to the edge, and r_0 and r

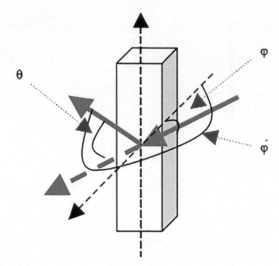

Figure 6.37   Abstract 3D view of diffraction from corners of a micro-metals defined with three related angles.

are the distances of transmitter and the receiver from the edge, respectively. The diffraction coefficient function $D_{G_0}(\theta)$ for a conducting screen is given by:

$$D_{G_0}(\theta) = \frac{-1}{\sqrt{2\pi k}} \left[\frac{1}{cos(\frac{\varphi - \varphi\prime}{2})} + \frac{\Gamma_{E,H}}{cos(\frac{\varphi + \varphi\prime}{2})} \right], \qquad (6.10b)$$

where $\Gamma_E = -1$ for E parallel to the edge and $\Gamma_H = 1$ for H parallel to the edge.

The transition (Fresnel) function, F(S), is given by:

$$F(S) = 2j\sqrt{S}e^{jS} \int_{\sqrt{S}}^{\infty} e^{-ju^2} du, \qquad (6.10c)$$

where S varies based on the characteristic of the wave. For the cases that we are addressing here, diffraction of waves in contact with two edges, S can be defined as:

$$S = 2kcos\psi \frac{rr_0}{r + r_0} sin^2(\theta/2) \qquad (6.10d)$$

An approximation to F(s) is given by:

$$\sqrt{2\pi s \left[f\left(\sqrt{2s\pi}\right) + jg\left(\sqrt{\frac{2s}{\pi}}\right) \right]}, \qquad (6.10e)$$

where

$$f(u) = \frac{1 + 0.926u}{2 + 1.792u + 3.104u}, \tag{6.10f}$$

$$g(u) = \frac{1}{2 + 4.14u + 3.492u^2 + 6.67u^3}. \tag{6.10g}$$

In [Ask11], the s from [Siw03] are used to analyze the effects of micro-metal on TOA ranging and validate these analyses with empirical UWB measurements.

6.8.3 Effects of Human Body on TOA Estimation

Another important application of unified theory of diffraction to positioning is the analysis of the effects of human body on TOA-based ranging. With the popularity of body-mounted sensors for health monitoring, application of localization using the RF signal transmitted from these sensors becomes appealing. In Section 6.6.2, we used empirical UWB measurements to model the behavior of TOA ranging for body-mounted sensors communicating with an access point mounted on a close by wall [Gen13]. That model is specific to a deployment scenario with specific location of the access point and the human body, which wears the body-mounted sensor. If we solve this problem analytically, we can extend the solution to more generalized scenarios. The unified theory of diffraction used for analysis of the effects of micro-metals is also instrumental to address this problem [Ask17a].

Human body for radio propagation is equivalent to a water tank and water is a conductor similar to metals. When the human body is located between a transmitter and a receiver, the diffracted signal survives the detection threshold of the receiver, but heavy attenuated (~70 dB) direct path is undetectable at the receiver. Figure 6.38(a) depicts the situation caused by human body blocking the direct path between a transmitter and a receiver and demonstrates how diffracted signal from corners of the body causes DME in short-distance TOA ranging. Figure 6.38(b) provides the methodology used in [Ask17a] to validate the results of unified theory of diffraction for the path between a body-mounted antenna and an external wall-mounted antenna. The human body is modeled as an elliptic cylinder, and the ray arriving on the body creeps over the body until it arrives at the body-mounted sensor. The amplitude and the length of the path is calculated and compared with the results of empirical measurements used for empirical modeling of the effects of human body [Gen16]. Figure 6.39 compares the results of analysis using unified theory of diffraction with the results of UWB measurements.

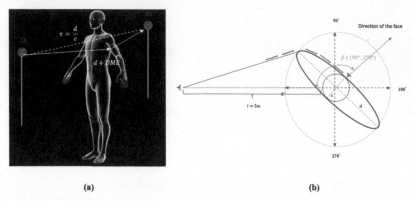

(a)   (b)

Figure 6.38   (a) DME caused by human body obstructing the direct path, (b) ray tracing model for the path between a chest mounted antenna and a wall mounted external antenna.

Results of TOA measurements (Figure 6.39(a)) shows two regions: one for LOS, when the angle of rotation of the human body is either less than 90°, and the other for more than 270°. In this situation, DME is theoretically zero and only a few centimeters in UWB measurements. As the body begins to form an OLOS situation, when the angle is between 90° and 270°, the DME jumps to high values on the orders of 30 cm. These results also show close agreement with the DME observed by measurements. Figure 6.39(b) shows the results of measured received power as a function of angle of orientation of the human body, and the same agreement is validated for path loss as well. The smoothness of the results of analysis and its generality for

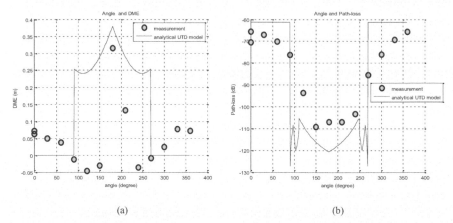

(a)   (b)

Figure 6.39   Comparison of the results of diffraction analysis with the results of UWB measurements, (a) comparison of DME, (b) comparison of path-loss in dB.

accommodating other scenarios makes this approach more appealing. Details of derivations of equation in [Ask17a] is beyond the scope of this book; however, these derivations and its associated MATLAB code are available in [Ask17b].

6.9 Computational Methods for Localization in Body Area Networks

The main challenge for the design of accurate algorithms for RF localization inside and around the human body is the lack of radio propagation channel models for localization applications. Since it is not practical to make RF measurements inside the human body in vivo, researchers resort to using phantoms, dead body animals, or computational techniques to measure the RF characteristics inside and around the human body. It is very difficult to emulate complex paths, such as those of inside the small intestine, in a phantom or a dead animal body and computational techniques may be perceived as less accurate and unrealistic. However, it is possible to use limited measurements on phantoms and on the surface of human volunteers to validate and calibrate software simulation. In Section 6.8.3, we described ray tracing method for effects of human body on TOA ranging and validated the results with empirical UWB measurements. However, ray tracing cannot be used for inside the human body because shapes of the organs are very complex and they do not form flat surfaces similar to the general indoor areas. Ray tracing was a method for approximate solution to Maxwell's equations using principles of geometric optics. An alternative approach to analyze radio propagation inside and around the human body is to resort to computation methods for direct solution of the Maxwell's equations. Since the mid-1960s, these methods have been introduced to solve a variety of RF propagation scenarios, and several popular software products for these purposes are commercially available.

In the literature on localization for body area networks, there are three software simulation tools reported for RF propagation inside the human body, the commercially available SemCAD X used in [Aoy09, Kur09], the Ansoft HFSS used in [Say10, Ask11], and a proprietary software on MATLAB developed at CWINS/WPI [Mak11]. The SemCAD X and Ansoft HFSS have fancy collection of waveforms, diversity of models for human bodies and organs, and simpler proprietary MATLAB software is a much faster solver [Kha11]. The SemCAD X and proprietary MATLAB tools use the finite difference time domain (FDTD) method to solve Maxwell's equations in time domain, and the HFSS software uses the more computationally complex frequency domain finite element method (FEM) to solve Maxwell's equations.

6.9.1 Validation of FDTD Methods for BAN Applications

Maxwell equations for electric (E_x, E_y, E_z) and magnetic fields (H_x, H_y, H_z) are a set of differential equations:

$$
\begin{cases}
\frac{\partial E_z}{\partial t} = \frac{1}{\varepsilon}\frac{\partial H_y}{\partial x} - \frac{1}{\varepsilon}\frac{\partial H_x}{\partial y} - \frac{\sigma}{\varepsilon}E_z \\[2mm]
\frac{\partial H_x}{\partial t} = -\frac{1}{\mu}\frac{\partial E_z}{\partial y} - \frac{\sigma^*}{\mu}H_x \\[2mm]
\frac{\partial H_y}{\partial t} = \frac{1}{\mu}\frac{\partial E_z}{\partial x} - \frac{\sigma^*}{\mu}H_y
\end{cases}
\tag{6.11}
$$

using permittivity, ε, conductivity, σ, and permeability, μ, of the medium, to relate the electric and magnetic fields.

The numerical solution of these differential equations over a designated area requires selection of a number of points at which the solution is to be determined iteratively. The Finite Difference Time Domain (FDTD) method is probably the most straightforward and most widely used method for numerical solution of Maxwell's equations. This method approximates Maxwell's equations by a set of finite-difference equations:

$$
\begin{cases}
\begin{aligned}
\frac{E_{zk,m}^{n+1} - E_{zk,m}^{n}}{\Delta t} &= \frac{1}{\varepsilon_{k,m}}\frac{H_{yk+1/2,m}^{n+1/2} - H_{yk-1/2,m}^{n+1/2}}{\Delta x} \\
&\quad - \frac{1}{\varepsilon_{k,m}}\frac{H_{xk,m+1/2}^{n+1/2} - H_{xk,m-1/2}^{n+1/2}}{\Delta y} - \frac{\sigma}{2\varepsilon_{k,m}}(E_{zk,m}^{n+1} + E_{zk,m}^{n}) \\[3mm]
\frac{H_{xk,m+1/2}^{n+3/2} - H_{xk,m+1/2}^{n+1/2}}{\Delta t} &= -\frac{1}{\mu_{k,m+1/2}}\frac{E_{zk,m+1}^{n+1} - E_{zk,m}^{n+1}}{\Delta y} \\
&\quad - \frac{\sigma^*}{2\mu_{k,m+1/2}}(H_{xk,m+1/2}^{n+3/2} + H_{xk,m+1/2}^{n+1/2}) \\[3mm]
\frac{H_{yk+1/2,m}^{n+3/2} - H_{yk+1/2,m}^{n+1/2}}{\Delta t} &= +\frac{1}{\mu_{k+1/2,m}}\frac{E_{zk+1,m}^{n+1} - E_{zk,m}^{n+1}}{\Delta x} \\
&\quad - \frac{\sigma^*}{2\mu_{k+1/2,m}}(H_{yk+1/2,m}^{n+3/2} + H_{yk+1/2,m}^{n+1/2})
\end{aligned}
\end{cases}
,
$$

where

$$
\begin{cases}
E_z(t,z) \to E_z\left((n-1)\cdot\Delta t,\ (k-1)\cdot\Delta x,\ (m-1)\cdot\Delta y\right) = E_{zk,m}^n \\[2mm]
H_x(t,z) \to H_x\left(\left(n-\tfrac{1}{2}\right)\cdot\Delta t,\ (k-1)\cdot\Delta x,\ \left(m-\tfrac{1}{2}\right)\cdot\Delta y\right) = H_{xk,m+1/2}^{n+1/2} \\[2mm]
H_y(t,z) \to H_y\left(\left(n-\tfrac{1}{2}\right)\cdot\Delta t,\ \left(k-\tfrac{1}{2}\right)\cdot\Delta x,\ (m-1)\cdot\Delta y\right) = H_{yk+1/2\ m}^{n+1/2} \\[2mm]
n = 1,2,...,N_t - 1 \quad k = 1,2,...,N_x + 1 \quad m = 1,2,...,N_y + 1
\end{cases}
\tag{6.12}
$$

By placing the electric and magnetic fields on a staggered grid, shown in Figure 6.40, and defining appropriate initial conditions, the FDTD algorithm employs the central differences to approximate both spatial and temporal derivatives, and it solves Maxwell's equations directly. The distribution of electric and magnetic fields over the whole grid is calculated incrementally in time; and when the simulation is finished, the propagation characteristics are known at every location in the area under study.

Programming for the FDTD on MATLAB is relatively simple and it is available in the appendix of [Kha17b]. To validate the accuracy of the FDTD simulation for wideband RF propagation in and around the human body,

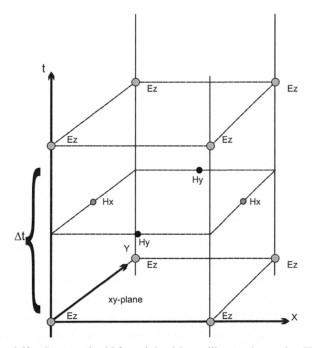

Figure 6.40 Staggered grid for solving Maxwell's equations using FDTD.

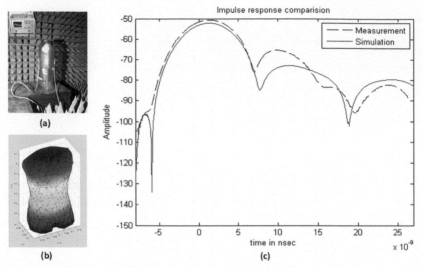

Figure 6.41   (a) Measurement set up inside the anechoic chamber, (b) simulation of the same environment using FDTD, (c) matching the waveforms from measurements and computation (simulation).

we need to compare its results with those obtained from empirical measurements. Figure 6.41 provides an overview of an experimental setup to compare the FDTD results with those of measurements from a VNA. The waveform measurements are taken by a network analyzer inside an anechoic chamber built with absorbing material covering the inside walls (Figure 6.41(a)). The simulated waveform of this environment uses the fast and simple FDTD software in MATLAB [Mak11]. The boundary conditions of the simulation are absorbing walls similar to those of the chamber. Figure 6.41(a) shows inside the chamber and the sample wideband measurements on the network analyzer on a hollow phantom filled with water inside the chamber. Figure 6.41(b) shows a sample snapshot of the results of FDTD simulation of the RF propagation using the CAD scan of the same phantom inside the chamber. The color chart indicates the strength of the electric field in each location inside the phantom, with red representing the strongest and blue the weakest signal strength.

Figure 6.41(c) shows the results of waveform simulation using FDTD and the actual wideband measurements inside the chamber for a transmitted signal in the ISM bands centered at 2.4 GHz and a bandwidth of 70 MHz. The overall waveform shape and the RSS and TOA obtained from measurement and simulation waveforms match closely. In [Swa12], this experiment is repeated for different distances between the transmitter and receiver antennas, narrower and wider bandwidths for transmission and other center frequencies for MedRadio and UWB spectrums. The accuracy of computational techniques to extract features of the waveforms pertinent to localization, RSS and TOA, have bandwidth limitations that are a function of the distance between the transmitter, the receiver, and the grid size used for discrete computation of Maxwell's equations. For practical distances of <2 m inside the human body and 2 mm grids and with reasonable computational time of around 10 minutes for each waveform simulation, accurate measurements were obtained of the localization features of the waveform for bandwidths up to 100 MHz. A more detailed quantitative relationship among the bandwidth, computational time, and accuracy of localization features of the waveform is reported in [Swa12].

6.9.2 Validation of FEM Method for BAN Applications

The result of research in radio propagation inside and around the human body that are presented in this book use either the time domain FDTD or frequency domain FEM. Computational complexity for solving Maxwell's equation using FDTD grows exponentially with the size of the grid, which is proportional to the wavelength of the carrier frequency. Wireless devices operate in frequencies above several hundred MHz and for that, we can rely on FDTD solutions with dimensions of a few meters. As a result, these tools have been popular in the analysis of radio propagation in and around the human body. The finite element method (FEM) is used to find approximate solution of partial differential or integral equations, and when it is used to solve Maxwell's equation, it allows adjustment of the grid to provide desirable precision in specific domains. The ANSYS HFSS software is the most popular commercially available software using FEM. In Section 2.2.2, we introduced the NIST path-loss model for inside the human body [Say09]. This model is based on the emulation of radio propagation inside the human body using the HFSS.

An example of modeling wideband channel characteristics using the FEM technique in Ansys HFSS has been presented in [Mak11, Kha18b]. Real-life experiments were first conducted, whereby surface-to-surface measurements were taken and then compared to their corresponding FEM simulations. Both in actual measurements and software simulations, two dipoles with 900 MHz as their center frequency were placed 50 cm apart (Figure 6.17(a)) and their channel parameters were plotted over a bandwidth of 100 MHz. This plot was then used to find the impulse response using the chirp z-transform function in MATLAB. Next, a person with height 172 cm and weight 156 lbs was positioned between the two antennas (Figure 6.17(b)). Additionally, a human body model with similar characteristics was placed in the HFSS simulation environment, between the two antennas (Figure 6.42). The impulse response for this new channel was also plotted in the same way. The comparative results are shown in Figure 6.43. The side faces of the radiation box in the HFSS simulations were assigned concrete as their material and the front and back faces were assigned the radiation boundary to imitate the environment of the lab.

From the measurement taken without the body, the TOA of the first path was calculated to be 1.70 ns, which roughly translates to about 51 cm – an error of 1 cm from the actual distance. The same value from the HFSS simulation equals 1.95 ns, which in turn translates to 58 cm, indicating an error of about 7 cm from the measurement. The TOA of the first path from measurements taken with the body computed as 2.07 ns, translating into 60 cm, which means that on average, the human body would slow down the signal enough to cause a 9 cm offset in the measurements. But the simulation

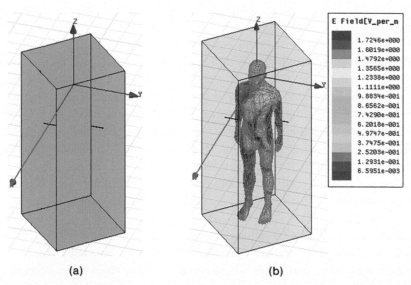

(a)   (b)

Figure 6.42   ANSYS HFSS TM simulation setup, (a) without body, (b) with body and electric field plot. The two horizontal black lines represent the dipoles.

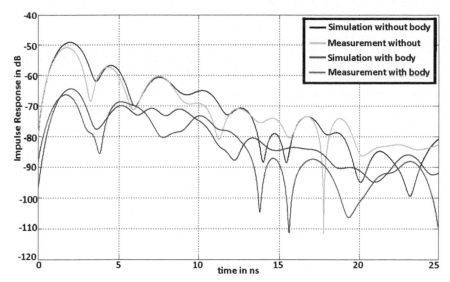

Figure 6.43 Comparison of channel time responses using HFSS simulations and channel measurement using a VNA inside a box with and without the human body.

with the body shows that the TOA of the first path is 2.00 ns, again an offset of about 9 cm from the measurements, showing a good correlation between measurements and simulations.

From Figure 6.43, the rms delay spread of the first three paths for the measurements without the body is 4.12 ns and the same value for the simulation without the body is 3.97 ns, a difference of just 0.15 ns. When the body was added to the measurement setup, the rms delay spread was 3.79 ns, the same value for the simulation with the body corresponds to 3.32 ns, an error of about 0.47 ns. Hence, it is shown using rms delay spread that the results obtained from the HFSS simulation are very close to those obtained from the actual measurements. This leads us to conclude that FEM is an effective means to simulate the wideband profile of a human body channel [Kha18a,b].

6.9.3 Modeling TOA from Inside to the Surface of the Human Body

In Section 5.6.2, we presented a model for the DME for TOA ranging inside the human body using a 3D ray tracing technique and the speed of travel of electromagnetic waves in different organs is given by $v = c/\sqrt{\varepsilon_r}$, where

ε_r is conductivity of an organ. Another approach to model the TOA behavior inside the human body is to use computational methods. In [Kha18c], using a similar setting as in Figure 6.42 for FEM simulation in HFSS, full body of a human with organs is used for this purpose. Figure 6.44 shows a sample impulse response waveform received at the sensor on the surface of the belly with the transmitter positioned inside the small intestine. Dipole antennas are modeled to act as both the pill and surface sensors for the simulation. This is for simplicity of design and to reduce simulation time. The downside of using these dipoles to represent the sensors is the capacitive dip we see before and after the received pulse in the waveform. The delay of arrival of the peak of the received pulse is the TOA of the pulse. In Figure 6.44, the transmitter and receiver are about 11 cm apart. As shown in the time domain response (TDR) plot in the figure, the pulse is received at 0.48 ns, which roughly translates to 14.4 cm, demonstrating a DME of 3.4 cm.

A number of such simulations were carried out with the transmitter kept in the intestine and the position of the receiver was rotated around the body

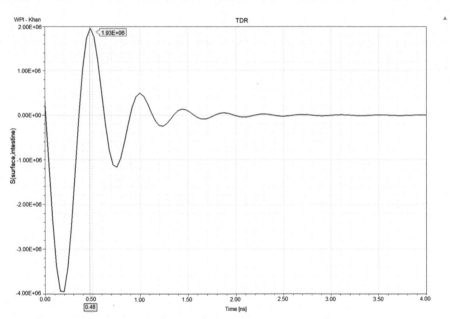

Figure 6.44   Impulse response between pill inside intestine and sensor on the surface of the belly.

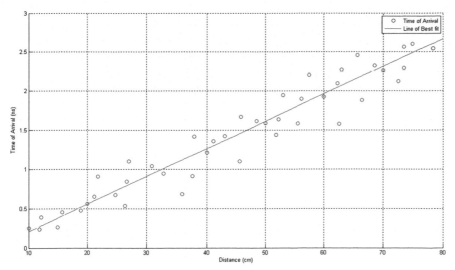

Figure 6.45 Results of HFSS simulation for measurement of TOA for different distance for a WVCE inside the small intestine and body mounted sensors in different distances on the surface of the human body.

at different positions at the same horizontal level, to account for different distances. All these simulations are then used to create a scattered plot of measured distance using TOA versus the physical distance (Figure 6.45). The best fit line to the scattered points provides for calculation of the average conductivity of the human being because

$$TOA = \frac{r}{v} = \frac{\sqrt{\varepsilon_r}}{c} \times r. \tag{6.13}$$

From the slope of the TOA versus distance line of Figure 6.45, we have $\varepsilon_r = 1.336$. Using this value, we can calculate the distance estimate and determine the DME. This value was also used to estimate the measured value for a distance of 5 cm between the sensors, mentioned in Section 4. Figure 6.35 shows the distance measurement error plot obtained from the simulations carried out. It can be seen that the distance measurement error (given in millimeters) increases linearly with distance.

Appendix 6.A

MATLAB code for calculation, storage in file "sample.raw), and plots (linear and in dB) of the channel impulse response between a transmitter at (xt, yt, zt) and a receiver at (xr, yr, zr) in linear and dB scales: 20 m × 50 m room with the ceiling height of 5 m (prepared by Julang Ying).

```
%% Example 6.6 3D In-room Ray Tracing Using Optical
Geometry
c=3e8;
Pt=1;Gr=1;Gt=1;
a=[1, -0.7, -0.7, -0.7, -0.7, -0.7, -0.7];
xt=10;yt=10;zt=10
xr=1;yr=20;zr=10
aa=20;bb=50;hh=5
d1=sqrt((xt-xr).^2+(yt-yr).^2+(zt-zr).^2);
d2=sqrt((-xt-xr).^2+(yt-yr).^2+(zt-zr).^2);
d3=sqrt((xt-xr).^2+(-yt-yr).^2+(zt-zr).^2);
d4=sqrt((xt-xr).^2+(yt-yr).^2+(-zt-zr).^2);
d5=sqrt((2*aa-xt-xr).^2+(yt-yr).^2+(zt-zr).^2);
d6=sqrt((xt-xr).^2+(2*bb-yt-yr).^2+(zt-zr).^2);
d7=sqrt((xt-xr).^2+(yt-yr).^2+(2*hh-zt-zr).^2);
%% Calculate phase, TOA and amplitude of each path
f1=2.4e9;
lambda1=c/f1;

phi1=-(2*pi*f1*d1)./c;
phi2=-(2*pi*f1*d2)./c;
phi3=-(2*pi*f1*d3)./c;
phi4=-(2*pi*f1*d4)./c;
phi5=-(2*pi*f1*d5)./c;
phi6=-(2*pi*f1*d6)./c;
phi7=-(2*pi*f1*d7)./c;
phi_all=rem([phi1(1);phi2(1);phi3(1);phi4(1);phi5(1);phi6(1);
phi7(1)],2*pi)+2*pi;


tao1=d1./c;
tao2=d2./c;
tao3=d3./c;
tao4=d4./c;
tao5=d5./c;
tao6=d6./c;
```

```
tao7=d7./c;
tao_all=[tao1(1);tao2(1);tao3(1);tao4(1);tao5(1);tao6(1);
tao7(1)];


Vr1=a(1)*exp(1i*phi1)./d1;
Vr2=a(2)*exp(1i*phi2)./d2;
Vr3=a(3)*exp(1i*phi3)./d3;
Vr4=a(4)*exp(1i*phi4)./d4;
Vr5=a(5)*exp(1i*phi5)./d5;
Vr6=a(6)*exp(1i*phi6)./d6;
Vr7=a(7)*exp(1i*phi7)./d7;


Po1=Pt*Gr*Gt*((lambda1/(4*pi))^2);
pr_all=Po1*abs([Vr1(1);Vr2(1);Vr3(1);Vr4(1);Vr5(1);Vr6(1);
Vr7(1)].^2);

W=[tao_all,pr_all,phi_all];
fid=fopen('sample.raw','w');

fprintf(fid,'%12s %12s %12s \r\n',
'Delay','Gain','Phase');
fprintf(fid,'%12s\r\n','-----------------------
------------------------');
fprintf(fid,'%12.10f %12.10f %12.6f\r\n',W');

fclose(fid);

%##############################################################
% I/O to enter the values from the .raw file
filename = 'sample.raw'; % Alternative to a fixed file name,
uncomment the text below:
% filename = input('Enter the filename => ','s');
%------ enter the whole file in one string ----------------
fid = fopen(filename, 'r');
s1 = fscanf(fid, '%c');
fclose(fid);
%------ remove the header characters ----------------------
k = 1;
while s1(k) ~= '_'
    k = k + 1;
end;
```

```
while s1(k) == '_'
    k = k + 1;
end;
j = 0;
for i=k:length(s1)
    j = j + 1;
    s2(j) = s1(i);
end;
%------- convert to numbers and sort ------------------------
s3 = str2num(s2);
rt = sortrows(s3,1);
tk  = rt(:,1);
% time values in seconds
ak  = rt(:,2);
% amplitude values (linear/no units)
phk = rt(:,3);
% phase values (in radians)
%###########################################################
npaths = length(tk);
% total number of taps
ck     = ak .* exp(-jay * phk); % construct the complex taps
ck_db  = 20 * log10(abs(ck)); % tap amplitude values in dB
% ---------------- Purely for making nice plots ------------
tmin = 10*(floor(1e9*tk(1)/10)-1)*1e-9; tmax = 10*ceil(1e9*tk
(npaths)/10)*1e-9;
min_ckdb = min(ck_db); min_db1  = 10*floor(min_ckdb/10);
min_db   = max(min_db1, -200);
max_ckdb = max(ck_db); max_db   = 10*ceil(max_ckdb/10);
%-----------------------------------------------------------
% -------------- Impulse Response dB Scale Plot -------------
figure(1)
for i=1:npaths
    hold on
    plot([tk(i) tk(i)] * 1e9, [min_db ck_db(i)]);
end
axp = [tmin*1e9 tmax*1e9 min_db max_db];
axis(axp);
title('Impulse response in dB scale')
xlabel('time in ns')
ylabel('dB')
% Note: axp, and axis(axp) are just for keeping the plot
within bounds
% defined by tmin, tmax, min_db and max_db
```

```
figure(2)
for i=1:npaths
    hold on
    plot([tk(i) tk(i)] * 1e9, [0 abs(ck(i))*1000]);
end
axp = [tmin*1e9 tmax*1e9 0 max(abs(ck))*1.1*1000];
axis(axp);
title('Impulse response in linear scale')
xlabel('time in ns')
ylabel('mU') % mU stands for milli Units
```

Appendix 6.B

MATLAB code for plotting the multipath profile for a given bandwidth, prepared by Julang Ying. This program reads the data file "sample.raw", which carries the magnitude, TOA, and phase of the paths (produced by code in Example 6.6 presented in Appendix 6.A) and uses the raicosfct function (Figure 6.24(c)) to generate the channel profile (6.2). Then, the program detects peaks of the pulses as the estimate of the arriving paths using a threshold in dB, which signifies the acceptable range of the signal peaks. The threshold is enforced to separate sidelobes of the raised cosine pulse from actual paths and it changes for different values of β. The shortest sidelobes (largest threshold) occurs for $\beta = 1$ (see Figure 6.24(c)). The program finally plots the channel profile with the detected peaks as well as channel profile with actual paths and calculates the DME associated with the file.

```
clear all; close all; clc; jay = sqrt(-1);
%###############################################################
% I/O to enter the values from the .raw file
filename = 'sample.raw'; % Alternative to a fixed file name,
uncomment the text below:
% filename = input('Enter the filename => ','s');
%------ enter the whole file in one string ----------------
fid = fopen(filename, 'r');
s1 = fscanf(fid, '%c');
fclose(fid);
%------ remove the header characters -------------------
k = 1;
while s1(k) ~= '_'
    k = k + 1;
end;
```

```
while s1(k) == '_'
    k = k + 1;
end;
j = 0;
for i=k:length(s1)
    j = j + 1;
    s2(j) = s1(i);
end;
%------ convert to numbers and sort -----------------------
s3 = str2num(s2);
rt = sortrows(s3,1);
tk  = rt(:,1);
% time values in seconds
ak  = rt(:,2);
% amplitude values (linear/no units)
phk = rt(:,3);
% phase values (in radians)
%#########################################################
npaths = length(tk);
% total number of taps
ck    = ak .* exp(-jay * phk); % construct the complex taps
ck_db = 20 * log10(abs(ck)); % tap amplitude values in dB
% ---------------------- Purely for making nice plots---------
tmin = 10*(floor(1e9*tk(1)/10)-1)*1e-9; tmax = 10*ceil(1e9*tk
(npaths)/10)*1e-9;
min_ckdb = min(ck_db); min_db1  = 10*floor(min_ckdb/10);
min_db   = max(min_db1, -200);
max_ckdb = max(ck_db); max_db   = 10*ceil(max_ckdb/10);
%-----------------------------------------------------
-------% -------------- Impulse Response Linear Scale Plot ----
--------------
figure(1)
for i=1:npaths
    hold on
    plot([tk(i) tk(i)] * 1e9, [0 abs(ck(i))*1000]);
end
axp = [tmin*1e9 tmax*1e9 0 max(abs(ck))*1.1*1000];
axis(axp);
title('Impulse response in linear scale')
xlabel('time in ns')
ylabel('mU') % mU stands for milli Units
```

```
%-------------------------------------------------------------
-------% ----------------------- Computing Power -------
--------------------
ak2 = ak .* ak;
norm power = ones(1, npaths)*ak2;
%-------------------------------------------------------------
-------% ----------- Computing Delay Spread -------------
tk2 = tk .* tk; rms1 = (tk2' * ak2) ./ norm_power; rms2 =
((tk' * ak2) ./ norm_power) .^ 2;
trms = (rms1 - rms2) .^ 0.5;
ex_delay = max(tk) - min(tk);
%-------------------------------------------------------------
----% ----- Computing Raised Cosine Filtered Output----------
npts = 1000;
t    = linspace(tmin, tmax, npts);
bw = 1e6*input('Enter the bandwidth in MHz => ');
for i=1:npts
    pnrt(i) = 0;
    for k=1:npaths
        pnrt(i) = pnrt(i) + ak(k)*exp(jay*phk(k))*raicosfct
        (bw, t(i) - tk(k));
    end;
end;
pul_db = 20*log10(abs(pnrt));
figure(2)
plot(t*1e9, abs(pnrt)*1000, 'g')
for i=1:npaths
     hold on
     plot([tk(i) tk(i)] * 1e9, [0 abs(ck(i))*1000]);
end;
axp = [tmin*1e9 tmax*1e9 0 max(abs(pnrt))*1.1*1000];
axis(axp);
s=['Filtered radio channel and ray tracing taps'];
title(s);
xlabel('time in ns')
ylabel('mU')
%-------------------------------------------------------------
% --------------- Peak Detection ----------------------------
thr       = 30;            % threshold for peak detection (dB)
timz      = abs(pnrt);
tz        = t;
Nt        = length(timz);
max1      = max(timz);
```

```
side_lobe = 10^(thr/20);
threshold = max1/side_lobe;
m = 0;
for k=2:Nt-1
    if timz(k)-timz(k-1)>0 & timz(k+1)-timz(k)<0 & timz(k)>
    threshold
            m = m + 1;
            peaks_amp(m)  = timz(k);
            peaks_time(m) = tz(k);
        end;
end;
npeaks = m;
figure(3)
plot(t*1e9, abs(pnrt)*1000, 'g')
for i=1:npeaks
    hold on
    plot([peaks_time(i) peaks_time(i)] * 1e9, [0 peaks_amp(i)*
    1000],'r');
end;
axis(axp);
s=['Filtered radio channel and peaks in the profile'];
title(s);
xlabel('time (in ns)');
ylabel('mU');
disp(' '); [s, err]=sprintf('Unfiltered radio channel (RT):
first TOA = %0.5g ns',tk(1)*1e9); disp(s);
[s, err]=sprintf('Filtered radio channel (Raised Cosine):
first TOA = %0.5g ns',peaks_time(1)*1e9); disp(s);
distance2 = abs((tk(1) - peaks_time(1)))*3e8; [s, err]=sprintf
('Calculated Real Distance = %0.5g meters',tk(1)*3e8);
disp(s);
[s, err]=sprintf('Measured Distance = %0.5g meters',
peaks_time(1)*3e8); disp(s);
[s, err]=sprintf('Distance Measurement Error = %0.5g meters',
distance2); disp(s);
```

Assignments for Chapter Six

Questions

1. Explain the three different classes of multipath conditions for TOA-based localization and describe how they affect localization precision?
2. What are the reasons for occurrence of undetected direct path conditions?

3. Why traditional multipath channel models designed for communications do not reflect localization aspects thoroughly?
4. Does increasing the bandwidth improve the TOA-based localization in a single path environment? Explain why.
5. Does increasing the bandwidth improve the TOA-based localization in a multipath environment? Explain why.
6. What is the effect of human body on TOA-based localization?

Problems

Problem 1:

a) Using [Ala06] model, determine the mean and variance of TOA-based ranging error for distances of 3, 5, and 15 m if we have a 1 GHz bandwidth and a LOS condition with no UDP condtion.
b) Repeat (a) for bandwidths of 100 MHz, 10 MHz, and 1 MHz at a distance of 5 m.
c) Repeat (a) and (b) when we have the UDP conditions with the same statistics as in the model.

Problem 2:

The bandwidth of a Wi-Fi device is 20 MHz and it has coverage of around 30m with a minimum acceptable SNR of 10 dB.

a) Determine the CRLB for the accuracy of RSS-based localization at the edge of the coverage of the system for a typical indoor area using the IEEE 802.11 path-loss model C from Chapter 2.
b) Repeat (a) for TOA-based localization.

Problem 3:

a) In a three-path channel, the distances travelled by the arriving three paths are 3, 5, and 8 m. Determine, sketch, and label the impulse response in this multipath medium if the frequency of operation is 2.4 GHz, and the transmission is only in air. Assume the received power of the signal along the first path is 0 dBm.
b) Repeat (a) if the experiment is conducted in water where the distance power gradient is 7, and the conductivity of the medium is approximately 85.

Problem 4:
Derive the TOA-based ranging error in a non-homogeneous environment in terms of conductivity and the length of the direct path segment in each medium.

Projects

Project 1 (UWB localization error using empirical data)
The objective of this project is to characterize localization errors versus distance from UWB empirical measurements of channel characteristics in a typical building. The National Institute of Science and Technology (NIST) has released a set of UWB channel measurements in a variety of buildings. These measurements are posted in a website at http://snad.ncsl.nist.gov/uwb. Go to this NIST website and identify UWB channel measurements and layout of measurement scenario in the North Building.

  a) Using individual channel profiles, visually estimate the TOA for each channel profile and record the estimate distance from that measurement.
  b) Use the layout map of measurement to calculate the actual distance for each measurement and by comparing that with results of part (a) calculate and record the DME for each location.
  c) Plot the CDF of the DME and calculate the median, mean and standard deviation of DME in all locations.
  d) Make a scatter plot of the DME versus the distance. How can you describe this relation under the light of CRLB analysis?

Project 2 (Raytracing for effects of multipath on TOA estimation)
This project uses Raytracing and the scenario shown in Fig. P1 to analyze the effects of multipath on TOA-based localization techniques.

Figure P1 shows floor plan of a 20 m × 50 m rectangular room and two mobile devices communicating in that room. The ceiling height is 5 m and two antennas are mounted on the devices at 1.5 m above the ground. Communication between the terminals is taking place through seven paths: the direct path, the path reflected from the ground, the path reflected from the ceiling, and four paths bounced through the four walls. Assume that paths with more than first-order reflections are negligible, the reflection coefficient from the walls, ground, and the ceiling is 0.7, and each reflection causes an additional 180° phase shift we can develop a simple Raytracing program to simulate the multipath arrivals in the room.

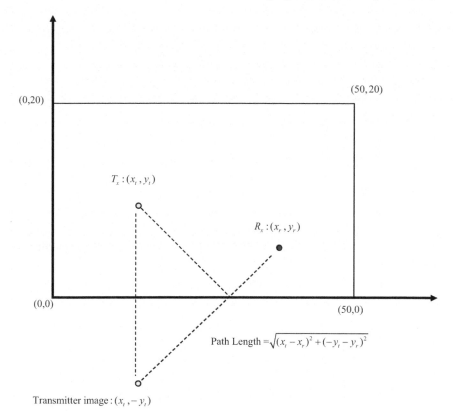

Figure P1 Geometry for calculation of path lengths in a rectangular room using a simple Raytracing techniques.

Figure P2 shows a phasor diagram that we can use for calculation of the amplitude, phase, and delay for each path, and they are calculated from:

$$i - th\,path\,phasor\,\alpha_i = a_i\frac{\sqrt{P_0}}{d_i}e^{j\phi_i} = a_i\frac{A_0}{d_i}e^{j\phi_i}$$

a_i : Reflection Coefficient

$$\tau_i = \frac{d_i}{c}$$

c : Speed of light

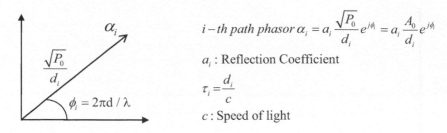

$$i-th\ path\ phasor\ \alpha_i = a_i \frac{\sqrt{P_0}}{d_i} e^{j\phi_i} = a_i \frac{A_0}{d_i} e^{j\phi_i}$$

a_i : Reflection Coefficient

$$\tau_i = \frac{d_i}{c}$$

c : Speed of light

Figure P2 Phasor diagram to calculate magnitude, delay, and phase of a path of length d.

a) For a carrier frequency of 2.4 GHz, when transmitter is at T_x : $(25, 10)$ and the receiver moves from R_x : $(1, 9)$ to R_x : $(49, 9)$ in 1 m steps, provide the code that calculates the complex channel impulse response defined as:

$$h(t) = \sum_{i=1}^{N} \alpha_i e^{j\phi i} \delta(t - \tau_i). \tag{6.14}$$

Provide the plot of the magnitude of the channel impulse response in the linear form for the first and last points on the path. (Hint: you can begin from codes in Appendix 6.A and 6.B for Example 6.6)

b) Using the impulse responses of (a), generate the channel profile for different bandwidths for a transmitted raised cosine pulses given by the following equation ($\beta = 0.5$):

$$f(t) = \frac{\sin(\pi W t)}{\pi W t} \frac{\cos(\beta \pi W t)}{1 - 4(\beta W t)^2}. \tag{6.15}$$

Write a peak detection algorithm to detect the TOA of the first peak of the profile and using that calculate the distance estimate and associated distance measurement error (DME). Give the sample channel profiles at the beginning and end of the path for bandwidths of 1, 10, and 100 MHz.

c) Provide the plot of DME on the route of motion, defined in (a), for the three bandwidths of 1, 10, and 100 MHz.

d) Repeat (b, c) assuming that a human body is blocking the direct path between the transmitter and the receiver causing a 50 dB reduction in the power of the direct path.

e) If we were using phase of the carrier (NB signal) for TOA estimation, what would have been the distance ambiguity?

f) If we were using two tones to estimate the distance using the phase of the receive carrier signal (NB measurement of TOA), what would be the frequency separation of the two tones to eliminate the phase ambiguity of measuring the phase of the direct path?

g) If we use the frequency separation recommended by (f), what would be the phase of the direct path and the estimate of the distance at the beginning and end of the path.

h) Repeat (g) if we use the effects of all multipath components on the phase of carrier signal (using a phasor diagram).

i) Write a short description of what you learned from this project from the effects of multipath on NB and WB measurement of the TOA in indoor multipath conditions.

7

Introduction to Positioning Algorithms

7.1 Introduction

In parts I and II of this book, we introduced the fundamentals of RF positioning using signals of opportunity, provided insights into the expected performance of systems using these signals, and described popular positioning systems using these technologies in indoor and urban areas. Part III of this book is devoted to the applied algorithms for RF localization in indoor and urban areas and it includes four chapters. Chapter 7 introduces the general concept of positioning algorithms. Chapters 8 and 9 are devoted to applied algorithms for RSS and TOA positioning, and Chapter 10 explains hybrid algorithms using RF signals as well as other positioning sensors.

In RF positioning, we can locate a target tag (TG) device either by using the signal it receives from known reference points (RPs) or by the received signal transmitted from the TG device at the RPs. For example, in GPS the TG device has a GPS chipset that measures the distance from the satellites to position its location. In TOA-based CPS systems, a module in the mobile switching center collects the signal from the TG device from different cell towers used as RPs to position the device. In either of these cases, RF positioning is based on multiple measurements of RSS, TOA, or DOA of a signal using location sensors in a TG device or at the RPs. Figure 7.1 shows the basic concept and parameters involved in positioning. We use measurements of features of the signal to estimate either the distance between the device and a RP, R, or the angle of arrival of the signal, α. Positioning algorithms use these estimated values of distances or angles and the known coordinates of the RPs, $(x_i\ y_i)$, to locate the coordinates of the TG, $(x\ y)$. Since both distance and angle are functions of the location of the TG and each measurement involves its own noise, the general statement of the problem in classical estimation theory is that we have N-observation of N-functions of two parameters, $(x\ y)$, in additive noise, which is described in Appendix Section A.5.2.

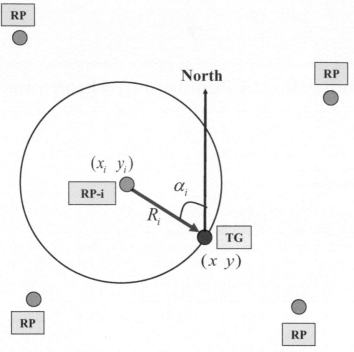

Figure 7.1   Basic concept of RF positioning of a target (TG), using N-known location of reference points (RP). A location sensor uses RSS, TOA or DOA to measure the distance or angle of arrival of the signal from RP.

In vector notation, the general multiple observations through different functions of two parameters become:

$$O_i = g_i(x\ y) + \eta_i \Rightarrow \mathbf{O} = \mathbf{G}(x\ y) + \boldsymbol{\eta}, \tag{7.1a}$$

where:

$$\begin{cases} \mathbf{O} = \begin{bmatrix} O_1 & O_2 & \cdot & O_N \end{bmatrix}^T \\ \mathbf{G}(\alpha, \beta) = \begin{bmatrix} g_1(x,y) & g_2(x,y) & \cdot & \cdot & g_N(x,y) \end{bmatrix}^T \cdot \\ \boldsymbol{\eta} = \begin{bmatrix} \eta_1(\sigma_1) & \eta_2(\sigma_2) & \cdot & \cdot & \eta_N(\sigma_N) \end{bmatrix}^T \end{cases}$$

The ML and the MMSE for the coordinates of this problem are obtained by solving:

$$\mathbf{O} = \mathbf{G}(x, y). \tag{7.1b}$$

Defining the Jacobian matrix, $\mathbf{H} = \nabla_{x,y}\mathbf{G}$, the FIM and the CRLB for this problem are given by:

$$\begin{cases} \mathbf{F} = \mathbf{H}^{\mathbf{T}}\mathbf{\Sigma}^{-1}\mathbf{H} \\ CRLB \geq Trace\left(\mathbf{F}^{-1}\right) = Trace\left(\mathbf{H}^{\mathbf{T}}\mathbf{\Sigma}^{-1}\mathbf{H}\right)^{-1} \end{cases}. \tag{7.1c}$$

To position the location of a TG, we need to estimate its coordinates, $(x\ y)$, and if we solve Equation (7.1b), we find the ML and MMSE estimates of these coordinates. Equation (7.1b) is a set of N geometric equations with two unknowns, and only for $N = 2$, it can provide unique analytical solutions. For $N > 2$, we need to define a cost function and solve the problem. Least square (LS) methods are commonly used to formulate this problem. The LS formulation can be solved either numerically or using the recursive least square (RLS) algorithm. In the remainder of this chapter, we begin with describing basic triangulation methods for positioning a TG with minimum RPs using different combinations of measurements of distance and angle of arrival. Then, we discuss LS formulation of the positioning problem and practical solutions to this problem using RLS algorithms and computational methods. We pay special attention to application of these algorithms to urban and indoor positioning problem.

7.2 Basic Triangulation Methods for Positioning

For positioning in indoor and urban areas, which are the focus of this book, we use 2D maps with a fix coordinate to visualize the location. We measure the distance, R, or the angle, α, to position the location on a typical 2D map of the area displayed on screen of a device. In 2D positioning, we need at least two measurements to solve Equation (7.1b) and find the $(x,\ y)$ coordinates of the TG device. There are three traditional options to form a set of geometric equations for positioning to find $(x\ y)$ using distance and angle, (R, α), (α, α), and (R, R). These three options need at least one, two, or three RPs to find a unique geometric location for a device, respectively. In practice, sometimes it is more convenient to measure the difference between distances or difference between the angles, rather than the actual value of the distance or angle. It is also possible to form geometric equations using the differences to find the location of the TG device. When we have the ability to measure the distance between different TGs, and some of the RPs do not cover both TGs, we can find the TG location using cooperative localization. In cooperative localization, geometric equations related to location of multiple TGs are used as a part of positioning algorithm.

7.2.1 Triangulation Using (R, α)

In this approach, we locate a TG using its distance, R, and its DOA angle, α, from an RP. One of the unique applications of this approach is in navigation of a mobile device to find an RF radiating TG. The mobile device uses the DOA of the signal to navigate motions toward the TG and uses the RSS or TOA to measure its distance from the TG. Figure 7.2a shows a classic example of this positioning method in a system designed for finding stolen vehicles using an RF device with a secure hidden location in the vehicle. The hidden RF device is activated by the service provider when the owner reports that the car is stolen [Duv99]. A police car searching for the stolen vehicle navigates towards the TG using, α, measured from activated RF device signal. The distance, R, is measured to guide the police in finding the stolen vehicle, when it is in visible sight. Figure 7.2b [Xu18] shows a similar application in an indoor environment. In this application, a first responder with sensors to measure (R, α) uses the angle to navigate itself towards a co-responder asking for help.

Figure 7.3 shows the basic principle of (R, α) positioning, where we can locate a TG by using its position information only from one RP. To formulate the geometric equations for the analysis, we assume RP is located in the

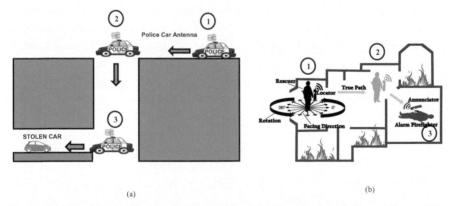

(a)   (b)

Figure 7.2   Examples of Basic principles of (R, α) localization, (a) a police car finds a stolen vehicle using the RSS and DOA of the transmitted RF signal from a device hidden in the car, (b) a first responder finds the location of a firefighter with an activated RF transmitter using the TOA and DOA of the received signal.

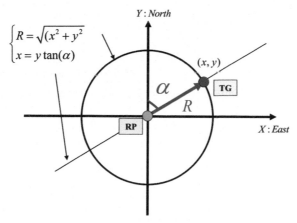

Figure 7.3   Basic principles of (R, α) localization, the distance of the target (TG) to and the direction of arrival of signal from the RF are used to position the TG in a 2D Cartesian coordinate.

origin of the coordinate system. Then, using classical formulation of Equation (7.1a), the observations are modeled as:

$$\begin{cases} R = \sqrt{x^2 + y^2} + \eta_r \\ \alpha = \tan^{-1}\left(\frac{y}{x}\right) + \eta_\alpha \end{cases}, \tag{7.2a}$$

where η_r, η_α are Gaussian measurement noise for measurement of the distance and direction of arrival, respectively. The vector, $G(x, y)$, is this formulation is given by:

$$\mathbf{G} = \left[\begin{array}{c} \sqrt{x^2 + y^2} \\ \tan^{-1}\left(\frac{y}{x}\right) \end{array} \right]. \tag{7.2b}$$

The ML estimates of $(x \; y)$ coordinates of the TG, from Equation (7.1b), are then obtained by solving:

$$\begin{cases} R = \sqrt{x^2 + y^2} \\ \alpha = \tan^{-1}\left(\frac{y}{x}\right) \end{cases} \Rightarrow \begin{cases} R = \sqrt{x^2 + y^2} \\ y = \tan(\alpha) \times x \end{cases}. \tag{7.3a}$$

Knowing the distance translates for knowing the TG being on a circle and knowing the angle translate to knowing the TG to be on a line. The intersect of the circle and the line is the location of the TG. Solution to the above set of two geometric equations is simple, and the ML estimate of the Cartesian coordinates of the TG is:

$$\begin{cases} x = R \sin(\alpha) \\ y = R \cos(\alpha) \end{cases}.$$ (7.3b)

To analyze the performance of this localization technique, we need to calculate the CRLB, and for that, we begin by calculating the Jacobian matrix:

$$\mathbf{H} = \nabla_{x,y}[\mathbf{G}] = \begin{bmatrix} \frac{2x}{x^2+y^2} & \frac{2y}{x^2+y^2} \\ -\frac{y}{x^2+y^2} & \frac{x}{x^2+y^2} \end{bmatrix},$$ (7.4a)

which we can use in Equation (7.1c) to calculate the CRLB:

$$CRLB \geq Trace(\mathbf{F}^{-1}) = Trace(\ ^{\mathbf{T}}\mathbf{\Sigma}^{-1}\mathbf{H})^{-1},$$ (7.4b)

with

$$\mathbf{\Sigma} = \begin{bmatrix} \sigma_R^2 & 0 \\ 0 & \sigma_\alpha^2 \end{bmatrix}.$$ (7.4c)

The variance of these measurement noises depends on the technology that we use for the measurement of distance and direction of arrival. For example, if we use a waveform with flat spectrum for TOA measurement and an antenna array for measurement of DOA, σ_R^2 is given by Equation (4.20a), and σ_α^2 by Equation (4.28c).

7.2.2 Triangulation Using (α, α)

A typical application for this approach is positioning a boat using a compass and location of two lighthouses on a map. Figure 7.4 shows this basic concept, a sailor uses a compass to measure the angle of observation of the two lighthouses and uses those angles to locate the boat on the map. Figure 7.5 shows a basic geometry for derivation of positioning equations for (α, α) positioning. RP-1 is located in the origin of the coordinate system, and RP-2 is located at the known location, $(x_2\ y_2)$. The observed angle between the TG and RP-1, RP-2, is represented by α_1, α_2, respectively. Similar to the last section, we have:

$$\begin{cases} \alpha_1 = \tan^{-1}\left(\frac{y}{x}\right) + \eta_\alpha \\ \alpha_2 = \tan^{-1}\left(\frac{y-y_2}{x-x_2}\right) + \eta_\alpha \end{cases},$$ (7.5a)

and the matrix $\mathbf{G}$ is:

$$\mathbf{G} = \begin{bmatrix} \tan^{-1}\left(\frac{y}{x}\right) \\ \tan^{-1}\left(\frac{y-y_2}{x-x_2}\right) \end{bmatrix}.$$ (7.5b)

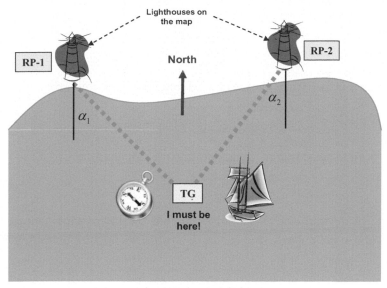

Figure 7.4   Locating a boat using (α, α) geometry, the sailor uses a compass to measure the angle of the two lighthouses and by using those angles position its boat on the map.

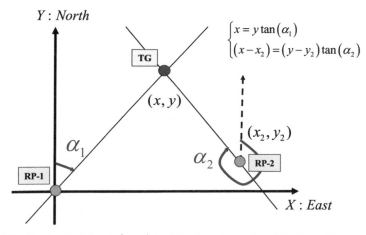

Figure 7.5   Basic principles of (α, α) localization, the angles of the TG with respect to two RPs are used to position the TG in a 2D Cartesian coordinate.

The ML estimate of the coordinates is found by solving the following two sets of geometric equations, representing two lines:

$$\begin{cases} \alpha_1 = \tan^{-1}\left(\frac{y}{x}\right) \\ \alpha_2 = \tan^{-1}\left(\frac{y-y_2}{x-x_2}\right) \end{cases} \Rightarrow \begin{cases} y = \tan(\alpha_1) \times x \\ y - y_2\alpha_2 = \tan(\alpha_2) \times (x - x_2) \end{cases}. \quad (7.6a)$$

The intersect of the two lines is unique and it is given by:

$$\begin{cases} y = \frac{y_2 \cdot \tan(\alpha_2) - x_2}{\tan(\alpha_2) - \tan(\alpha_1)} \\ x = y \tan(\alpha_1) \end{cases}. \qquad (7.6b)$$

The Jacobian matrix for calculation of FIM and CRLB for this case is:

$$H = \nabla_{x,y}[\mathbf{G}] = \begin{bmatrix} -\frac{y}{x^2+y^2} & \frac{x}{x^2+y^2} \\ -\frac{y(x-x_2)^2}{(x-x_2)^2+(y-y_2)^2} & \frac{(x-x_2)^2}{x[(x-x_2)^2+(y-y_2)^2]} \end{bmatrix}, \qquad (7.73a)$$

which we need to substitute in Equation (7.4b) to calculate the CRLB, with:

$$\Sigma = \begin{bmatrix} \sigma_\alpha^2 & 0 \\ 0 & \sigma_\alpha^2 \end{bmatrix}. \qquad (7.7b)$$

Calculation of the variance of the noise depends on the measurement technique. If we use a compass, we need to determine its angular measurement error.

7.2.3 Triangulation Using (R, R)

Locating a TG from its distance from RPs is one of the most popular positioning techniques. Distance-based positioning is the choice of all popular TOA-based positioning systems, and in Section 4.3.6, we derived the CRLB for these systems. Sensor networks use the distance estimates from RSS measurements for short-range positioning, and in Section 3.2.4, we derived CRLB for the distance-based RSS localization. Here, we review these materials when we have only two RPs to compare analysis of (R, R) positioning with (R, α) and (α, α) positioning provided in Sections 7.2.1 and 7.2.2, respectively. Figure 7.6 shows the basic principles of (R, R) localization, where we locate a TG by knowing its distance from at least two RPs. For simplicity of calculations, we assume RP-1 is located in the origin of the coordination and the RP-2 is located at $(x_2 \ y_2)$. With these locations, the geometric positioning equations from two observations are given by:

$$\begin{cases} R_1 = \sqrt{x^2 + y^2} + \eta_R \\ R_2 = \sqrt{(x - x_2)^2 + (y - y_2)^2} + \eta_R \end{cases}, \qquad (7.8a)$$

where η_R is the Gaussian measurement noise for measurement of the distance. The associated **G** matrix is then given by:

$$\mathbf{G} = \begin{bmatrix} \sqrt{x^2 + y^2} \\ \sqrt{(x - x_2)^2 + (y - y_2)^2} \end{bmatrix}. \qquad (7.8b)$$

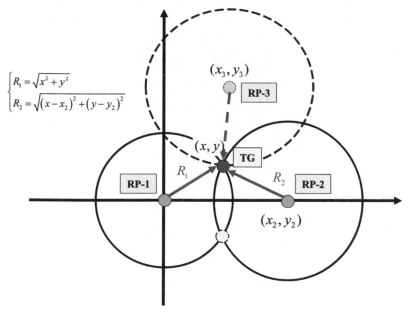

Figure 7.6 Basic principles of (R, R) localization, the distances of the TG from the RPs are used to position the TG in a 2D Cartesian coordinate.

The ML estimate of the location is obtained by solving the following two quadratic functions:

$$\begin{cases} R_1 = \sqrt{x^2 + y^2} \\ R_2 = \sqrt{(x - x_2)^2 + (y - y_2)^2} \end{cases}. \tag{7.8c}$$

To find a solution to the above two sets of quadratic equations geometrically, we need to draw two circles with the radius of measured distances from the RPs and determine the intersect between the two circles. Since these are quadratic functions, we have two answers for the solution and we need additional information to distinguish the correct answer. For example, if we are locating a flying object using two RPs mounted on the ground, one of the answers is above the ground and one under the ground. We can easily select the correct answer because a flying object cannot be under the ground. If we are tracing an object as it moves, the answer closest to the last location is the correct answer. In case none of these guiding information is available to select the correct answer for the two solutions, we need a third RP, shown by dashed lines in Figure 7.6, to identify the correct answer. However, with

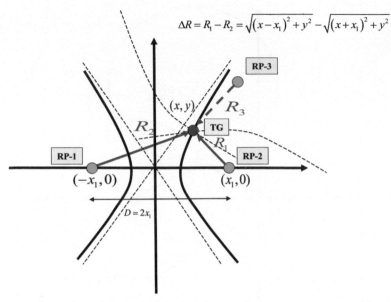

Figure 7.7   Basic principles of positioning using difference of distances, each measured difference maps to a hyperbola shape. Using at least two hyperbola we can locate the target, with three hyperbolas we have a unique solution.

noisy estimates of the distance, all three circles do not intersect and we need to resort to least square solution that we discuss in Section 7.3.2.

The Jacobian matrix for two observations of the distance in Figure 7.5 is:

$$H = \nabla_{x,y}[\mathbf{G}] = \begin{bmatrix} \dfrac{x}{\sqrt{x^2+y^2}} & \dfrac{y}{\sqrt{x^2+y^2}} \\ \dfrac{x-x_2}{\sqrt{(x-x_2)^2+(y-y_2)^2}} & \dfrac{y-y_2}{\sqrt{(x-x_2)^2+(y-y_2)^2}} \end{bmatrix}, \qquad (7.9a)$$

which we can use to calculate the CRLB from Equation (7.4b) with:

$$\Sigma = \begin{bmatrix} \sigma_R^2 & 0 \\ 0 & \sigma_R^2 \end{bmatrix}. \qquad (7.9b)$$

This variance of measurement noise depends on the technology that we use for the measurement of distance.

7.2.4 Differential Triangulation Using (ΔR, ΔR)

Some *(R, R)* systems use the difference of distances for localization. For example, in U-TDOA positioning systems for cellular networks (see Section 5.5.3), the signal from mobile phone at different cell towers is processed for positioning using the time difference of arrival. In these systems, $\Delta R = R_1 - R_2 = c \, \Delta\tau$, where $\Delta\tau$ is the difference in time of arrival at two different cell towers and c is the speed of light. Figure 7.7 shows the basic geometry of localization using the differences of distances. Each measured distance difference from two RPs results in a hyperbola for possible location of the TG. For a y-axis in the middle of the two RPs, the relation between the coordinates and the measurement is described by the following quadratic equation:

$$\frac{x^2}{a^2} - \frac{y^2}{b^2} = 1, \tag{7.10a}$$

where:

$$\begin{cases} D = 2x_1 \\ a = \frac{R_1 - R_2}{2} = \frac{\Delta R}{2} \\ b = \sqrt{(D/2)^2 - a^2} \end{cases} \tag{7.10b}$$

with $D = 2x_1$, the physical distance between the two RPs. On a hyperbola, ΔR, the difference between the measured distances of any point from the two RPs, is the same. In general, any measured difference in distance generates a parabolic equation given by:

$$\Delta R_i = R_i - R_j = \sqrt{(x - x_i)^2 + (y - y_i)^2} - \sqrt{(x - x_j)^2 + (y - y_j)^2} + \eta_{\Delta R}, \tag{7.10c}$$

where the variance of observation noise is the sum of the variances of the two measurement noises associated with the two measured distances. With only two RPs, we can only measure one set of equations; therefore, to have two or three hyperbolas, we need at least three RPs. Similar to solutions to circles for (R, R), if we use two time differences, we may have two solutions and we need to eliminate one. If we use three of them, ambiguity is resolved by imperfect measurements, and we need to resort to LS solution. Calculation of CRLB is very similar (R, R) and we leave that as an exercise.

7.2.5 Cooperative Triangulation

Imagine we have a large sensors network with two classes of sensors. For the first class, we know the location of the sensor in a known coordinate. For the second class, we do not know the location of the sensor, but we can measure their distances from neighboring sensors. In a 2D coordinate, each sensor with unknown location has two unknown coordinates, and for that, we need two distance measurements to form two quadratic equations to solve for the location parameters. Therefore, for N unknown sensor locations, we need to have a minimum of $2N$ distances to locate them. Figure 7.8 shows an example of such a network with three known RPs and two TGs with unknown locations. TG-1 with coordinates of $(\alpha_1 \ \beta_1)$ can measure R_{11}, its distance from RP-1. TG-2 can measure $(R_{22}, R_{32}$, its distance from two other RPs, RF-2 and RF-3, respectively. In addition, the two TGs can measure the distance between themselves, ρ_{12}. In this scenario, we have four unknown parameters, $(\alpha_1, \beta_1, \alpha_2, \beta_2)$, associated with the coordinates of the two TGs and four measured distances, $(R_{11}, \rho_{12}, R_{21}, R_{31})$, offering four quadratic equations to solve for the four ML estimates of the unknown locations:

$$\begin{cases} R_{11} = \sqrt{(\alpha_1 - x_1)^2 + (\beta_1 - y_1)^2} \\ \rho_{12} = \sqrt{(\alpha_1 - \alpha_2)^2 + (\beta_1 - \beta_2)^2} \\ R_{22} = \sqrt{(\alpha_2 - x_2)^2 + (\beta_2 - y_2)^2} \\ R_{32} = \sqrt{(\alpha_2 - x_3)^2 + (\beta_2 - y_3)^2} \end{cases} \quad . \tag{7.11a}$$

Circles shown in Figure 7.8 help visualization of the geometric solutions of these quadratic equations to find the location of the two TGs. This type of localization is referred to as cooperative localization, because TGs in unknown locations cooperate with the RPs in known locations, to position themselves.

Following the same terminology of Equation (7.1) for classical estimation theory, in the presence of measurement noise for calculation of distance, the matrix **G** is:

$$\mathbf{G} = \begin{bmatrix} \sqrt{(\alpha_1 - x_1)^2 + (\beta_1 - y_1)^2} \\ \sqrt{(\alpha_1 - \alpha_2)^2 + (\beta_1 - \beta_2)^2} \\ \sqrt{(\alpha_2 - x_2)^2 + (\beta_2 - y_2)^2} \\ \sqrt{(\alpha_2 - x_3)^2 + (\beta_2 - y_3)^2} \end{bmatrix} . \tag{7.11b}$$

Then, the Jacobian matrix is given by:

$$\mathbf{H} = \nabla[\mathbf{G}] = 2 \begin{bmatrix} \frac{\alpha_1 - x_1}{R_{11}} & \frac{\beta_1 - y_1}{R_{11}} & 0 & 0 \\ \frac{\alpha_1 - x_2}{\rho_{12}} & \frac{\beta_1 - y_2}{\rho_{12}} & \frac{\alpha_2 - x_1}{\rho_{12}} & \frac{\beta_2 - y_1}{\rho_{12}} \\ 0 & 0 & \frac{\alpha_2 - x_2}{R_{22}} & \frac{\beta_1 - y_2}{R_{22}} \\ 0 & 0 & \frac{\alpha_2 - x_3}{R_{23}} & \frac{\beta_2 - y_3}{R_{23}} \end{bmatrix}, \tag{7.11c}$$

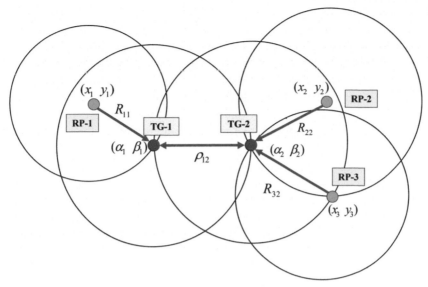

Figure 7.8 Example of a cooperative triangulation, locating two TGs using the distances between three RPs and three of the TGs, and the distance between the two TGs.

and we can calculate the CRLB from $CRLB \geq Trace\left(\ ^{T}\boldsymbol{\Sigma}^{-1} \ \right)^{-1}$, where:

$$\boldsymbol{\Sigma} = \begin{bmatrix} \sigma_{R_{11}}^2 & 0 & 0 & 0 \\ 0 & \sigma_{\rho_{12}}^2 & 0 & 0 \\ 0 & 0 & \sigma_{R_{21}}^2 & 0 \\ 0 & 0 & 0 & \sigma_{R_{31}}^2 \end{bmatrix}. \quad (7.11d)$$

7.3 LS Method for Positioning

In Section 7.2, we introduced geometric algorithms to locate a TG using distance and angle of arrival of the signal. We showed that angle of arrival, α, defines a geometric line, distance, R, defines a geometric circle, and distance difference, ΔR, defines a geometric hyperbola. When we have two of any combination of these, we can locate a TG. If we have N TGs, we showed that using cooperative localization we can locate them using $2N$ measurements of distance, angle, or distance difference to locate them. In addition, using CRLB, we showed how we can measure the performance of these techniques versus one another. In this section, we consider situations with more than enough measurements to localize. That means we have more than two measurements of distance or angle per each TG with unknown location. If all measurements were noiseless, all of these lines or circles intersect in the same location. Figure 7.6 shows an example of this situation where we have used three distances to find the location of the TG on the x- and y-axis without any ambiguity. If we had many more exact measurements of distances or angels from RPs, all the circles and lines associated with them would intersect in the same location and extra ones are redundant. When measurements are noisy, this situation changes and we have multiple solutions by solving different sets of equations. To avoid multiple solutions and define a unique solution, we need to define a cost function and the least square (LS) solution to the problem. In Section 2.2.1, we used the LS algorithm to estimate the three parameters of the path-loss model using a large number of measured RSS. In the remainder of this section, we provide the LS solution to localization problems using DOA estimate of the angles, TOA and RSS estimated distances and for cooperative localization.

7.3.1 LS Method for Positioning with Angles

Figure 7.9 shows an example scenario for a system measuring the noisy DOA angles of a TG from three RPs using the RF signal transmitted from the TG. Associated with each DOA, we have a line and the three lines intersect in three different locations all different from the actual location of the TG. To find a unique answer to this problem, we need to define a cost function and find the location that optimizes the cost function. In the LS approach, we use summation of the distance squares of the TG location from the three lines as the cost function. As a refresher from basic analytical geometry, the distance between a point with the coordinates $(x, \ y)$ and a line identified by equation $a_i x + b_i y - c_i = 0$ is given by:

$$r_i = \frac{a_i x + b_i y - c_i}{\sqrt{a_i^2 + b_i^2}} \tag{7.12a}$$

If we define the cost function as the average distance square of the location from the DOA line equations, the cost function is given by:

$$\varepsilon(x, y) = \frac{1}{N} \sum_{i=1}^{N} \left[\frac{a_i x + b_i y - c_i}{\sqrt{a_i^2 + b_i^2}} \right]^2 . \tag{7.12b}$$

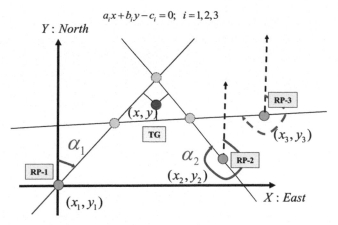

Figure 7.9 Basic principles of LS localization using DOA. Three DOA angles each produce a geometric line resulting in three answer. The LS algorithm finds a unique solution with minimum distance square from the lines.

To find the optimum location, we take the derivative of the cost function with respect to our two variables, x and y:

$$\begin{cases} \frac{\partial \varepsilon}{\partial x} = \frac{1}{N} \sum_{i=1}^{N} \left[\frac{2a_i(a_i x + b_i y - c_i)}{a_i^2 + b_i^2} \right] = 0 \\ \frac{\partial \varepsilon}{\partial y} = \frac{1}{N} \sum_{i=1}^{N} \left[\frac{2b_i(a_i x + b_i y - c_i)}{a_i^2 + b_i^2} \right] = 0 \end{cases} . \qquad (7.12c)$$

Equation (7.12c) is a set of two linear equations with two unknowns and it has a unique solution, which we refer to it as the LS estimation of the localization using multiple DOA angle measurements. If we substitute the LS values of the coordinate into Equation (7.12b), we have the minimum cost function, which is the average of distance squares of the location from the DOA lines. The value of this minimum mean square error reflects the accuracy of position estimation, and it is referred to as the residual error.

7.3.2 LS Method for Positioning with Distances

Figure 7.9 shows an example scenario for a system measuring the noisy TOA of a TG from three RPs using the RF signal transmitted from the TG. Associated with each TOA, we have a circle and the three circles intersect in three different locations all different from the actual location of the TG. To find a unique answer to this problem, we define a cost function and find a unique location coordinate, (x, y) that optimize the cost function. In the LS approach, we use summation of the distance squares of the TG location from the three circles as the cost function. As a refresher from basic analytical geometry, the distance between a point with the coordinate, (x, y), and a circle defined by equation, $\sqrt{(x - x_i)^2 + (y - y_i)^2} - R_i = 0$ is:

$$d_i = \sqrt{(x - x_i)^2 + (y - y_i)^2} - R_i. \tag{7.13a}$$

If we define the cost function as the average distance square of the location from the TOA circles, the cost function is:

$$\varepsilon(x, y) = \frac{1}{N} \sum_{i=1}^{N} \left[\sqrt{(x - x_i)^2 + (y - y_i)^2} - R_i \right]^2 \tag{7.13b}$$

To find the optimum location coordinates, (x, y), we take the derivative of the cost function with respect to our two variables, x and y:

$$\begin{cases} \frac{\partial \varepsilon}{\partial x} = \frac{1}{N} \sum_{i=1}^{N} \frac{2(x-x_i)\left[\sqrt{(x-x_i)^2+(y-y_i)^2}-R_i\right]}{\sqrt{(x-x_i)^2+(y-y_i)^2}} = 0 \\ \frac{\partial \varepsilon}{\partial y} = \frac{1}{N} \sum_{i=1}^{N} \frac{2(y-y_i)\left[\sqrt{(x-x_i)^2+(y-y_i)^2}-R_i\right]}{\sqrt{(x-x_i)^2+(y-y_i)^2}} = 0 \end{cases}. \tag{7.13c}$$

Equation (7.13c) is a set of two nonlinear equations with two unknowns that is very complex. To simplify the solution, we may redefine our cost function in (7.14) as:

$$\varepsilon(x, y) = \frac{1}{N} \sum_{i=1}^{N} \left[(x - x_i)^2 + (y - y_i)^2 - R_i^2 \right]^2 \tag{7.14a}$$

The set of equations for calculation of the optimum location for this new cost function is:

$$\begin{cases} \frac{\partial \varepsilon}{\partial x} = \frac{1}{N} \sum_{i=1}^{N} 2(x - x_i)\left[(x - x_i)^2 + (y - y_i)^2 - R_i^2\right] = 0 \\ \frac{\partial \varepsilon}{\partial y} = \frac{1}{N} \sum_{i=1}^{N} 2(y - y_i)\left[(x - x_i)^2 + (y - y_i)^2 - R_i^2\right] = 0 \end{cases}. \tag{7.14b}$$

Presentation of Equation (7.14b) is less complex than Equation (7.13c), but in both cases, we are involved in solving two sets of nonlinear equations with multiple answers. Equation (7.12c) for multiple angles of arrival is simple because it is a set of linear equations. The difference between the two cases is that if we measure distance from an RF, the TG is somewhere on a circle around the RP. If we measure the angle from an RP, the TG is somewhere on a line passing through the RP. Finding intersect of line is much simpler and unique, while intersect of circles are two and they provide ambiguity. The problem of solving the LS problem has been around for centuries and recursive least square (RLS) algorithms have been the transitional solution for solving this problem.

7.4 Practical Solutions to LS Problem

The obstacle in solving the LS problem is that the solution becomes involved in tedious calculations, e.g. Equation (7.13c) or (7.14b). Iterative LS (RLS) algorithms using the Newton–Gauss method offer a practical solution and they are commonly used in a variety of positioning applications. In these methods, we assume TG is in an arbitrary location and then we adjust the location iteratively until we get very close to the solution of the equation.

$$\left(x-x_i\right)^2+\left(y-y_i\right)^2-R_i^2=0; \quad i=1,2,3$$

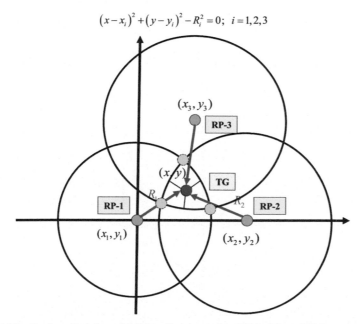

Figure 7.10 Basic principles of LS localization using TOA. Three TOA each produce a geometric circle resulting in three answer. The LS algorithm finds a unique solution with minimum distance square from the circles.

We introduce the Newton–Gauss method in solving a quadratic equation with a simple example.

Example 7.1: (Newton–Gauss Method to Solve a Quadratic Equation)
Figure 7.11 shows a quadratic parabola function $f(x) = (x-1)^2 - 4$ and its intersections with the x-axis, where $f(x) = (x-1)^2 - 4 = 0$. The algebraic solution to this problem is simple, and we have two answers:

$$(x-1)^2 - 4 = 0 \Rightarrow x = -1, 3.$$

For the iterative solution to this problem, we begin from an arbitrary point on the parabola, at $x_n = 5$ and $f(x_n) = 12$. Then, using the fact that

$$\frac{x_n - x_{n+1}}{f(x_n)} = \frac{dx}{dy} = \frac{1}{f'(x_n)} \Rightarrow x(n+1) = x_n - \frac{f(x_n)}{f'(x_n)}, \qquad (7.15)$$

we can iteratively calculate the location. For our specific example:

$$x_{n+1} = x_n - \frac{f(x_n)}{f'(x_n)} = x_n - \frac{(x_n - 1)^2 - 4}{2(x_n - 1)}.$$

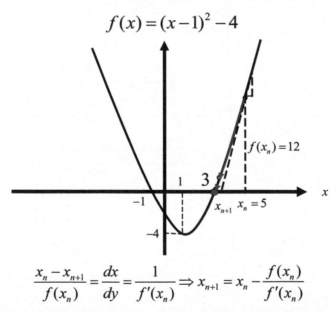

$$\frac{x_n - x_{n+1}}{f(x_n)} = \frac{dx}{dy} = \frac{1}{f'(x_n)} \Rightarrow x_{n+1} = x_n - \frac{f(x_n)}{f'(x_n)}$$

Figure 7.11   Basic principles of the iterative Newton-Gauss method to solve a quadratic equation.

In this equation, $f'(x)$ is the derivative of our original quadratic function. Then, beginning with our arbitrary point, we have:

$$x_n = 5$$

$$x_{n+1} = 5 - \frac{16 - 4}{2(4)} = 5 - 1.5 = 3.5$$

$$x_{n+2} = 3.5 - \frac{6.25 - 4}{2(2.5)} = 3.5 - 0.45 = 3.05$$

. . . .

At the third iteration, the solution is already close to intersect at 3.05. More iterations get the solution closer to the actual solution 3. If we had started from the left side of the curve, we would converge to the other solution at -1. In positioning applications, we usually have some a priori information of location of the TG that helps in finding the right solution.

7.4.1 RLS Algorithm for Distance-Based Positioning

Similar to Example 7.1, complex geometrical triangulation equations for TOA localization using the LS approach can be solved iteratively with the Newton–Gauss method in vector notation. As shown in Figure 7.10, due to estimation errors of distances from the landmark RPs caused by inaccurate measurement of the distance between the RPs and the TG device, the geometrical triangulation technique can only provide a *region* for the location of the TG device with uncertainty, instead of a single position fix. To obtain an estimate of location coordinates in the presence of errors in the measurement of location metrics, a variety of direct and iterative statistical positioning algorithms have been developed to solve the problem by formulating it as a set of nonlinear iterative equations and using different criteria for LS solutions (Section 7.2.2). The simplest iterative algorithms are the iterative least square (LS) algorithms. The general statement of the problem in 2D localization is that we have a set of N nonlinear equations defining the distances from N circles associated with the RPs:

$$f_i(x, y) = (x_i - x)^2 + (y_i - y)^2 - R_i^2; \quad i = 1, 2, \ldots N \qquad (7.16a)$$

In these N nonlinear equations, R_i is the measured distance from the i-th RP, and (x, y) and (x_i, y_i) are the location of the TG device and RPs, respectively. The function $f_i(x, y)$ reflects the ranging error for the distance of the device from the circle associated with the i-th RP. Defining the distance vector, $\mathbf{F}$, as:

$$\mathbf{F} = [f_1(x, y),\ f_2(x, y),\ \ldots\ldots,\ f_N(x, y)]^T, \tag{7.16b}$$

the cost function of Equation (7.14a) becomes:

$$\varepsilon(x, y) = \frac{1}{N} \sum_{i=1}^{N} \left[(x - x_i)^2 + (y - y_i)^2 - R_i^2 \right]^2 = \frac{1}{N} \mathbf{F}^T \mathbf{F}. \tag{7.16c}$$

We call $\mathbf{F}$ a distance vector because each element of the vector represents the distance from one of the circles around an RP. The Jacobian matrix of the F, representing differential increments in location, is defined as the gradient of the function:

$$\mathbf{J} = \nabla_{x,y}[\mathbf{F}] = \begin{bmatrix} \frac{\partial f_1(x,y)}{\partial x} & \frac{\partial f_1(x,y)}{\partial y} \\ \cdots & \cdots \\ \frac{\partial f_N(x,y)}{\partial x} & \frac{\partial f_N(x,y)}{\partial y} \end{bmatrix} \tag{7.16d}$$

If we begin from an arbitrary location

$$\mathbf{U}_n = [x_n\ y_n]^T \tag{7.17a}$$

similar to Equation (7.15), we can apply the iterative Newton–Gauss method to solve the problem iteratively.

$$\mathbf{U}_{n+1} = \mathbf{U}_n - \mathbf{E}_n = \mathbf{U}_n - \frac{\mathbf{F}_n}{\nabla_{x,y}[\mathbf{F}_n]} = \mathbf{U}_n - \mathbf{J}_n^{-1} \mathbf{F}_n. \tag{7.17b}$$

Since J is not a square matrix, its inverse is calculated as:

$$\mathbf{J}\mathbf{J}^{-1} = \mathbf{I} \Rightarrow \mathbf{J}^{-1} = \left(\mathbf{J}^T \mathbf{J}\right)^{-1} \mathbf{J}^T. \tag{7.16}$$

Substituting this inverse value in Equation (7.17b), we have the final RLS algorithm using the Newton–Gauss iterative solution:

$$\mathbf{U}_{n+1} = \mathbf{U}_n - \left(\mathbf{J}_n^T \mathbf{J}_n\right)^{-1} \mathbf{J}_n^T \mathbf{F}_n. \tag{7.17c}$$

This algorithm begins from an arbitrary location and solves the quadric set of equations iteratively. Figure 7.12 shows a typical path that algorithm may take to update the location from an arbitrary location to a location in the region of intersection among all circles.

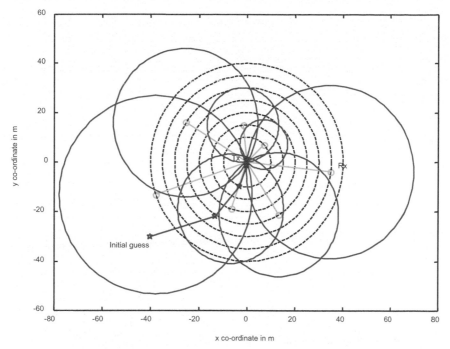

Figure 7.12 A typical path of updates for the RLS algorithm to approach the region of intersection among all circles.

Example 7.2: (MATLAB Code for RLS algorithm in a room)

Assume that we have three RPs at $(10, 10)$, $(0, 15)$, and $(-5, 5)$ and a device measures its location from these distances to be 15, 10, and 5 m, respectively. To solve this problem, we need to solve the following three sets of quadratic equations:

$$
\mathbf{F} = \left[\begin{array}{c} f_1(x, y) \\ f_2(x, y) \\ f_3(x, y) \end{array} \right] = \left[\begin{array}{c} (x_1 - 10)^2 + (y_1 - 10)^2 - 15^2 \\ (x_2 - 0)^2 + (y_2 - 15)^2 - 10^2 \\ (x_3 + 5)^2 + (y_3 - 5)^2 - 5^2 \end{array} \right]
$$

The Jacobian matrix for the above matrix is:

$$
\nabla_{x,y}\mathbf{F} = \left[\begin{array}{cc} 2(x_1 - 10) & 2(y_1 - 10) \\ 2(x_2 - 0) & 2(y_2 - 15) \\ 2(x_3 + 5) & 2(y_3 - 5) \end{array} \right]
$$

The MATLAB code provided in Appendix 7.A finds the RLS solution of Equation (7.22) starting from location (5, 2). The first eight iterations, shown in Figure 7.13 and their associated errors are:

```
The initial location estimation is: 5 2
The 1th estimated location is: [-1.9174,1.2254] with an
    error of 6.9606
The 2th estimated location is: [-3.8817,4.2477] with an
    error of 3.6045
The 3th estimated location is: [-4.3327,5.6743] with an
    error of 1.4962
The 4th estimated location is: [-4.4744,6.1228] with an
    error of 0.47038
The 5th estimated location is: [-4.5271,6.2505] with an
    error of 0.13809
The 6th estimated location is: [-4.5444,6.2876] with an
    error of 0.040933
The 7th estimated location is: [-4.5497,6.2985] with an
    error of 0.012182
The 8th estimated location is: [-4.5514,6.3018] with an
    error of 0.0036286
```

The final solution is somewhere around $(-4.5, 6.3)$.

RLS algorithm is very useful in navigation of mobile TGs. As a mobile moves along a path location, sensors update their distances from the RPs and these updates will be used in Equation (7.16a,b) for calculation of distance matrix and Jacobian matrix, and for location update Equation (7.17c). This feature makes the algorithm suitable for TOA-based CPS or GPS systems using the distance from RPs for positioning. The RLS algorithm is also applicable to any other geometric localization algorithm for continual positioning using updated measurements from a mobile using distance or angle of arrival. For any combination of R, α, ΔR or for cooperative localization, we can apply RLS algorithm. All we need is to define the set of nonlinear equations that describe our measurements of these features to form **F**, the characteristic function of the measurements we use for positioning.

7.4.2 Residual Weighted RLS Algorithm for Indoor and Urban Areas

The RLS algorithm described in the previous section is a simple traditional gradient algorithm used in basic GPS systems, where the multipath and shadow fading is limited and the noise level associated with different measured distances from the RPs are similar. In many dense urban and

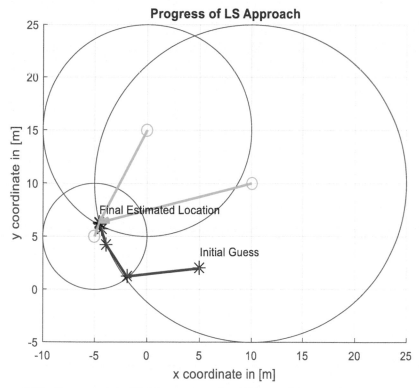

Figure 7.13   Progress of the RLS for the scenario of Example 7.2 produced by the MATLAB code in Appendix 7.A.

indoor area application scenarios, behaviors of the distance measurement error in LOS and OLOS are widely apart and the RLS or other algorithms need to adjust to the situation. In outdoor urban areas, when the LOS path between the transmitter and the receiver is blocked by a building, we have an OLOS condition with a blocked direct path. In this situation, the TOA-based distance measurement will measure a path that is not the direct path and that mistake causes large errors. To handle these situations, variations of the RLS algorithm have been proposed for localization in urban and indoor areas.

The RLS algorithm in 2D applications needs a minimum of three RPs. With more RPs, the algorithm is expected to provide a more accurate positioning. The residual weighted RLS (RW-RLS), originally proposed for cellular positioning in [Che99] and later on used for UWB indoor positioning in [Kan04a], is another version of the RLS algorithm that fits better for indoor and urban areas, characterized by bipolar behavior in LOS and OLOS TOA ranging. This algorithm operates under the condition that we have more than three RPs, which is very typical in urban and indoor positioning scenarios. RW-RLS algorithm makes a location estimate based on three and more combination of the RPs and calculates the residual error of each combination. The final location is estimated by weighted averaging all combinations, based on the inverse of the residual value of each of them. This way, those combinations with large errors will have a smaller contribution in the final position estimate. The indoor scenario used in [Kan04a] assumes a grid indoor installation of RPs with four RPs installed in corners of a square area covering everywhere inside them. In this scenario, we have all possible 3-RPs combinations from the four and one combination with all four-RPs. Using a random number generator, a uniformly distributed location in the room is selected. Using the UWB distance measurement error model for TOA-based ranging provided in Section 6.6.1, the estimated distances from the four RPs are determined for a variety of system bandwidths. These distances are then used with the RLS or RW-RLS algorithms to determine the estimated location of the terminal in the area. The difference between the estimated location and the actual randomly selected location is used as the positioning error. By repeating this experiment, the statistics of the positioning error associated with each of the algorithms is determined for different bandwidths of the system. Figure 7.14 [Kan04a] shows the comparison of the performance of the two algorithms for a variety of system bandwidths. The RW-RLS algorithm provides a slightly better performance than RLS in lower bandwidths.

7.4.3 Closest Neighbor LS Grid Algorithm

LS solution turns the positioning problem to an optimization problem. We define an error function, similar to Equation (7.13c) or (7.14b), and we find the location of the TG, (x, y), which minimizes that error function for the given location of RPs, (x_i, y_i). The error function is a function related to the distance between the TG and a circle representing the trace of location of the TG from the location of RPs. Section 7.4.1 discusses the RTS iterative solution to LS problem. The advantage of RLS solution to direct solution of LS problem is that RLS begins in an arbitrary location and the solution emerges in a few steps and after that, it updates the solution from the last estimated location. Therefore, it suites well for tracking mobile devices. To solve a set of nonlinear equations, however, the iterative LS algorithm may converge to a local optimum rather than the universal optimum location.

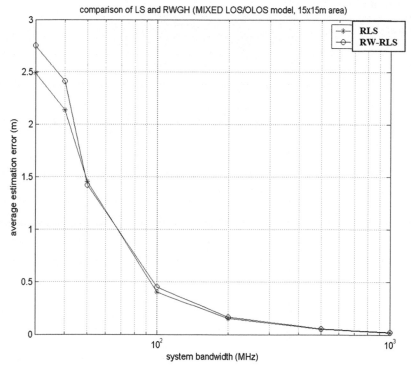

Figure 7.14 The average estimated positioning error versus bandwidth of a TOA system for RW-RLS algorithm in a 15×15m room with mixed LOS and OLOS conditions.

In certain situation among the circles with respect to one another, the algorithm may even diverge from the area of expected location. For example, if we have an indoor geolocation problem, solution may diverge to the outdoors.

Numerical solution to the localization problem offers itself as an alternative with more robust positioning of a location. Principles of numerical solution are very simple, we calculate the value of the cost function, for example, those defined in Equation (7.13c) or (7.14b), over a grid set of locations and we declare the location with minimal value of the cost function as the solution to the problem. Figure 7.15 shows the basic concept behind the numerical solution to the LS function using the nearest neighbor algorithm. We form an L × M grid around the area of the positioning scenario, where all the RPs and the TG are located. The dimension of the grid can be the physical boundaries of the solution, for example, dimension of a room if we are locating an object inside a room. The size of the grid lattice, the distance

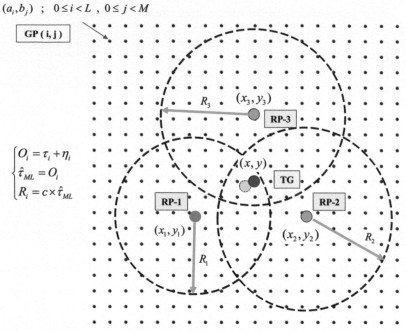

Figure 7.15   Closest neighbor LS (CNLS) algorithm to solve the LS problem numerically for TOA-based positioning.

between the consecutive points on the grid, should be on the order of the standard deviation of the DME obtained from the CRLB. If we represent the location of each point in the grid by $\{(a_j, b_k);\ 0 \leq j < L,\ 0 \leq k < M\}$ and use the metric in Equation (7.13c) as the definition of the error, the optimization process reduces to finding

$$\min \{\varepsilon(a_j, b_k)\}_{\substack{0 \leq j < L \\ 0 \leq k < M}}$$

$$= \min \left\{ \frac{1}{N} \sum_{i=1}^{N} \left[\sqrt{(a_j - x_i)^2 + (b_k - y_i)^2} - R_i \right]^2 \right\}_{\substack{0 \leq j < L \\ 0 \leq k < M}} \tag{10.18}$$

In other words, this algorithm calculates the distance of the location with all locations on the grid to find the nearest neighboring location for the LS solution. For that reason, we may refer to it as the closest neighbor LS algorithm (CNLS) [Kan04B].

As compared with the RLS, the CNLS always provides the global solution to numerical solution of the LS problem and this solution always converges.

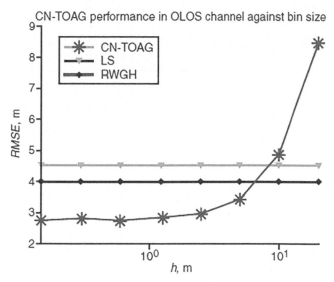

Figure 7.16   Performance of an UWB TOA-based localization system using nearest neighbor algorithm with a bandwidth of 500 MHZ in a 20 × 20 m room as a function of grid size. Performance is compared with the LS and the residually weighted LS algorithm.

The computational complexity of the algorithm is in the order of square of the number of grid points on one side of the grid. This complexity increases to the third order for 3D localization. If we apply that to cooperative localization with two unknown TGs in 2D, the complexity goes to fourth order of the one dimension and so on. As a result, this approach is suitable for cases where the ratio of the dimension to standard deviation of range measurement error is not too large. Figure 7.16 [Kan04B] shows performance of the CNLS algorithm as a function of grid size, h, used for UWB localization with 500 MHz of bandwidth in a 20 m × 20 m room. Performances of LS and RW-RLS algorithms are provided as a reference for comparison. As the grid size increases, performance of the CNLS gets closer to the LS and RW-RLS algorithms. The channel model introduced in Section 6.5.1 is used to emulate the measured DME for localization of the TG point in test locations.

Appendix 7.A: MATLAB Code for RLS of Example 7.2

```
clc;clear all;close all;
%% This Matlab code solve Example 7.2 in textbook
known_references = [10,10;0,15;-5,5];
```

```matlab
initial_guess = [5,2];
distances = [15,10,5];

if size(known_references,2) ~= 2
    error('location of known RPs should be entered as Nx2
    matrix');
end

figure(1);
hold on
grid on

% Draw Circles
theta = 0:pi/360:2*pi;
circle1 =
[known_references(1,1)+distances(1)*sin(theta'),known_
references(1,2)+distances(1)*cos(theta')];
plot(circle1(:,1),circle1(:,2),'b')
circle2 =
[known_references(2,1)+distances(2)*sin(theta'),known_
references(2,2)+distances(2)*cos(theta')];
plot(circle2(:,1),circle2(:,2),'b')
circle3 =
[known_references(3,1)+distances(3)*sin(theta'),known_
references(3,2)+distances(3)*cos(theta')];
plot(circle3(:,1),circle3(:,2),'b')

i=1;
temp_location(i,:) = initial_guess ;
temp_error = 0 ;

for j = 1 : size(known_references,1)
    temp_error = temp_error + abs((known_references(j,1) -
    temp_location(i,1))^2 + (known_references(j,2) - temp_
    location(i,2))^2 - distances(j)^2) ;
end

estimated_error = temp_error ;
plot(temp_location(i,1),temp_location(i,2),'k*','MarkerSize',
10) ;
%plot
text(temp_location(i,1), temp_location(i,2)*(1 + 0.8),
'Initial Guess');
disp(['The initial location estimation is:
', num2str([temp_location(i,1),temp_location(i,2)])]);
% new_matrix = [ ];
```

```
while norm(estimated_error) > 1e-2 %iterative process for
LS algorithm

    for j = 1 : size(known_references,1)   %Jacobian has been
    calculated in advance
        jacobian_matrix(j,:) = -2*(known_references(j,:) -
temp_location(i,:)) ;  %partial derivative is i.e. -2(x_1-x)
        f(j) = (known_references(j,1) - temp_location(i,1))^2 +
(known_references(j,2) - temp_location(i,2))^2 - distances(j)
^2 ;
    end

    estimated_error = -inv(jacobian_matrix' * jacobian_matrix)
* (jacobian_matrix') * f' ; %update the U and E

    temp_location(i+1,:) = temp_location(i,:) + estimated_error'
    ;

plot(temp_location(i+1,1),temp_location(i+1,2),'k*',
'MarkerSize',10) ;
% plot
    dp = temp_location(i+1,:)-temp_location(i,:);

quiver(temp_location(i,1),temp_location(i,2),dp(1),dp(2),0,
'Color','r','LineWidth',2);

    %text(temp_location(i+1,1), temp_location(i+1,2)*(1 +
    0.005), num2str(i));

    i = i + 1;
lx=num2str(temp_location(i,1));ly=num2str(temp_location(i,2)
);err=sqrt(estimated_error(1)^2+estimated_error(2)^2);
disp(['The ',num2str(i-1), 'th estimated location is:
',' [',lx,',',ly,']',' with an error of ', num2str(err)]);
end

plot(known_references(:,1),known_references(:,2),'go',
'MarkerSize',10);
for i=1:length(known_references)
    dp = temp_location(end,:) - known_references(i,:);

quiver(known_references(i,1),known_references(i,2),dp(1),dp(2)
,0,'Color','g','LineWidth',2);
end
text(temp_location(end,1), temp_location(end,2)*(1 + 0.2),
'Final Estimated Location');
```

```
%axis([1.1*min(temp_location(:,1)),1.1*max(temp_location(:,1)),
0.9*min(temp_location(:,2)), 1.1*max(temp_location(:,2))]);
title('Progress of LS Approach')
xlabel('x coordinate in [m]');
ylabel('y coordinate in [m]');
```

Assignments for Chapter Seven

Questions

1. What are different alternatives to locate a target device from its distance and angle from reference points? What are the minimum number of reference point needed for each technique?
2. How can we locate a device using a single reference point?
3. Can we find a unique location for a target in the sky by measuring its distance from two reference points on the earth on a 2D map the plane connecting all three of these locations?
4. What are the differences among cooperative and traditional localization techniques?
5. What are the benefits of differential TOA localization and how this method relates to synchronization problem?
6. Compare is the RLS algorithm and what are its advantages and disadvantages over numerical solution of the LS algorithm using computational techniques.
7. How can we compare the results of CRLB for 2D positioning with the results obtained from solving LS solution to positioning?

Problems

Problem 1:
Two base stations located at (x, y) coordinates (500, 150) and (200, 200) are measuring the angle of arrival of the signal from a mobile terminal with respect to the x-axis. The first base station measures this angle as $45°$ and the second as $75°$. What are the coordinates of the mobile terminal?

Problem 2:
In Problem 1, what happens if the first base station incorrectly measures the AOA from the mobile terminal as $50°$ or $30°$?

Problem 3:
Base stations A, B, and C located at (50, 50), (300, 0), and (0, 134) are found to be at distances 90, 200, and 100 m from a mobile terminal. Draw circles corresponding to these values and try to determine the location of the mobile terminal.

Problem 4:
In Problem 3, what happens if the mobile incorrectly measures the distance from base station B as 100 m or 300 m?

Problem 5:

a) Using Figure 7.5 show that for AOA localization using two RPs, we have:

$$y = \frac{y_2 \cdot \tan(\theta_2) - x_2}{\tan(\theta_2) - \tan(\theta_1)}$$
$$x = r \tan^{-1}(\theta_1)$$

b) Using Figure 7P.1, show that for TDOA localization we have:

$$a = \frac{d_1 - d_2}{2}$$
$$b = \sqrt{(D/2)^2 - a^2}$$

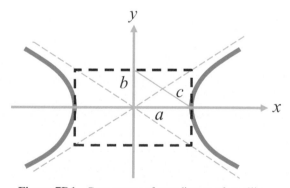

Figure 7P.1 Parameters of coordinates of an ellipse.

Problem 6:

a) For three RPs at $(x_i \ y_i)$, $i = 1, 2, 3$, give the three set of quadratic functions needed to find the location of a terminal with the TOA estimate of τ_i; $i = 1, 2, 3$ from the RPs.

b) Repeat (a) and if we use the TDOA

$$\Delta\tau_i; \quad i = 1, 2, 3$$
$$\Delta\tau_1 = \tau_1 - \tau_2$$
$$\Delta\tau_2 = \tau_1 - \tau_3$$
$$\Delta\tau_3 = \tau_2 - \tau_3$$

Problem 7:

Let us suppose that there are N estimates d_i of the distance of an MS from N known locations with coordinates (x_i, y_i) for $i = 1,2,3, \ldots, N$. We then have N equations of the form:

$$f_i(x, y) = (x_i - x)^2 + (y_i - y)^2 - d_i^2 = 0$$

where (x, y) is the unknown location of the MS. The *least squares* technique provides a method of estimating x and y when there are errors in the estimates d_i. The technique works as follows. Let $\mathbf{F} = [f_1(x, y) \, f_2(x, y) \ldots f_N(x, y)]^T$. First, we have to construct the *Jacobian* matrix given by:

$$\mathbf{J} = \begin{bmatrix} \frac{\partial f_1(x,y)}{\partial x} & \frac{\partial f_1(x,y)}{\partial y} \\ \cdots & \cdots \\ \frac{\partial f_N(x,y)}{\partial x} & \frac{\partial f_N(x,y)}{\partial y} \end{bmatrix}$$

Next we pick an estimate of the solution $\mathbf{U} = [x^* \, y^*]$ and we determine the error in the solution as $\mathbf{E} = -(\mathbf{J}^T\mathbf{J})^{-1}\mathbf{J}^T \, \mathbf{F}$ evaluated at the estimate $\mathbf{U}$. The new solution is $\mathbf{U} + \mathbf{E}$. Iteratively, the error in the solution is reduced by computing a new error that is added to the previous solution to obtain a new solution till a point is reached where the solution does not change. Let the known locations of RPs be (10, 10), (0, 15), and (−5, 5). The measured distances from these RPs are 15, 16, and 5 m. Use the least squares approach to determine the estimate of the location by using the location (2, 2) as the initial estimate.

Problem 8:

Repeat Problem 7 for LS estimation of the location if the cost function is defined as:

$$\varepsilon(x, y) = \|F\|^2 = F^T F = \sum_{i=1}^{N} \left[\sqrt{(x - x_i)^2 + (y - y_i)^2} - d_i \right]^2,$$

where

$$f_i(x, y) = \sqrt{(x - x_i)^2 + (y - y_i)^2} - d_i; \quad i = 1, ...N.$$

Instead of

$$f_i(x, y) = (x - x_i)^2 + (y - y_i)^2 - d_i^2; \quad i = 1, ...N.$$

8

RSS-Based Positioning Algorithms

8.1 Introduction

Selection of RF positioning algorithms depends on density of deployment of reference points, multipath characteristics of the environment, and precision requirement for the application. RSS-based localization is based on measurement of the received signal strength (RSS), which is very simple but unreliable. Measurement of RSS is simple because it can be done without any coordination between the transmitter and the receiver. Using RSS for ranging is unreliable because, in theory, its CRLB is proportional to the square of the distance making this method impractical for applications involved in long distances. In addition to this theoretical fact, there are several practical issues in reliability of models for RSS measurements at the receiver. The general model for calculation of the RSS at the receiver, $P_r = P_0 - 10\alpha + X(\sigma)$, has three parameters, (P_0, α, σ). The received power at a reference point, P_0, needs the knowledge of the transmitted power and antenna gains. Standard organizations, such as IEEE 802.11, specify a highest transmitted power and manufacturers design their products with adjustable power levels and different antenna gains. Besides, mobile devices use batteries to power up the device, power drops in time with the life of the battery and power amplifier. These practical issues affect the reliability of calculation of P_0. Value of distance power gradient, α, and the variance of the shadow fading, σ, are also unreliable because they are dependent on the environment of operation. As a result, because of simplicity of implementation of RSS-based positioning, a variety of popular applications use RSS, and because of complexity of RSS behavior, a variety of algorithms are used in RSS-based localization applications for different environments.

This chapter is devoted to RSS-based positioning algorithms, which includes a wide variety of algorithms used in numerous applications. In Chapter 3, we divided positioning systems into systems using RSS directly and the systems using fingerprinting. To proceed with our discussions in this chapter, we follow the same division to discuss RSS-based positioning algorithms. Short-range positioning applications use the RSS directly because the ratio of the coverage of the reference point to the required precision is low. These applications include localization using RFID, localization in wireless sensor networks (WSN), positioning in local IoT [Yin17], and localization inside the human body using body-mounted sensors [Li12]. When the ratio of coverage to precision is high, we need to resort to fingerprinting. Finger-printing is used for Wi-Fi localization in RTLS and WPS [Hat06, Pah09], and RSS-based CPS [Akg09] positioning systems. In these systems, we resort to fingerprinting to increase the precision to an acceptable level for applications in urban and indoor areas. In this chapter, we further divide fingerprinting algorithms into outdoor and indoor finger printing algorithms. In outdoor, fingerprinting is based on the GPS readings and creation of a big database of RSS reading from a huge number of transmitters. In indoor, fingerprinting is based on manual selection of the location for measurement of the RSS and it produces a smaller database in a specific indoor area from a limited number of transmitters. The algorithms used for these techniques are different and we discuss them separately.

8.2 RSS-Based Algorithms for Short-Range Positioning

In Section 3.3, we introduced RSS-based positioning systems using the RSS directly to triangulate a target with a few reference points. These are low-power short-range systems such as RFID systems with a range of approximately 1 m, sensors Bluetooth, LEB (iBeacon), or ZigBee wireless sensor technologies with a range of approximately 10 m, wireless devices operating inside the human body such as wireless endoscopy capsule with an approximated range of a few tens of centimeter. When the ratio of precision to coverage is close to 1, we can localize using one RP and the positioning algorithm is very simple; when the target (TG) reads a certain reference point (RP), its ML location is the location of the RP. This algorithm is an RF sign positing algorithm, because when we see the RP, we know we are there. For larger and smaller ratios of precision to coverage, we need triangulation and several RPs. The simplest approach to solve the problem is to use the LS method. As we discussed in Section 8.4, we can use either RLS or computational methods to solve the problem. The numerical solution introduced in Section 8.4.3 solves the geometric positioning equations using minimum mean square error (MMSE) criterion. Another approach to solve the problem is to use a probabilistic grid, which assigns probabilities to each point on the grid and solves the problem using these probabilities.

8.2.1 Physical Grid Closest Neighbor Algorithm

In Section 3.3, we introduced RSS-based positioning systems using RFID technology from the architectural and performance assessment point of view. In this section, we discuss these systems from the point of view of their positioning algorithms. RFID positioning systems consist of two elements: the RFID tag and the RFID reader, which are used for location monitoring in a variety of popular applications. Figure 8.1(a) shows an RFID reader monitoring vehicles carrying an externally visible RFID tag, while the vehicle passes through a certain highways post. Figure 8.1(b) shows the RFID reader posted on the entrance door of a hospital nursery room to monitor newborns traffic with an RFID tag attached to the child. Figure 8.2(a) shows the general block diagram of an RFID monitoring system. The positioning algorithm is very simple: when the reader reads the RFID tag, it associates its location to the location of the child carrying the RFID at that time.

RFID systems are also used for one-dimensional (1D) and two-dimensional positioning and navigations in indoor areas. Figure 8.2(b) shows an example of a 1D RFID positioning system inside a tunnel. RFID readers

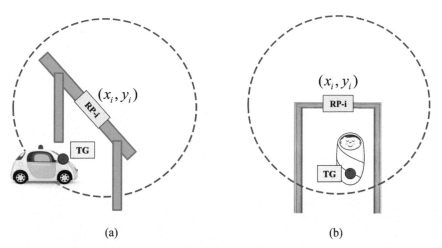

(a)   (b)

Figure 8.1 RFID for sign posting applications, an RFID reader in a known location is utilized as RP-i to read the RFID of the TG, (a) reader on a pay toll post and RFID in a car, (b) reader on a newborn entrance room in a hospital and RFID on the newborn.

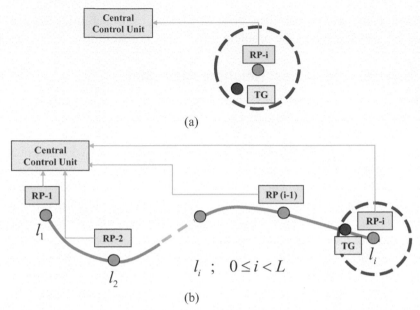

(a)

(b)

Figure 8.2   (a) RFID application in RF sign posting to localize an object carrying an RFID, (b) 1D localization along the path of movement using RFID technology.

deployed in every few meters along the motion path of the tunnel read the RFID tags on the mineworker to find how deep the minor has traveled inside the tunnel. Defining L_i; $0 \leq i < L$, the positioning algorithm associates the location of the reader that reads the RFID with the distance that mineworker has travelled. The closest neighboring RP identifies the location of the TG.

Another interesting application of RFID positioning is 2D positioning of a robot on the floor of a warehouse. For the three RFID positioning applications, which we discussed earlier in this section, the RFID readers are installed on fix RP locations and the RFID TG is installed on the moving objects. For 2D robotics positioning in a large warehouse, situation is in reverse, RFID tags are installed on a grid on the floor, for example, in the middle of the floor tiles, and the RFID readers is on a robot operating on the floor (Figure 8.3). In Figure 8.3, we have RPs physically installed on a grid in known locations (x_i, y_j); $0 \leq i < L$, $0 \leq j < M$. The algorithm positions the robot by associating the RFID reader input to the known locations of the RFID tag. If RFID reader reads more than one RFID, it can triangulate using simple algorithms. For example, the reader can select the location of the RFID with the highest power as the position estimate, or it

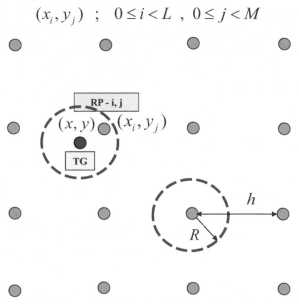

Figure 8.3 Physical Grid Closest Neighbor algorithm for 2D localization of an RFID reader using a dense deployment of RFID tags.

can use the average of (x_j, y_j) coordinates of RFIDs as the location estimate. Since RFIDs are deployed on a grid, we can consider this algorithm a grid-based algorithm and since it decides on the location based on the closest neighbors to the reader, we can also call it a closest neighbor algorithm. Therefore, we refer to this algorithm as Physical Grid Closest Neighbor (PGCN) algorithm.

8.2.2 CNLS for RSS-Based Localization

In Section 8.4.3, we describe another grid algorithm, the closest neighbor LS (CNLS) grid algorithm. The difference between the CNLS and PGCN algorithms is that in CNLS algorithm grid points, $(a_j, b_k); \quad 0 \le j < L$, $0 \le k < M$ are made based on assumption for location of the TG, in PGCN grid points, $(x_i, y_j); \quad 0 \le i < L, \ 0 \le j < M$, which are the actual location of physical RPs. The PGCN was introduced for RSS-based dense grid deployment (Figure 8.3), where dimension of the grid is close to coverage of the RP, $R \sim h$. Such a dense deployment is only possible for inexpensive and small RFID tags. For other short-range sensor devices using BLE or ZigBee technologies, the CNLS algorithm is practical. Our presentation of CNLS in Section 8.4.3 was for UWB TOA-based localization, where we observe the TOA in noise:

$$\begin{cases} O_i = \tau_i + \eta_i \\ \hat{\tau}_{ML} = O_i \\ R_i = c \times \hat{\tau}_{ML} \end{cases} \qquad (8.1a)$$

Then, we find the minimum of the LS cost function using, R_i, location of RPs, $(x_i \ y_i)$, and the grid locations, (a_j, b_k); $0 \le j < L$, $0 \le k < M$:

$$\min \left\{ \varepsilon(a_j, b_k) \right\}_{\substack{0 \le j < L \\ 0 \le k < M}}$$

$$= \min \left\{ \frac{1}{N} \sum_{i=1}^{N} \left[\sqrt{(a_j - x_i)^2 + (b_k - y_i)^2} - R_i \right]^2 \right\}_{\substack{0 \le j < L \\ 0 \le k < M}} . \qquad (8.1b)$$

The CNLS is also applicable to RSS-based positioning with minor changes (Figure 8.4). In RSS positioning with N measured power, P_r^i; $0 \le i < N$,

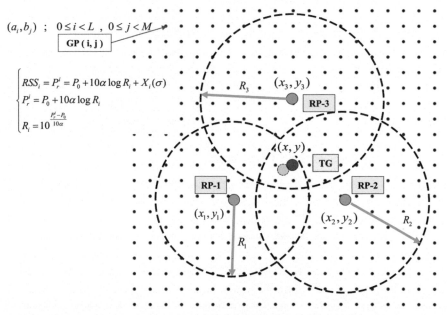

Figure 8.4   Closest neighbor LS (CNLS) algorithm for RSS-based localization.

we calculate the distance from:

$$\begin{cases} P_r^i = P_0 + 10\alpha \log R_i + X_i(\sigma) \\ P_r^i = P_0 + 10\alpha \log R_i \\ R_i = 10^{\frac{P_r^i - P_0}{10\alpha}} \end{cases} \qquad (8.1c)$$

Then, we use Equation (8.1b) to find the grid point that provides the minimum of the cost function. The measurement noise for the TOA is complex and varies in LOS and OLOS multipath conditions and availability of the direct path. Measurement of distance from RSS is simple but involves large measurement noise from shadow fading and uncertainty on estimation of parameters, (P_0, α), which are environment-dependent.

8.2.3 Probabilistic Grid Closest Neighbor Algorithms

The CNLS algorithms, described in Sections 8.4.3 and 8.2.2, for TOA- and RSS-based localization, respectively, are numerical solutions to ML geometric triangulation equations using the minimum mean square error (MMSE) criteria. We solve ML equations for multiple observations of positioning parameters, $(x\ y)$, and using computational methods, we find a unique estimation of the location. We can use maximum likelihood formulation directly to assign probabilities to the grid points and select the grid point with highest probability to triangulate the TG. We refer to this algorithm the Probabilistic Grid Closest Neighbor (PRGCN) algorithm.

Figure 8.5 shows the basic idea behind the PRGCN algorithm. We have *N-RPs* located at (x_i, y_j); $0 \leq i < N$, and we have measured N values of RSS, P_r^i; $0 \leq i < N$. Our model defining the relation between RSS measurements and the distance is given by the path loss model, $P_r = P_0 - 10\alpha \log r + X(\sigma)$. Using this model, for any given value of power measurement, P_r^i; $0 \leq i < N$, we can assign a probability to each location on the grid, specified by its coordinate, (a_j, b_k); $0 \leq j < L$, $0 \leq k < M$. The distance between a grid location and RP-i is:

$$r_i(j, k) = \sqrt{(a_j - x_i)^2 + (b_k - y_i)^2}; \quad 0 < i \leq N, 0 < j \leq L, 0 < k \leq M, \qquad (8.2a)$$

and probability that this measurement belongs to a given location on the grid is:

$$p_i(j, k) = 1 - \frac{1}{2} erfc \left(\frac{P_0 - 10\alpha \log [r_i(j, k)] - P_r^i}{\sqrt{2}\sigma} \right). \qquad (8.2b)$$

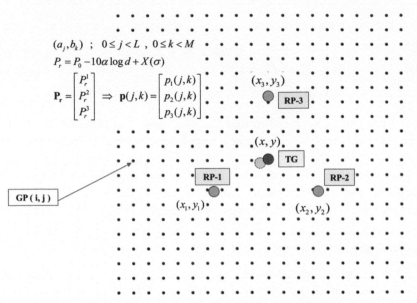

Figure 8.5 Probabilistic grid closest neighbor (PRGCN) algorithm for RSS localization using physical reference points (RP). Using the vector of RSS measurements by the TAG we calculate the probability of that point to be the solution.

Then, associated with every grid point with location,(a_j, b_k); $0 \leq l < L$, $0 \leq m < M$, we have a probability vector:

$$\mathbf{p}(j,k) = \begin{bmatrix} p_1(j,k) & p_2(j,k) & \ldots\ldots & p_N(j,k) \end{bmatrix}^T, \tag{8.2c}$$

associated with N-RSS measurements of the TG from the N-RPs. Each element of the vector for a grid point represents a probability for that grid point to be the answer, given the measurement of RSS from an RP.

Association of grid locations to a probability vector generated from the measurements of the RSS from a set of reference points opens the horizon for a class of algorithms using these probabilities with different statistical interpretations or cost functions. We can simply define the cost function for optimization as the average of probabilities of being in a location:

$$\varepsilon(a_j, b_k)|_{\substack{0 \leq j < L \\ 0 \leq k < M}} = \frac{1}{N} \sum_{j=1}^{N} p_i(j,k). \tag{8.3a}$$

Then, the value of (a_j, b_k) which maximizes the above cost function is the estimated location of the tag:

$$\max\{\varepsilon(a_j, b_k)\}_{\substack{0 \leq j < L \\ 0 \leq k < M}} = \max\left\{\frac{1}{N} \sum_{i=1}^{N} p_i(j, k)\right\}_{\substack{0 \leq j < L \\ 0 \leq k < M}}. \qquad (8.3b)$$

Weakness of the cost function defined in Equation (8.3a) is that it does not perform well when the TG is very close to an RP. In this situation, probability associated with that RP is close to 1 and probabilities for others are very low. Then, the cost function defined by Equation (8.3a) will have an average probability of close to 1/N and the locations in the middle will be detected as the estimation of the position, causing large errors.

One approach to remedy this weakness of the average probability algorithm of Equation (8.3a,b) is to define a threshold p_{th} (e.g., 90%) to probabilities that we include in the averaging. That means, if M is the number of elements of the vector, $p(j, k)$, with $p_1(j, k) > p_{th}$, we find the optimum location from:

$$\max\{\varepsilon(a_j, b_k)\}_{\substack{0 \leq j < L \\ 0 \leq k < M}} = \max\left\{\frac{1}{M} \sum_{i=1}^{N} p_i(j, k)\right\}_{\substack{0 \leq j < L \\ 0 \leq k < M}}. \qquad (8.3c)$$

Using this modification, when we are close to an RP, the average probability cost function gets close to 1. Another similar approach is to adopt a binary scoring system for the probabilities by defining:

$$\begin{cases} s_i(j, k) = 1 & ; if \, p_i(j, k) \geq p_{th} \\ s_i(j, k) = 0 & ; if \, p_i(j, k) < p_{th} \end{cases} \qquad (8.4a)$$

This way, we divide the reference points in effective and ineffective and we select the gird points with the highest score:

$$\max\{\varepsilon(a_j, b_k)\}_{\substack{0 \leq j < L \\ 0 \leq k < M}} = \max\left\{\sum_{i=1}^{N} S_i(j, k)\right\}_{\substack{0 \leq j < L \\ 0 \leq k < M}}. \qquad (8.4b)$$

If we have K grid points with the same score, position estimate is the centroid of those points given by:

$$\begin{cases} \hat{x} = \frac{1}{K} \sum_{i=1}^{K} a_i \\ \hat{y} = \frac{1}{K} \sum_{i=1}^{K} b_i \end{cases}. \qquad (8.4c)$$

8.2.4 Maximum Likelihood Grid Triangulation Algorithm

In Section 2.4.2, we used the path loss model $P_r = P_0 - 10\alpha \log r + X(\sigma)$ to define a rim shape certainty region for the RSS-based ranging. Figure 8.6 shows the summary of the method we used for calculation. For a given certainty, γ (e.g., 90%), we first calculate the fade margin, F_σ, from variance of shadow fading, σ, and the required certainty from:

$$F_\sigma = \sqrt{2}\sigma \times erfc^{-1}(1 - \gamma). \tag{8.5a}$$

Then, we use the fade margin and the measured RSS, P_M, to calculate the radius of two circles, defining inner and outer boundaries from:

$$\begin{cases} r_1 = 10^{\frac{P_0 - F_\sigma - P_M}{10\alpha}} \\ r_2 = 10^{\frac{P_0 + F_\sigma - P_M}{10\alpha}} \end{cases} . \tag{8.5b}$$

The area between the two circles designates a region for the location of the tag with the desirable certainly.

The maximum likelihood grid triangulation (MLGT) is a grid-based computational algorithm that extends the idea of rim shape certainty region to triangulation of a TG using multiple RPs. Figure 8.7 shows the basic principles of MLGT, and the algorithm forms a grid and plots regions of confidence associated with measurements from different RPs. Then, it finds the grid points in the overlap of all regions of confidences. For K grid points

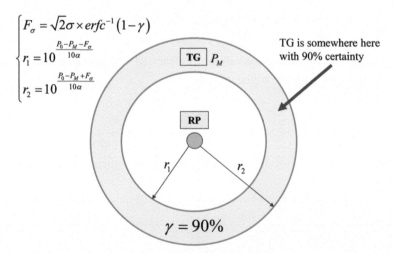

Figure 8.6   Calculation of certainty region from the path-loss model and fade margin.

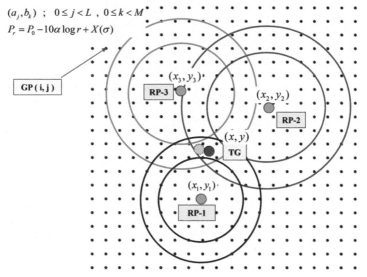

Figure 8.7 Principles of Maximum Likelihood Grid Triangulation (MLGT) algorithm for RSS based localization.

in the overlap area, the location estimate is the centroid of those points given by:

$$\begin{cases} \hat{x} = \frac{1}{K} \sum_{i=1}^{K} a_i \\ \hat{y} = \frac{1}{K} \sum_{i=1}^{K} b_i \end{cases}.$$

(8.6)

Example 8.1: (MATLAB Code for 2D MLGT)

We have three RPs at (10, 10), (0, 15), and (−5, 5), and a TG measures the three received power in dBm from them, $P_{M-i} = (-53, -54, -51)$. The radii of the circles are given by:

$$\begin{cases} r_{1-i} = 10^{\frac{P_0 - F_\sigma - P_{M-i}}{10\alpha}} \\ r_{2-i} = 10^{\frac{P_0 + F_\sigma - P_{M-i}}{10\alpha}} \end{cases}.$$

Assuming that propagation parameters of the environment are ($P_0 = -30dBm$, $\alpha = 2, 3.5, \sigma = 2$), plot the circles for $\gamma = 0.9$ and overlay a grid over the area with the dimensions of 0.5 to show the position of desired points in the intersect. The MATLAB code provided in Appendix 8.A finds the MLGT solution to this problem. Figure 8.8 shows the results of MLGT using the MATLAB code in Appendix 8.A.

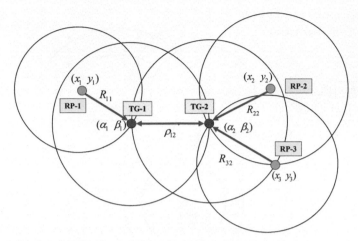

Figure 8.8   Results of Maximum Likelihood Grid Triangulation algorithm for Example 8.1 with three reference point using the MATLAB code in Appendix 8.A.

Example 8.2: 3D MLGT for Localization inside Human Body

Figure 8.9 [Li12] shows the basic concept and results of performance evaluation for the 3D MLGT algorithm in small intestine of human body using four body-mounted sensors. The RSS at the sensors are used for localization. Figure 8.9(a) shows the basic concept of likelihood area in 3D and the shape of the likelihood volume. Figure 8.9(b) shows the results of performance evaluation of the algorithm against the 3D CRLB. Performance is very close to that of the CRLB. The NIST channel model described in Table 2.1 is used for modeling the behavior of RSS inside the human body.

The MLGT algorithm is a class of PRGCN algorithms with a binary decision on selecting candidate grid points, similar to Equation (8.4b). Everything above the threshold probability is assumed to have a probability of 1, and everything below the threshold is assumed to have a probability of 0. In MLGT, we use rim-shaped geometry and focus on the rim areas for different APs to find the optimum solution. In PRGCN of Equation (8.4.b), we calculate the cost function for optimization over all grid points.

In this section, we introduced probabilistic algorithms for RSS-based ranging, and these concepts can be extended to other combinations of localization algorithm using DOA, as well as cooperative localization algorithms.

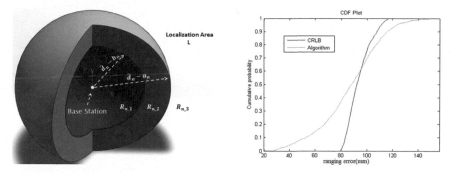

Figure 8.9 3D MLGT inside the small intestine, (a) likelihood volume in 3D (b) performance of the MLGT against the CRLB.

8.3 Pattern Recognition Algorithms for WPS

In this section, we describe RSS-based localization algorithms using outdoor fingerprinting. This includes WPS and CPS systems using GPS for war driving in the streets of the cities to collect their RSS signature database. Fingerprinting systems are opportunistic positioning systems taking advantage of an existing network deployed for wireless communications. WPS uses the RSS of Wi-Fi access points and CPS takes advantage of RSS from cell towers. Wi-Fi has a coverage of tens of meters and cell towers cover up to a few kilometers. For these long-distance wide-area coverage applications, the density of the deployment is not adequate to use the RSS directly and we need to resort to fingerprinting methods and algorithms.

Figure 8.10 shows the basic concept behind the fingerprint positioning algorithms. From RF signal radiated from Access Points (APs) located at

$$\{(x_j\ y_j);\ j = 1, 2, \ldots, N\}, \tag{8.7a}$$

we have formed a database of RSS measurements:

$$\left\{ \mathbf{P}_{r-i} = \begin{bmatrix} p_{r-i}^1 & p_{r-i}^2 & \cdots & p_{r-i}^M \end{bmatrix}^T;\ i = 1, 2, .., M \right\}, \tag{8.7b}$$

at Reference Points (RPs) located in:

$$\{(\alpha_i\ \beta_i);\ i = 1, 2, .., M\}. \tag{8.7c}$$

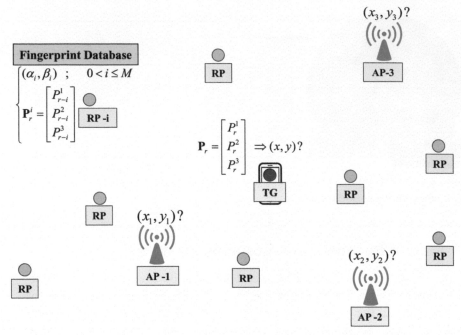

Figure 8.10   Basic concept behind positioning using fingerprints. Access Points (APs) radiate RF signals, and we have results of measurements in RPs, we want to locate a target (TG), which measures power from the APs using the similarities between RSS in RPs and RSS in location of the TG.

We want to position a target (TG), which measures RSS from the APs:

$$\mathbf{P}_r = \begin{bmatrix} P_r^1 & P_r^2 & .... & P_r^N \end{bmatrix} \Rightarrow (x, y) \qquad (8.7d)$$

We have two approaches to solve the problem. In the first approach, we locate the APs by triangulation among RPs, which read the RSS from a specific AP to form a database of location of the APs. Then, we position the TG and RSS values that it reads and the location of their APs in the AP location database. In the second approach, we locate the TG directly by comparing its RSS readings from a set of APs with those measured from the same APs in the RP fingerprint database. The first approach is utilized for big databases, created from war driving outdoors for fingerprinting in metropolitan area in WPS and CPS. The first approach is utilized for smaller databases collected for indoor fingerprinting in specific buildings in RTLS.

8.3.1 Centroid Algorithm and QoE for AP Database

WPS systems collect a big database by war driving in the streets of the cities and using GPS to tag the location for measurement of RSS from different Wi-Fi access points. WPS systems generally use the simple centroid algorithms to map the database of RSS fingerprints to a database of AP locations, and they assign a practical quality of estimate (QoE) parameter to each AP location in the database. To localize a device using its RSS reading from different APs, they use the weighted centroid algorithm and the AP location database [Pah09].

In analytical geometry, centroid of a set of locations is a point that its coordinates are the average of each of the coordinates of the set of locations. In Section 8.2.3, we used the centroid algorithm, in Equation (8.6), to find the center of K possible locations as the answer to the MLGT algorithm. WPS systems, described in Section 3.4.2, use the centroid algorithm to locate the Wi-Fi access points from their RSS fingerprints in known locations. Figure 8.11 shows the basic formulation of the centroid algorithm for 1, 2, and 3 reference points. With one RP, the algorithm is the same as sign posting; if a TG reads the RSS from an AP in a known location, the location of the TG is estimated as the location of the AP. With two RPs, estimate of the TG

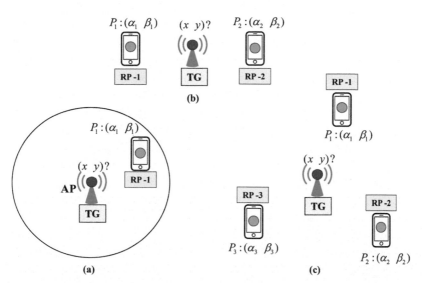

Figure 8.11 Basic formulation of the centroid algorithm utilizing an AP localization and RSS measurements from it, in known reference locations (RPs), with, (a) one RP, (b) two RPs, (c) three RPs.

location is in the middle of the two RPs. With three or more RPs, the location of TG is in the middle of all of them. For M-RSS measurements from an AP at locations:

$$\{(\alpha_i, \beta_i) \; ; \; i = 1, 2, \dots M\},$$ (8.8a)

the estimate of the location of the AP is given by:

$$\begin{cases} \hat{x} = \frac{1}{M} \sum_{i=1}^{M} \alpha_i \\ \hat{y} = \frac{1}{M} \sum_{i=1}^{M} \beta_i \end{cases}.$$ (8.8b)

The centroid algorithm always estimates the location of TG as the middle of all RPs. Basic properties of the algorithm are simplicity and stability. In addition, centroid algorithm works for any number of RPs and it does not rely on the accuracy of measured RSS.

WPS systems use the simple centroid algorithm for locating the APs because of the pattern of the RSS measurements obtained from war driving in the streets. Wi-Fi APs are commonly deployed in indoor areas with the objective of optimization of coverage inside the buildings. Therefore, the RSS measured outside in the streets associates with the locations with weak RSS on the fringe of the coverage of the APs. The range of RSS readings from a typical Wi-Fi AP is between -20 dBm and -90 dBm. When we drive in the streets, the range reduces to -60 dBm to -90 dBm (see Figure 3.17(c)). The centroid algorithm is a good choice for this situation because variations of the RSS are low and RSS readings are close to each other on a straight line, following the path of the street. The RLS algorithm will have convergence problem and weighted centroid (Section 8.3.3) does not show improvement over centroid.

In Wi-Fi fingerprinting by war driving in the streets, the number of RPs used for locating each AP are quite different and they range from a few up to hundreds or even more. Figure 8.12 provides an abstract visualization of WPS fingerprints for APs installed in different locations in a building. In Figure 8.12(a), the AP is deployed in the middle of the building covering parts of a street crossing the side of the building. In Figure 8.12(b), the AP is installed in a corner of a building converging parts of two crossing streets, and in Figure 8.12(c), the AP is installed deep inside the building with no outside coverage. In addition to Wi-Fis inside the building, sometimes we have APs installed outside on traffic lights or street light posts, with a large

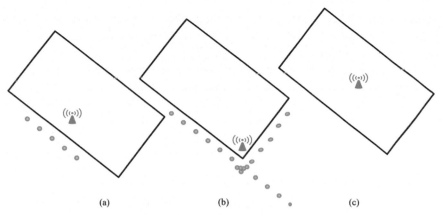

(a)   (b)   (c)

Figure 8.12   Abstract visualization of WPS fingerprints for an AP installed, (a) in the middle of a building covering one street, (b) in a corner of a building coverage two crossing streets, (c) installed deep inside the building with no outside coverage.

coverage area. As a result, the quality of estimate (QoE) for each access point in the AP location database is different. Quantifying these differences is useful when we want to use the AP locations database to position a device, TG. Two quantitative parameters affect the QoE for the location of an AP, the number of RP locations that are used for estimating the location of the AP, and size of geographical coverage of these APs exposed to war driving in the streets.

To map these two phenomena to numerical parameters reflecting the QoE of an AP, one may use the measured RSS values used for estimation of the location of an AP to calculate a path-loss model of the AP (see Example 2.1). Path loss parameters, (α, σ, P_0), are the distance power gradient, variance of the shadow fading, and the received power at a reference distance to develop the general model $P_r = P_0 - 10\alpha + X(\sigma)$ for each AP from the exposed data to war driving. These path-loss models can be used to determine the coverage. For the number of points, one may ignore multiple measurements in one points, for example, at stop lights, and just count them all as one and come up with a number representing total number of locations in which we have measurements rather than total number of measurements. Variations of combining the path-loss model parameters and the number of locations available in the database can be used to define a QoE for each AP. A practical example for formulation of QoE using path-loss model parameters and number of locations in the database is available in [Ali09a,b].

In this patent, a formula is presented to assign a relative probability to the QoE of an AP and using the Skyhook database, practical parameters of the formula are calibrated. The quality of estimate is one if we know the exact location of an AP, which happens only for a very small fraction of APs for which the user or a service provider has reported the location for installation of the AP. Other alternatives to define algorithms for similar purposes are open to innovations. But defining a QoE is useful for localization algorithms for positioning a device based on the AP location database and the RSS readings of the device.

8.3.2 Weighted Centroid Algorithm for Device Localization

In the previous section, we introduced the centroid algorithms and showed that how WPS systems transfer their big data of RF fingerprints of APs collected by war driving to another more trimmed and uniform database of AP locations and their associated QoE. Using the AP database, a device that reads a vector of RSS values from its surrounding APs:

$$\left\{ \mathbf{O} = [P_r^1, ..., P_r^M]/(x, y) \right\} \tag{8.9a}$$

and sends these data to the central server of the location service provider (Figure 3.22). The service provider algorithm uses the RSS values along with stored AP locations in the database and their QoE:

$$\left\{ (x_i, y_i) , \ QoE_i; \ i = 1, 2, ... M \right\} \tag{8.9b}$$

to position the location of the device. The general format of RSS readings for collection of fingerprint database and the data reported by the device readings are the same. But there are two fundamental differences between the contents of the two. The fingerprint database is collected by war driving in the streets and the value of RSS readings are always weak and between -60 and -90 dBm. Devices read RSS values most of the times in indoor areas, and they are sometimes very close to specific APs. Therefore, the RSS readings of the device include all possible values from -20 to -90 dBm. Besides, the QoE of the APs that the device reads from are different. Considering the wide range of RSS readings and variations in QoE of the APs, we have a better estimate of location of the device if we weighted the location of APs used for positioning the device. WPS systems generally use the weighted centroid algorithms for positioning the location of a device.

In a WPS weighted centroid algorithm, the RSS readings from *M-APs,* with estimated locations, $(x_i , y_i); i = 1, 2 ..., L$, the locations of the device,

(x, y), is estimated by:

$$\begin{cases} \hat{x} = \sum\limits_{i=1}^{M} p_i^x x_i \\ \hat{y} = \sum\limits_{i=1}^{M} p_i^y y_i \end{cases}, \qquad (8.10a)$$

where P_k^x and P_k^y are the *weights* of the averaging along the X and Y axis for localization. These are the probabilities of having the device in location of a specific AP. This probability is a function of RSS reading from the AP and the QoE of the AP. Figure 8.13 shows the basic scenarios for positioning a device using weighted centroid algorithm and location and QoE of APs. The simple centroid algorithm defined by Equation (8.8b) is a special case of weighted centroid defined by Equation (8.10a), which assigns the same probability to all reference points:

$$p_i^x = p_i^y = \frac{1}{M}. \qquad (8.10b)$$

Unlike centroid algorithm, the weighted centroid algorithm does not estimate the location of TG as the middle of all RPs. The estimated location of the target device is tilted toward the AP with higher power and QoE. However, the basic properties of the weighted centroid algorithm remain the same as

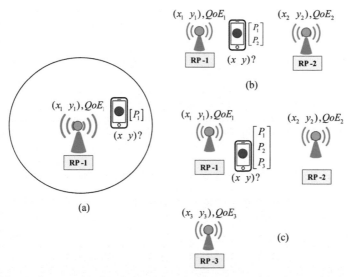

Figure 8.13 Basic scenarios for positioning a device using weighted centroid algorithm and location as well as QoE of APs used as the RPs, for (a) one AP, (b) two APs, (c) three APs.

those of centroid. The main attraction is stability and the fact that algorithm provides a unique solution for any number of RPs. As shown in Figure 8.13, for one AP weighted centroid, similar to centroid, is the same as sign posting and with two RPs it provides a unique solution.

Defining the weight is a practical problem open to innovation and different WPS systems use different approaches. For example, with M values of RSS in dBm, $\left\{ P_r^i; \ i = 1, 2, ..., M \right\}$ from M of the APs in the database, each with a QoE of $QoE_i; \ i = 1, 2, ..., M$, we may define the weight as:

$$w_i = \frac{QoE_i}{|P_r^i|}, \tag{8.11a}$$

arguing that weight is proportional to QoE and absolute value of the RSS in dBm, which is always a negative number. Equivalently, we may calculate the distance from the path-loss model for any AP using:

$$\begin{cases} P_r^i = P_0 - 10\alpha \log d_i + X(\sigma) \\ d_i = 10^{\frac{P_0 - P_i}{10\alpha}} \end{cases}$$

then, define the weight as:

$$w_i = \frac{QoE_i}{d_i^\alpha}. \tag{8.11b}$$

These weights need to be replaced by probabilities using:

$$p_i = p_i^x = p_i^y = \frac{w_i}{\sum_{i=1}^{M} w_i}. \tag{8.11c}$$

Earlier, we can use them in Equation (8.10a). Experience over the large database has shown that these algorithms do not make decisive differences in the results. WPS location service providers use their own proprietary algorithms based, which optimizes their own fingerprint databases.

8.3.3 Organic Data and Positioning the Hidden APs

In WPS systems, each time a device sends a request with a set of M-measurements of the RSS to the location server, indeed the device adds another point to the fingerprint database. As we discussed in Section 3.4.4, we refer to this type of data as organic data. If the organic data include the GPS estimate of the location, the reading is the same as the readings of a device used for the war driving. But this happened only when the device allows the WPS algorithm access to the GPS data and when the device is operating outdoor and it is able to obtain a GPS fix. Regardless of the availability of GPS, localization system estimates the location of the device and we can use that as a fingerprint. Statistically, as we showed in Section 3.4.3, location estimates of WPS are comparable to those of GPS. Therefore, organic data collection expands the fingerprinting database without any cost for war driving. In war driving, we systematically collect data on specific streets to cover populated areas and organic data is arriving from popular indoor and urban areas in a random manner. War driving is performed by specific cars driving with schedules and organic data arrives continually at a rate of approximately one billion measurements per day for each of the Wi-Fi localization service providers.

Similar to additional war driving fingerprint data, organic data can also enhance the QoE of existing APs in the database. However, one of the distinct features of organic data is that they allow discovery of the APs deep inside the building with no outside visibility (Figure 8.12(c)). The method to discover a hidden AP is illustrated in Figure 8.14. A device reads M powers from surrounding APs and sends that for localization at the server. The server notices that the MAC address associated with one of the APs is not in its database, a hidden AP. Server uses M-1 of the RSS values to locate both the

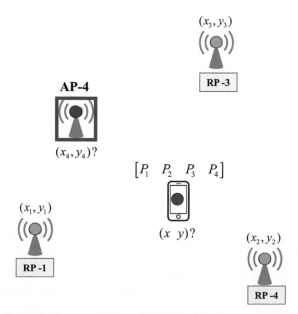

Figure 8.14 Localization of a hidden AP using organic data.

device and the hidden AP. The location of the device is sent to the user, and the location of the hidden AP is added to the AP database. The localization algorithm for the hidden AP could be a simple centroid or a more complex weighted centroid using the differences in received powers and QoE of the existing APs. In practice, except for hidden APs, we have APs installed after war driving or APs that have been relocated, which can be identified using organic data. The expansion and collection of AP databases depends of the service provider access to data. Skyhook AP database claims up to 40% increase in the number of APs using organic data.

8.4 Pattern Recognition Algorithms for RTLS

Both RTLS and WPS systems are opportunistic Wi-Fi localization systems using pattern recognition algorithms. But they use different methods for fingerprinting database collection and different localization algorithms. WPS mainly relies on automated RF fingerprinting in the city streets and RTLS relies on a database collected by site survey and manual data collection in specific indoor areas. The RF fingerprint database for RTLS is much smaller, but it contains much higher variations of the RSS signal because it is recorded inside the building in locations that can be very close to the location of the APs. In addition, RTLS is focused on specific floors of buildings to attain accuracies on the order of a few meters as compared with general expectation of a few tens of meters for WPS. In RTLS, we are collecting a few samples of the RSS by rotating around in the same location (Figure 8.13) as compared with WPS in which we use the fingerprints collected while we are driving a vehicle (Figure 3.16). Since the size of the fingerprint database, the area of coverage, structure of the signature points, and the performance expectations for RTLS and WPS are different, the pattern recognition algorithms used for the two technologies have evolved differently.

The fundamental difference between the pattern recognition algorithms for WPS and RTLS is that in WPS, the database is huge and diversified, so we process positioning in two steps. First, we associate the fingerprint database to the location and QoE of APs, then we use those values to localize a device reading RSS from APs. Our main objective is to position the location of a device and locating the APs is redundant. We adopt that path in WPS as a practical intermediate step to facilitate the large and disorganized fingerprint of WPS systems. In pattern recognition for RTLS applications, the RF fingerprint database is much smaller and manageable. Therefore, we can eliminate the intermediate step and use the signature database to directly locate the device using a pattern recognition algorithm. The simplest of the pattern recognition algorithms is the closest neighbor, and more sophisticated algorithms can be designed with a probabilistic approach.

8.4.1 Closest Neighbor Power Difference Algorithm

The Closest Neighbor Power Difference (CNPD) algorithm was the first algorithm used in the simple experiment at Microsoft to demonstrate the capability of RSS-based opportunistic Wi-Fi localization in indoor areas [Bah00], what we refer to as RTLS. The general statement of this problem is described by Equation set (8.7) and it is illustrated in Figure 8.10. We have RSS Wi-Fi fingerprint of N access points (APs) at M known locations:

$$\begin{cases} (\alpha_i, \beta_i); \quad i = 1, 2, \ldots \ldots M \\ \mathbf{P_i} = [P_i^1, P_i^2 \ldots \ldots \ldots P_i^M]^T \end{cases} \quad (8.11a)$$

A device at an unknown location, (x, y), reads another set of RSS values:

$$\mathbf{P} = [P_1, P_2 \ldots \ldots, P_M]^T , \quad (8.11b)$$

and we want to estimate its location.

Similar to other closest neighbor algorithms we discussed before, for example, CNLS in Section 8.2.2, the CNPD algorithm defines a cost function related to the distance between the device and the RPs and selects the location of the RP with minimal cost function as the estimate of location of the device. The cost function in CNPD algorithm is defined by:

$$\min \left\{ \varepsilon \left(\alpha_i, \beta_i \right) \right\}_{0 \leq i < N}$$

$$= \min \left\{ \frac{1}{M} \sum_{j=1}^{M} \left(P_r^j - P_i^j \right)^2 \right\}_{0 \leq i < N}$$

$$= \min \left\{ \frac{1}{M} |\mathbf{P_r} - \mathbf{P_i}|^2 \right\}_{0 \leq i < N}. \quad (8.11c)$$

We select the location of the RP with minimum of the above cost function as the estimate of the location of the device.

The advantage of this CNPD method over distance-based localization techniques, such as CNLS, is that we do not need to know the location of the APs to calculate radius of the circles. In opportunistic localization, we cannot control the number of APs, but we can increase the number of RP to increase the accuracy of the algorithm.

Example 8.3: (CNPD performance in a typical laboratory environment)

Figure 8.15(a) [Pah02a] presents a partial layout of the Telecommunications Laboratory at the Centre for Wireless Communications at University of Oulu, Finland. The locations of four Access Points (AP) and 31 measurement locations along a long corridor, with about 2 m separation between adjacent points, are illustrated in the figure. A mobile terminal is carried along the corridor and the RSS values are measured at each location. Figure 8.15(b) shows the measured RSS at all four APs as the terminal travels from the right corner close to the AP-I to the end of the vertical corridor after AP-IV. When the CNPD algorithm is applied to the measurement data, the standard deviation of the positioning error is 2.4 m and at about 80% of locations, the position error is less than 3 m. The accuracy of the positioning algorithm is on the orders of separation between the RPs.

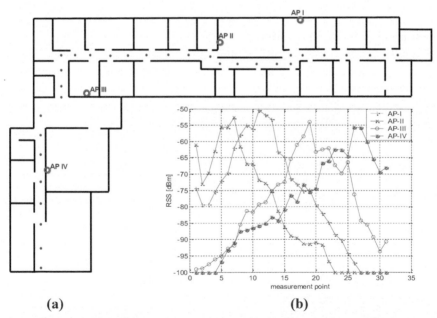

(a)   (b)

Figure 8.15   (a) The building layout and location of WLAN APs at the first floor of the CWC Laboratory at the University of Oulu, Finland, (b) RSS signature in different locations [Pah02].

8.4.2 K-Closest Neighbor Power Difference Algorithm

A simple modification to the CNPD algorithm is the K-CNPD algorithm in which we select K locations with the lowest cost functions from the N calculation of the cost function in Equation (8.11c):

$$(\alpha_k, \beta_k) \; ; \; k = 1, .., K. \tag{8.12a}$$

Then, we declare the location of the device as the centroid of these locations with lowest cost functions:

$$\begin{cases} \hat{x} = \frac{1}{K} \sum_{k=1}^{K} \alpha_k \\ \hat{y} = \frac{1}{K} \sum_{k=1}^{K} \beta_k \end{cases} . \tag{8.12b}$$

We have seen these similar approaches for other algorithms (e.g., in Section 8.2.3) in which rather than finding the sole absolute location with minimal cost function, we select the top K cost functions and average their locations using the centroid algorithm. This approach is a spatial filtering approach and, like any other filter, smoothens fluctuations in the DME at the expense of reduction in minimum error.

One of the characteristics of fingerprinting for RTLS systems is that we make multiple measurements in a single location by turning around and sampling the RSS (Figure 3.13). This process allows inclusion of more diversified samples of RSS measurements affected by shadow fading. In these situations, RSS measurements in each location becomes a matrix:

$$\begin{cases} \mathbf{O}_i(\alpha_i, \beta_i) = \left[\mathbf{P}_i^1, \mathbf{P}_i^2 \ldots\ldots \mathbf{P}_i^M \right] \; ; \; i = 1, 2, \ldots.. N \\ \mathbf{P}_i^j = \left[P_{i,1}^j, P_{i,2}^j, \ldots P_{i,L}^j \right]^T \; ; \; j = 1, 2, \ldots M \end{cases} . \tag{8.13a}$$

A device at an unknown location, (x, y), reads another set of M-RSS values given by Equation (8.11b). Since we measure L samples of the RSS in each location, the cost function (8.11c) becomes:

$$\min \left\{ \varepsilon \left(\alpha_i, \beta_i \right) \right\}_{0 \leq i < N} = \min \left\{ \frac{1}{ML} \sum_{j=1}^{M} \sum_{k=1}^{L} \left(P_j - P_{i,k}^j \right)^2 \right\}_{0 \leq i < N} . \tag{8.13b}$$

With this cost function, as definition of the distance, we can apply CNPD or K-CNPD algorithms by either selecting the location associated with the minimum distance as the location estimate or selecting the K locations with minimal distances and use the centroid algorithm (8.12b) to estimate the location of the device.

The general characteristic of CNPD and K-CNPD algorithms is that they use the measured RSS in known locations directly to calculate the location of a device. An alternative to this approach is to use RSS readings and models for RSS to assign probabilities to each location in the fingerprint database and then proceed with the numerical minimization of the cost function.

8.4.3 Maximum Likelihood Kernel Algorithms

In Section 8.2.3, we showed how we can map the RSS readings in a location to the probabilities of being in a location. Using similar approaches, one can use the difference between the RSS readings by a device in an unknown location and the RSS readings in known locations in a signature database to assign probability of being in any of the reference points in the database.

One of the pioneering works for using probabilistic approaches for indoor positioning is reported in [Roo02a,b]. In this approach, referred to as the Kernel method, the measured database is collected from N locations from M reference points with L measurements per location as it is described in Equation (8.13.a). A device in an unknown location, (x, y), reads another set of RSS values from the M access points given by $P = [P_1, P_2 \ldots \ldots, P_M]^T$.

The maximum likelihood kernel algorithm (MLKA) introduced in [Roo02a] assumes that the difference in RSS readings from an AP forms a zero mean Gaussian distribution with variance σ. Then, the probability distribution function of the difference between a measured power, k, from AP number j in location number i is:

$$f\left(P_k,\ P_{i,j}^k\right) = \frac{1}{\sqrt{2\pi}\sigma}e^{-\frac{1}{2\sigma^2}(P_k,\, P_{i,j}^k)^2}. \qquad (8.14a)$$

Defining a mass probability distribution function based on L differences in measured RSS values by the devices and the L-samples of RSS measurement in a specific reference location in the signature database, it defines a kernel function as:

$$K(\mathbf{O},\mathbf{P}_i^j) = \frac{1}{(\sqrt{2\pi}\sigma)^L}e^{-\frac{1}{2\sigma^2}\sum_{k=1}^L(P_k,\, P_{i,j}^k)^2} \qquad (8.14b)$$

Figure 8.16 [Roo02a] shows the sensitivity of the kernel function to the value of σ parameter. For larger values of the parameter, we have a smoother curve.

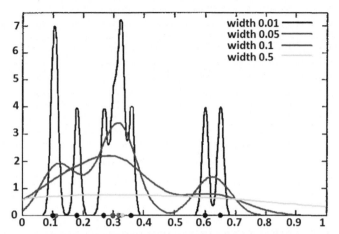

Figure 8.16 Examples for the shape of the kernel function difference values of 0.1, 0.11, 0.18, 0.27, 0.3, 0.32, 0.33, 0.36, 0.6, 0.65 and different values of σ (0.01, 0.05, 0.1, 0.5) [Ro02].

For using kernel algorithm, we need to find a suitable value for this parameter to optimize the error. Then, the probability of our RSS observations being in a given reference location, i, in the signature database is given by average of all kernels from M measurements from APs:

$$p(\mathbf{O}|l_i) = \frac{1}{M} \sum_{j=1}^{M} K(\mathbf{O}, \mathbf{P}_i^j), \qquad (8.14c)$$

where $l_i(x\,y)$ represent the location of i-th signature RP. Using Bayes' theorem, we can calculate the probability of being in location, n, given that we had the specific M measurements of the RSS by the device:

$$p(l_i|\mathbf{O}) = \frac{p(\mathbf{O}|l_i).p(l_i)}{p(\mathbf{O})} = \eta \cdot p(\mathbf{O}|l_i).$$

Since the probability of being in location, $p(l_i)$, and the probability of having an observation, $p(\mathbf{O})$, are fixed values, we represent their ratio as the fixed parameter, η. The estimated location in this approach is the expected value of the location given the observed valued of the power:

$$l(\hat{x}, \hat{y}) = E[l|\mathbf{O}] = \sum_{i=1}^{N} l(x_i, y_i)p(l_i|\mathbf{O}) = \eta \sum_{i=1}^{N} l(x_i, y_i)p(\mathbf{O}|l_i). \quad (8.14d)$$

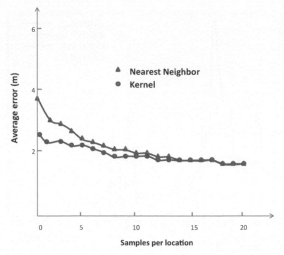

Figure 8.17   Comparison of the performance of nearest neighbor and Kernel methods.

In this equation, η is determined so that the total probability is normalized, which means:

$$\eta = 1 / \sum_{i=1}^{N} p(\mathbf{O}|l_i).$$

Equation (8.14d) is a basic weighted centroid algorithm in which the weight of each location is calculated probabilistically assuming that the difference in powers forms a Gaussian distribution. Figure 8.17 provides a comparison among nearest neighbor and kernel algorithms [Roo02a]. The test area is a 16×40 m office with concrete, wood, and glass structures with 10 access points (8 around the perimeter and 2 in the center). The training data is 155 points using a 2-m grid, and at each grid (*calibration*) point, 40 RSS observations are recorded. The test data is collected independently by using a similar 2-m grid, but by selecting the test points to be as far as possible.

Example 8.4: (MATLAB code for MLKA)

Appendix 8.B provides a sample MATLAB code for implementation of maximum likelihood Kernel algorithm in a 20 m $\times$ 20 m area shown in Figure 8.18(a). We have two APs and four RPs. The RF signature database collects three independent RSS values in each reference point. The IEEE 802.11 path-loss model before the breakpoint with a distance power gradient of two is used to emulate the power readings. The standard deviation of the shadow fading to generate the three independent RSS readings is 8. The test

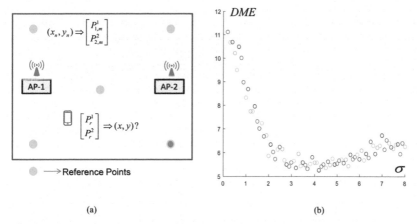

(a)   **(b)**

Figure 8.18 (a) Performance measurement scenario for Kernel Algorithm with 2 AP and 5 RP, (b) the DME as a function of parameter σ.

point for localization is located at (5,5). Figure 8.18(b) shows the DME as a function of the parameter σ of the Kernel algorithm. A value in the range of (3–6) provides good results for the algorithm.

8.5 Alternatives to Manual Indoor Fingerprinting

There are two alternatives to manual fingerprinting in indoor areas to form the database more efficiently. We can use measurement calibrated channel models to produce the fingerprint database or we may roll a robot in indoor areas to collect the RF fingerprint database. In this section, we explain these alternative method to expensive manual fingerprinting used for Wi-Fi localization.

If we have a reliable channel model and we know the location of APs, we can avoid measurements and use the fingerprinting algorithms with predicted value of the RSS calculated from the channel model. We have building specific ray tracing models and statistical path-loss models for the behavior of the radio propagation. These models, however, need to know the RSS at a reference point, P_0, to calculate the RSS in other locations. Since we have uncertainties on the transmitted power and antenna gains used by different manufacturers, we may calibrate these models with empirical measurement of the P_0. If we use this approach, we only need a few training points around each AP to estimate its P_0. Then, we can use the measurement calibrated models to efficiently generate a large database of RSS fingerprints.

With the popularity of inexpensive robots operating in indoor areas, such as low-cost vacuum cleaner robots, one may think of using them for massive fingerprint data collection for RTLS. For example, if we want to survey a Stop and Shop or a Home Depot building, we can drop a vacuum cleaner robot in the store and equip that with a smart device collecting RSS signature from the APs overnight, when the store is closed.

8.5.1 Ray tracing for Fingerprinting Algorithms

Ray tracing algorithms are building specific models and they provide reliable estimates of the RSS. Figure 8.19 shows a sample result of RSS measurements versus measurement calibrated ray tracing around a square route in a typical office building at the third floor of the Atwater Kent Laboratory, Worcester Polytechnic Institute (Figure 3.15). Results of measurements and calibrated ray tracing are very close. To show the impact of using ray tracing, instead of actual measurements of the RSS, on performance of fingerprinting positioning algorithms we resort to an example.

Example 8.5: (Ray tracing for Fingerprinting Algorithms)

Figure 8.20 [Hat06a] shows a summary of results for comparing the performance two algorithms, the CNPD and the maximum likelihood kernel algorithms in a typical office building at the Worcester Polytechnic Institute (Figure 3.15). The objective is to compare the performance of two typical algorithms for RTLS applications when we use the actual empirical measurement and when we use calibrated ray tracing software. Figure 8.20(a) shows the results when we use the empirical data, and Figure 8.20(b) presents results obtained from calibrated ray tracing. Results are presented as a function of grid resolution. Results for both cases closely follow each other.

Ray tracing is an analytical tool for comparative performance evaluation of positioning algorithms. We can use this tool to answer questions for different building layouts.

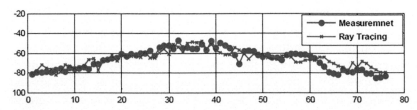

Figure 8.19   A sample results of RSS measurements versus measurement calibrated raytracing.

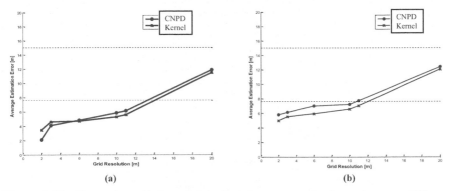

Figure 8.20 Comparison of the results of performance evaluation using two different algorithms, (a) results using empirical data, (b) results obtained from calibrated raytracing software.

Example 8.6: (Performance Analysis with Ray tracing)

In [Hat06b], a 2D ray tracing software is used for positioning applications in the first floor of the Atwater Kent Laboratories at the Worcester Polytechnic Institute. Figure 8.21 shows the layout of the building and the surrounding walls around the building, the route used for performance analysis, the locations of five access points, and the area covered by the grid used as the reference points (RP). The ray tracing software generates a vector of five RSSs from the access points (AP) for every node of the grid. Using the CNPD algorithm, the location of the mobile host is calculated. The estimated location is used to calculate the distance measurement error (DME). Figure 8.22(a) shows the cumulative distribution function of the DME for different sizes of grid. For the grid spacing of 1 m, in 60% of the locations, the distance measurement errors are less than 5 m. Figure 8.22(b) shows the effects of the number of APs. If we reduce the number of APs from five to two, for 60% of locations, we have increases in DME from 5 m to around 13 m.

8.5.2 Modified IEEE 802.11 Model for Fingerprinting

Statistical channel models for the RSS are based on three parameters: RSS in a reference point, P_0, distance-power gradient, α, and standard deviation of shadow fading, σ. The first parameter, the received power at first meters distance, P_0, which depends on manufacturers transmitted power specification and the type of antenna, and as we explained in the previous section,

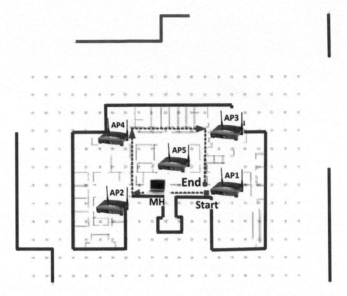

Figure 8.21   Layout of the first floor of AKL with outside walls from other buildings, location of the APs, and the grid used for raytracing. The dashed line is used for positioning performance evaluation.

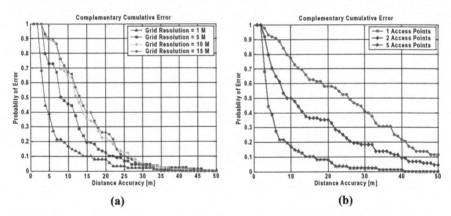

Figure 8.22   Statistics of the distance measurement errors for positioning using 2D ray tracing, (a) effects of grid size, and (b) effects of number of APs [Hat04].

in positioning applications, it is better to measure that empirically. For positioning applications, statistical models provide the other two parameters.

The most popular statistical model for indoor radio propagation is the IEEE 802.11 model, described in Section 2.2.3. The model divides the indoor propagation conditions into six different categories, and for each category, it defines a break point and a standard deviation for the shadow fading (Table 2.2). The model assumes the distance–power gradient before the breakpoint as $\alpha = 2$, and after the breakpoint, it becomes $\alpha = 3.5$. We can generalize this model for radio propagation in site-specific applications using the layout of a building. The idea behind the generalized building specific model is that in short distances, we always have an LOS connection between the transmitter and the receiver and the distance–power gradient follows the free space propagation, for which $\alpha = 2$. When the receiver passes through the first wall of the building, the OLOS region begins, the distance–power gradient is the same as in IEEE 802.11 model, $\alpha = 3.5$. This way, we define the intersection of the direct path and the first wall as the breakpoint of the IEEE 802.11 model.

If we interpret the IEEE 802.11 model for the standard deviation of shadow fading, σ, in a building-specific context, we can say, it assumes that in LOS $\sigma = 5dB$, in the OLOS, for distances between 5 and 20 m, $\sigma = 8dB$, and for distances more than 20 m $\sigma = 10dB$. The model does not pay attention to sharp reduction of the σ as we get very close to antenna because it is designed for calculation of the coverage and long distances. As we explained in Section 2.2.5, in short distance, the standard deviation of shadow fading reduces exponentially, and in [Liu15], an empirical model is derived for this behavior. Figure 8.23 shows a model for the standard deviation of the shadow fading for *positioning applications* based on discussions in Section 2.2.5 and the interpretation of shadow fading values of the

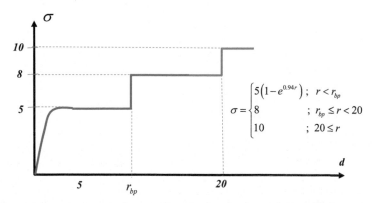

Figure 8.23 Model for standard deviation of shadow fading for building specific positioning applications.

IEEE 802.11 models. The mathematical expression for the model is given by:

$$\sigma = \begin{cases} 5\left(1 - e^{0.94r}\right) & ; \quad r < r_{bp} \\ 8 & ; \quad r_{bp} \le r < 20 \\ 10 & ; \quad 20 \le r \end{cases} \tag{8.15a}$$

where r_{bp} is the distance of the transmitter from the intersect of the first wall and direct path. The standard deviation of the shadow fading close to the antenna is zero and it increases exponentially up to 5 dBm at around 5 m distance using an approximation to Equation (2.6):

$$\sigma = 4.31 - 4.28e^{-0.9372r} \sim 5\left(1 - e^{-0.94r}\right). \tag{8.15b}$$

The standard deviation remains at approximately 5 dBm and continues until the break point created by the first wall. After the break point, standard deviation turns to 8 until close to 20 m, where the standard deviation elevates to its final value at 10 dBm.

Example 8.7: Performance of the Calibrated IEEE 802.11 Model

A comparison of performance of CNPD and maximum likelihood kernel algorithms using fingerprint database, measurement calibrate ray tracing, and measurement calibrated IEEE 802.11 models is available at [Hat06a]. Figure 8.24(a) shows comparison of RSS from measurements, calibrated ray tracing, and calibrated IEEE 802.11 models. Figure 8.24(b) shows the result

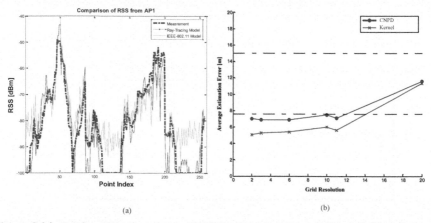

(a)   (b)

Figure 8.24 (a) Sample comparison of RSS from measurement, calibrated raytracing, and calibrated IEEE 802.11 model, (b) results of performance evaluation using IEEE 802.11 model.

of performance evaluation of the two algorithms, the CNPD and kernel, when we use IEEE 802.11 model to generate the fingerprint database. These results should be compared with results associated with the same algorithms, when we use actual measurements and, when we use a measurement calibrated ray tracing (Figure 8.20(a,b)).

The conclusion of this discussion is that if we know the location of the APs, model-based localization either using the more sophisticated ray tracing software or simpler path-loss model suggested in this section, we can provide comparable results with those obtained by physical collection of the finger-print database. The model-based approach relies on smaller measurement sets close to the location of the APs. More results in comparing RT with the cali-brated path-loss models in a controlled environment are available in [Ass07].

8.5.3 Robots to Collect RF Fingerprints

In WPS systems, we drive in the streets and use the GPS as a tag to identify the location of the car when we read the RSS from the APs. This way, we collect a large database to locate billions of the APs randomly deployed all over the world. Since GPS does not work indoors, in RTLS systems, we resort to manual data collection. With popularity of inexpensive robots covering indoor areas, we can use them as an alternative to expensive manual fingerprint database collection in indoor areas. Robots use their cameras and other sensors for simultaneous localization and mapping (SLAM) to localize and navigate themselves through an indoor area. We can use the location estimate from the SLAM to tag the RSS readings from the APs and build a large automated database. The role of location estimate by SLAM for collecting the fingerprint for the RTLS database is similar to the role of GPS location estimate in collecting fingerprint for the WPS database. As a result, the format of the robot-assisted fingerprint database is similar to that of WPS database. We have one set of RSS measurements in each location tagged by its SLAM estimate of location. In traditional manual signature database collection for RTLS, we often measure multiple sets of RSS by rotating around in each location (Figure 3.13).

To evaluate the performance of RTLS using manual and robot collected fingerprints, a group of students at WPI carried out an experience reported in [Niu16]. They collected two databases, one by a human using a cell phone to collect multiple set of fingerprints in manually identified locations, and the second one with a robot measuring a set of fingerprints as it moves and

Table 8.1   Comparison of average error for different algorithms and RPs

Number of RP	Database Collected by Human					Database Collected by Robot				
	Kernal	NN	2-NN	3-NN	4-NN	Kernal	NN	2-NN	3-NN	4-NN
4 corner	2.49	4.85	2.90	4.66	6.99	4.06	6.11	3.23	4.13	5.49
4 center	5.36	5.40	5.84	7.51	9.10	4.67	5.89	4.65	5.96	7.03
8RP	4.25	5.70	3.47	5.15	6.40	5.54	5.77	5.14	4.68	4.19
12RP	2.80	3.30	3.30	3.13	3.39	2.98	3.76	3.07	2.98	3.10
16RP	NA	NA	NA	NA	NA	2.92	3.30	3.42	3.11	2.95
20RP	NA	NA	NA	NA	NA	2.88	3.31	3.19	3.39	3.04
24RP	NA	NA	NA	NA	NA	3.05	3.57	3.13	3.34	3.19
28RP	NA	NA	NA	NA	NA	3.06	3.22	3.11	3.29	3.34
32RP	NA	NA	NA	NA	NA	2.99	3.23	3.11	3.10	3.16
40RP	NA	NA	NA	NA	NA	2.96	3.20	2.98	2.95	3.01
48RP	NA	NA	NA	NA	NA	3.03	3.12	2.91	3.04	3.04

tagging them with the SLAM identified location of the robot at the time of measurement. Table 8.1 shows the results of average error for different positioning algorithms and a variety of RPs. The algorithms are the kernel, CNPD (NN), and K-CNPD. Measurements are taken on a square route in the third floor of Atwater Kent Laboratory (Figure 3.15). The number of RPs range from 4 to 48, 4-RPs are either located in the corners or in the middle of each side. Taking manual RPs for more than 12 locations was not practical, while robot could take measurements in up to 48 locations. This experiment shows that robot-based collection of database can result in slight performance improvement with minimal human interference.

8.6 Pattern Recognition Algorithms for the CPS

In cellular networks, we can also use RSS for cell tower localization. Indeed, the original iPhone was using RSS-based cell tower localization as a backup for WPS positioning in areas such as highways, where Wi-Fi localization does not work. In Section 3.3.5, we provided an overview of these systems. The RSS-based CPS systems work based on signature database and fingerprinting the same way as WPS systems work. However, in CPS, antennas are deploys outdoor for comprehensive coverage of the streets in a large area. As a result, during war driving, we can visually detect the location of the antennas and we measure a wide variety of signal levels along large distances. Knowing the location of the antenna and having a large fingerprinting database allow us to consider algorithms that are used for both WPS and RTLS.

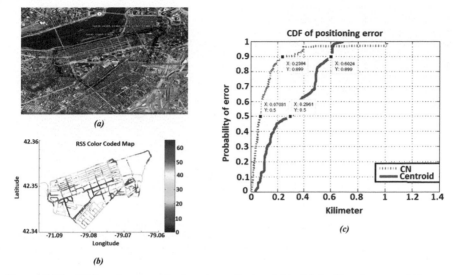

Figure 8.25 CPS performance in downtown Boston, MA, (a) the Google map of the area with the test route, (b) colored map of the signal strengths on the route, and (c) performance of two algorithms.

The following example uses empirical measurements to compare the performance of centroid algorithm with CNPD.

The following example provides some test results for performance of CPS.

Example 8.8: (Performance of centroid and CNPD algorithms for CPS)

Figure 8.25 shows the CPS performance in downtown Boston, MA [Akg09]. Figure 8.25(a) shows the Google map of the area with the test route and location of cell towers. Figure 8.25(b) shows the colored map of the signal strengths on the route, the range of variations of the RSS is over 60dB. Figure 8.25(c) shows the performance of CPS in this area using the centroid and CNPD algorithms. The centroid algorithm uses the center of all BS locations that the devices read as the estimate of the device location (Section 8.3.1). The CNPD algorithm uses the GPS location tagged fingerprint of RSS readings during war driving and measurements of RSS measurements to position the device (Section 8.4.1). The median accuracy of the centroid is approximately 296 m, and using the signature, database reduces it to 70 m. In the downtown areas, we have micro-cellular deployment with a separation distance of approximately 500 m, which is consistent with the estimate using

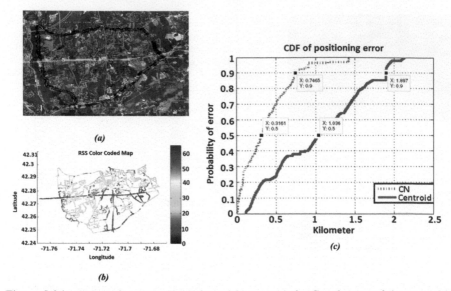

Figure 8.26   CPS performance Shrewsbury, MA area (a) the Google map of the area with the test route, (b) colored map of the signal strengths on the route and (c) performance of two algorithms.

centroid algorithm. Using the signature database improves the performance more than four times. The accuracy of the CNPD should be compared with that of WPS, which provided an accuracy close to 10 m in downtown areas. Figure 8.26 provides CPS performance in Shrewsbury, MA, in the suburb of Worcester. Figure 8.26(a) illustrates the Google map of the area with the test route and location of the cell towers in the area. Figure 8.26(b) shows the colored map of the signal strengths on the route. Figure 8.26(c) shows the result of performance evaluation using centroid and CNPD algorithms. In this area, the centroid algorithm results in a median error of 1 km, which reflects the increase in separation of the cell towers in that area. The CNPD provides a median accuracy of 32 m, which is close to the performance of WPS in the suburb of the Boston.

Knowing the location of antenna allows us to use channel model-based pattern recognition algorithms, similar to Section 8.5.2, and using organic data, we can improve the accuracy of these systems beyond the existing U-TDOA systems. This area is open for further research in cell tower localization for E-911 type of applications.

Appendix 8.A: MATLAB Code for ML Triangulation Grid Algorithm

```
(prepared by Julang Ming)
clc;clear all;close all;
%% This experiment is based on Maximum Likelihood
Triangulation to solve
%% the problem
CertainRange=0.9; % Certainty Range
% Use 802.11 channel model C to find bound for certainty range
sigma=2;dbp=5;
Pt=13;
fc=2.4e9;
c=3e8;
L0=-20*log10(c/4/pi/fc);
known_references = [10,10;0,15;-5,5];
    % Known Reference Locations
RSS = [-53, -54, -51];
Lp = Pt - RSS;
FadeMargin=sigma*sqrt(2)*erfcinv(1+CertainRange);
RSS_bp = Pt - L0 - 20*log10(dbp);
for i=1:length(RSS)
    if RSS(i)>RSS_bp
        Bound(:,i)=[dbp.*10.^((Lp(i)+FadeMargin-L0)./20);
dbp.*10.^((Lp(i)-FadeMargin-L0)./20)]
    else
        Bound(:,i)=[dbp.*10.^((Lp(i)+FadeMargin-L0-20*log10
(dbp))./35); dbp.*10.^((Lp(i)-FadeMargin-L0-20*log10(dbp))
    ./35)]
    end
end
% grid the space
pace=0.2;
x=-100:pace:100;y=-100:pace:100;
l1=length(x);
k=0;
for i=1:l1
    for j=1:l1
        circle1 = (x(i)-known_references(1,1))^2+(y(j)-
known_references(1,2))^2 >= (Bound(1,1))^2;
        circle2 = (x(i)-known_references(1,1))^2+(y(j)-
known_references(1,2))^2 <= (Bound(2,1))^2;
        circle3 = (x(i)-known_references(2,1))^2+(y(j)-
known_references(2,2))^2 >= (Bound(1,2))^2;
        circle4 = (x(i)-known_references(2,1))^2+(y(j)-
```

```
known_references(2,2))^2 <= (Bound(2,2))^2;
     circle5 = (x(i)-known_references(3,1))^2+(y(j)-
known_references(3,2))^2 >= (Bound(1,3))^2;
     circle6 = (x(i)-known_references(3,1))^2+(y(j)-
known_references(3,2))^2 <= (Bound(2,3))^2;
     if circle1 && circle2 && circle3 && circle4 && circle5
&&circle6
          k=k+1;
          sol(k,1)=x(i);
          sol(k,2)=y(j);
       end
    end
end
figure(1)
angle=0:2:360;
hold on
plot(Bound(1,1).*cosd(angle)+known_references(1,1),Bound(1,1).
*sind(angle)+known_references(1,2),'r.')
plot(Bound(2,1).*cosd(angle)+known_references(1,1),Bound(2,1).
*sind(angle)+known_references(1,2),'r.')
plot(Bound(1,2).*cosd(angle)+known_references(2,1),Bound(1,2).
*sind(angle)+known_references(2,2),'b.')
plot(Bound(2,2).*cosd(angle)+known_references(2,1),Bound(2,2).
*sind(angle)+known_references(2,2),'b.')
plot(Bound(1,3).*cosd(angle)+known_references(3,1),Bound(1,3).
*sind(angle)+known_references(3,2),'g.')
plot(Bound(2,3).*cosd(angle)+known_references(3,1),Bound(2,3).
*sind(angle)+known_references(3,2),'g.')
text(10, 10*(1 + 0.2) , 'Known Reference 1');
plot(10, 10 ,'r.','MarkerSize',20);
text(0, 15*(1 + 0.1) , 'Known Reference 2');
plot(0, 15 ,'b.','MarkerSize',20);
text(-5, 5*(1 + 0.3) , 'Known Reference 3');
plot(-5, 5 ,'g.','MarkerSize',20);
for i=1:k
    plot(sol(i,1),sol(i,2),'c.');
end
ex=mean(sol(:,1));ey=mean(sol(:,2));
text(ex, ey*(1 + 0.3) , 'Estimated Location');
plot(ex,ey,'m.','MarkerSize',20);
hold off
disp(['The estimated location is:
','[',num2str(ex),',',num2str(ey),']']);
tag = [0, 5]; % Exact Location of the Target
DME = sqrt((tag(1)-ex)^2+(tag(2)-ey)^2);
disp(['The Distance Measurement Error is: ',num2str(DME),
```

```
'm']);
xlabel('[meter]');ylabel('[meter]');
line1=['Maximum Likelihood Algorithm
with Certainty Range of ',
num2str(CertainRange)];
line2=['and Standard Deviation of ',num2str(sigma)];
title({line1;line2})
```

Appendix 8.B: MATLAB Code for a Simple Kernel Algorithm

```
clc;clear all;
% Space: 20*20
x_i=5;y_i=5;
x=3; % number of trainning points
y=2; % number of access points
z=5; % number of reference locations
s=20;
ref=[0,0;s,0;s,s;0,s;s/2,s/2];
ap=[0 s/2;s s/2];
sigma=0.1;
eta=1;
m=1;
figure
hold on
while sigma<=8
    for n=1:1:100
    d1=s/2;d2=sqrt(s^2+(s/2)^2);
    d3=sqrt((ap(1,1)-x_i)^2+(ap(1,2)-y_i)^2);
d4=sqrt((ap(2,1)-x_i)^2+(ap(2,2)-y_i)^2);
    % use path-loss model for free space
    pt=10;
    p1=pt-40-20*log10(d1);
    p2=pt-40-20*log10(d2);
    p3=pt-40-20*log10(d3);
    p4=pt-40-20*log10(d4);
    p=[p1 p2;p2 p1;p2 p1;p1 p2;p1 p1];
    p_i=[p3 p4];
    t=zeros(x,y,z);
    for i=1:1:x
        for j=1:1:z
            t(i,1,j)=p(j,1)+sigma*randn(1);
            t(i,2,j)=p(j,2)+sigma*randn(1);
        end
    end
    knorm=zeros(1,z);
```

```
        kgauss=zeros(x,z);
        for k=1:1:z
            for i=1:1:x
                sse=(p_i(1)-t(i,1,k))^2+(p_i(2)-t(i,2,k))^2;
                kgauss(i,k)= (exp(-
sse/(2*(sigma)^2)))/((sqrt(2*pi)*sigma)^y);
            end
    end
    for k=1:1:z
        for b=1:1:x
            knorm(k)=knorm(k)+kgauss(b,k);
        end
    end
    knorm=knorm/3;
    kmax=max(knorm);
    knorm=knorm/kmax;
    ksum=sum(knorm);
    prob=eta*knorm/ksum;
    probsum=sum(prob);
    x_hat=sum(ref(:,1).*prob');
    y_hat=sum(ref(:,2).*prob');
    dme(n)=sqrt((x_hat-x_i)^2+(y\_hat-y\_i)^2);
    end
    plot(sigma,mean(dme),'o');
    sigma=sigma+0.1;
end
```

Assignments for Chapter Eight

Questions

1. What are advantages and disadvantages of RFID localization?
2. If you were designing the following Robot application, what would be your algorithm https://www.youtube.com/watch?v=5r7Mu-wtxjo&nohtml5=False ?
3. If we use channel model (such as IEEE 802.11 or RT) to find the power in different locations, why do we need finger printing?
4. In data collection for WPI, we drive in the street and we sample the RSS of all readable APs with a location tag produced by GPS. Can you explain why simple centroid algorithm is used to process the database for finding AP locations?
5. What are the differences among database collection techniques for WPS and RTLS in terms of number of samples per location and the method used to tag the location?
6. What are the differences among the algorithms used in RTLS and WPS?

Problems

Problem 8.1:
Assume we have four reference point installed on the four corners of a rectangular 20×30 room $(x_i, y_i) = (0,0), (0,20), (30,20), (30,0)$ if we have a tag at $(x, y) = (5,5)$ and we measure the distance from reference points to be $d_i = (8, 16, 27, 30)$, determine the location estimate and the DME

a) Using the RLS algorithm described in Problem 15.2 and starting at (2,2).
b) Using the ML algorithm with certainties ranging from 10 to 95% in steps of 10%.
c) Plot the DME as a function of certainty and compare that with the results of RLS algorithm.
d) Repeat (a) for centroid and weighted centroid algorithms and compare the results with those of part (a).
e) Compare LS with Maximum Likelihood Triangulation in term of accuracy, convergence, and computational complexity.

Projects

Project 8.1 (Wi-Fi Localization):
In this project, we build on Project 2.1 by mapping the APs and collecting a database in a typical office building. We then use the database to compare the performance of RSS-based localization algorithms. These algorithms are either using the location of the APs or the database that is collected in different locations.

a) Walk on a specified closed route in corridors of the office (e.g. in the second floor of the Atwater Kent laboratory, shown in Figure P8.1), and observe the RSS and the number of MAC addresses along the path. Go to each of the 5-points identified by circles in Figure P8.1 and log 20 RSS reading in each location while you stay in the same location with a normal laptop holding posture, while turning around slowly. We refer to this database as the "training or fingerprint database". Give a sample print of your logged data showing the MAC addresses and the associated RSS readings. Give the number of APs and the CDF of the RSS from different APs for your sample logs.

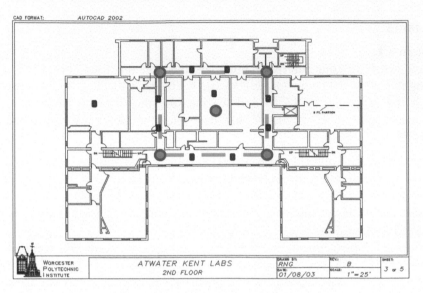

Figure P8.1 Layout of the second floor of the Atwater Kent Laboratory.

b) Do war driving to find the location of the APs in that floor of the building. Identify these locations in the layout of the floor map. Give the log for all AP readings under each AP and identify the RSS of the APs in the log of that location. Note that you might have several APs in one physical location. Give the number of APs and the range of RSS from different APs for your sample logs.

c) Go to the 10 locations identified by small squares and create another database with one RSS reading log per location. Let us call this the "test database".

A: Algorithms and bounds using the location of APs

1) Use the AP powers from "test database" and an appropriate IEEE 802.11 model to calculate the estimated distance of each test point from the APs in the that floor.

2) Use the distances from (A.1) and the RLS iterative algorithm described in Section 7.4.2 to calculate the estimated location based on power readings in all 10 test locations. Provide a table with actual distance, estimated distance and the DME for all of the 10 points. Note that in some locations algorithm may not converge; if you observe such a point, make a note in the table and explain why it has happened.

3) Repeat (A.2) using the ML triangulation algorithm

4) Repeat (A.2) using the centroid algorithm described in Section 8.3.1.

5) Design a weighted centroid algorithm of your choice and repeat (A.4).

6) Determine the CRLB of the RSS signals for all 10 test locations.

7) Give a table for the mean and variance as well as a graph for the CDF of the DME for the three algorithms and the CRLB.

B: Pattern recognition algorithms

1) Use the "training/fingerprint database" in Part A and the Closest Neighbor algorithm described in Section 8.4.1 to calculate the location and DME of the 10 test-points and provide a table similar to (A.2).

2) Use the "training/signature database" and the Kernel algorithm described in Section 8.4.3 to determine the location and DME of the 10 test-points and associated table.

2) Give a table for the mean and variance as well as a graph for the CDF of the DME for the two algorithms.

C: Analysis

How do you compare the two classes of algorithms? What are your advices for the location of the training points and the location of the APs to improve the performance of the algorithms?

9

TOA-Based Positioning Algorithms

9.1 Introduction

The fundamental benefit of RSS-based localization is that it is easy to measure the RSS. If we take samples of the received signal at the front end of the receiver, after passing through a filter centered at carrier frequency, we have a measurement of the RSS. There is no need for synchronization between the transmitter and the receiver, and we do not need any details of the transmitted waveform. The RSS is calculated by the receivers for power management anyways, so it is available for reading through software. However, RSS is an unreliable measure of the distance and it is disturbed by short-range multipath and long-term shadow fading. Averaging the received power will eliminate the effects of short-term multipath fading, which leaves shadow fading as the main source of noise for RSS-based positioning. Since variance of the RSS-based ranging is smaller in short distances, we use traditional LS or ML algorithms for in-room positioning applications using RFID or iBeacon technologies. For larger distances in indoor and urban area RSS-based positioning, we resort to fingerprinting pattern recognition algorithms.

The fundamental benefit of TOA-based positioning is the reliability of TOA to estimate the distance, which makes it the choice for long-range positioning using basic RLS like algorithms. However, measurement of the TOA requires synchronization between the transmitter and the receiver and detection process is affected by the design of the transmitted waveform. More importantly, measurement of the TOA is highly affected by the multipath details, and we need signal processing techniques to mitigate the effects of multipath. As a result, algorithms used for TOA-based localization are more relevant to signal processing aspects. We need signal processing algorithms to design the systems specification, to establish synchronization, and find methods to handle diversified multipath conditions in different application

315

environments. In the absence of multipath, precision of TOA-based localization is affected by the bandwidth of the transmitted waveform. In multipath conditions, this becomes a complex problem depending on the nature of the multipath. The nature of multipath depends on the environment and the deployment of the RP antennas, which varies substantially in open space, sub-urban, dense urban, indoor, and inside the human body.

In this chapter, we present useful signal processing algorithms and methods for TOA-based positioning systems. We begin by a review of traditional signal processing techniques for ranging, using the carrier frequency or envelope of a DSSS signal. Then, we describe technologies and algorithms used for ranging in extensive multipath, which includes UWB, multi-carrier transmission, and super-resolution algorithms. Finally, we address localization algorithms in the absence of direct path, where we discuss localization using multipath diversity and cooperative localization.

9.2 Basic Algorithms for Measurement of TOA

Measurement of TOA requires time synchronization between the transmitter and the receiver. In Section 4.2.1, we discussed how we can measure the TOA using NB and WB signals, and in Section 5.5, we discussed methods to establish timing for the measurement of the TOA. To synchronize a transmitter and a receiver, we need a method to create a replica of the transmitted signal at the receiver so that we can cross-correlate the transmitted and the received signal and extract the TOA. As we discussed in Section 5.5, knowing the transmitted waveform, we can either duplicate it at the receiver when we have established synchronization between the two devices or by using a transponder, which returns the transmitted signal from the receiver to the transmitter. The first approach is used for long distances in more sophisticated systems, such as GPS, and the second approach is used for short-range sensor positioning, using ZigBee or UWB technologies. A third alternative used in U-TDOA CPS systems, introduced in Section 5.5.3, uses the TDOA of the received signals in different base stations to positioning a device from its transmitted signal. This approach relies on synchronization among receivers, and it by passes the need for absolute synchronization between the transmitter and the receiver.

9.2.1 NB TOA Estimation Algorithms and Multipath

From the algorithm point of view, calculation of the range relies on the calculation of the cross-correlation function between the replica of transmitted signal and the received signals. Figure 9.1 provides a summary of application of the cross-correlation to calculate the distance in a traditional positioning systems using DSSS technology. In practice, we can either use the carrier frequency of the transmitted signal for NB measurement of the TOA or we use the WB envelope of the transmitted signal, for measurement of the distance. As shown in the top part of Figure 9.1, the DSSS transmitted signal has a carrier frequency represented by $\cos \omega t$ and an envelope represented by $p(t)$. The carrier frequency is a NB signal, $\cos \omega t$, which arrives at the receiver with a phase shift, $\cos(\omega t - \phi_0)$ (Figure 9.1(a)). We measure the phase by cross-correlating the receives carrier and the replica of the transmitted carrier signal:

$$\begin{cases} R_{xx}(\phi_0) \simeq \frac{1}{T} \int_0^T \cos \omega t \cos(\omega t - \phi_0) dt = \frac{1}{2} \cos \phi_0 \\ \hat{\phi} = \cos^{-1}[2R_{xx}(\phi_0)] \end{cases} \qquad (9.1a)$$

Since phase and delay are related by $\phi = \omega \times \tau$, we can determine the distance from:

$$d = \frac{c \times \hat{\phi}}{\omega} = \frac{\lambda \hat{\phi}}{2\pi}. \qquad (9.1b)$$

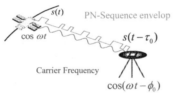

$$\begin{cases} R_{xx}(\phi_0) = \frac{1}{T} \int_0^T \cos \omega t \cos(\omega t - \phi_0) dt = \frac{1}{2} \cos \phi_0 \\ \hat{\phi} = \cos^{-1}[2R_{xx}(\phi_0)] \end{cases}$$

$$p(\tau - \tau_0) = R_{ss}(\tau - \tau_0) = \frac{1}{T} \int_0^T s(t - \tau_0) s(t - \tau) dt$$

$$\begin{cases} \hat{\phi} = \omega \times \hat{\tau} \\ d = c \times \hat{\tau} = c \times \hat{\phi} / \omega = \lambda \hat{\phi} / 2\pi \end{cases}$$

$$\begin{cases} \hat{\tau} = \tau_0 \\ d = \hat{\tau} c \end{cases}$$

Distance measurement ambiguity is: λ

Distance measurement ambiguity is: $T \times c$

(a)

(b)

Figure 9.1 Summary of NB vs WB measurement of the range using DSSS, (a) NB measurement using carrier signal, (b) WB measurement using PN-sequence of the DSSS envelope of the received signal.

This is indeed the ML estimate of the noise and variance of the noise in the absence of multipath can be calculated from the CRLB calculated in Section 4.3.2, Equation (4.16c). Measurement noise for NB measurement of the distance using the carrier signal is small, resulting in good precision for the measurement. However, if the distance goes beyond the value of λ (e.g., 6 cm at 5 GHz), the NB measurement of the distance becomes ambiguous, but it measures the distance precisely. To extend the ambiguity range, we may down convert the received signal to lower center frequencies or resort to multiple tones (Sections 4.2.3 and 4.2.4).

We need to analyze the effects of multipath to understand techniques and algorithms used to mitigate the effects of multipath in TOA-based localization. Figure 9.2 analyzes the effects of multipath on NB measurement of distance using TOA. If we represent the normalized received reference signal by:

$$x(t) = \cos(\omega_c t) = \text{Real}\left[ae^{j\omega_c t}\right], \tag{9.2a}$$

then, the received signal in the absence of multipath is given by:

$$y(t) = \beta_0 \cos(\omega t - \phi_0) = \text{Real}\left[\beta_0 e^{j(\omega_c t - \phi_0)}\right], \tag{9.2b}$$

where (β_0, ϕ_0) are the magnitude and phase of the direct path between the transmitter and the receiver, respectively. The ML estimate of the phase of

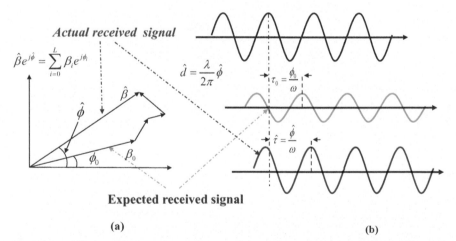

Figure 9.2 NB carrier phase measurement in multipath, (a) phasor diagram for calculation of magnitude and phase of the received carrier, (b) the expected received carrier signal and the actual received carrier signal in multipath.

the received signal is $\hat{\phi} = \phi_0$ and if we use it in Equation (9.1b), we have the ML estimate of the distance between the transmitter and the receiver. If we have L multipath arrivals, the received signal will become a summation of L signals arriving from all paths:

$$
\begin{aligned}
y(t) &= \sum_{i=0}^{L} \beta_i \cos(\omega t - \phi_i) = \text{Real}\left[\sum_{i=0}^{L} \beta_i e^{j(\omega t - \phi_i)}\right] \\
&= \text{Real}\left[e^{j\omega t} \sum_{i=1}^{L} \beta_i e^{j\phi_i}\right] = \text{Real}\left[\hat{\beta} e^{j(\omega t - \hat{\phi})}\right] = \hat{\beta}\cos(\omega_c t - \hat{\phi})
\end{aligned}
$$

(9.2c)

Figure 9.2(a) shows the phasor diagram for calculation of magnitude and phase of the received carrier using magnitude and phase of different multipath components. Figure 9.2(b) shows the expected received carrier signal and the actual received carrier signal in multipath. If we use the phase of the direct path, ϕ_0, in Equation (9.1b), we measure the actual distance. However, what we measure is the phase of the complex addition of all paths, $\hat{\phi}$, and substituting that in Equation (9.1b) results in a large distance measurement error, given by:

$$
DME = \frac{\lambda\left(\hat{\phi} - \phi_0\right)}{2\pi}.
$$

(9.2d)

The geometric relation between the expected phase and the measured phase in multipath can be illustrated using a phasor diagram (see Figure 6.10(b)). The measured phase in multipath is subject to the value of amplitude and phase of all multipath components (9.3c), which means we need the magnitude and phase of all multipath components to calculate the exact value of the phase of the received signal at the receiver. Since phase of the multipath components changes rapidly, on the order of the wavelength of the signal, the measured phase is subject to fast multipath fading. The DME in Equation (9.3d) and the ambiguity range of the measurements are both proportional to the wavelength. An increase in the wavelength to increase the ambiguity range will increase the DME as well. Therefore, if we do not apply a signal processing techniques to reduce the effects of multipath, measurement of phase using a carrier signal is only practical where we have very light multipath condition, such as application in open outdoor areas for GPS, or short-range LOS indoor areas for RFID localization. Emerging wireless communication networks using beamforming and multiple streaming over different paths are also subject to limited multipath for each stream and they may consider using phase of the carrier for measuring the distance between the transmitter and the receiver.

Using carrier phase for opportunistic TOA positioning is suitable for existing NB communication systems, such as those used in wireless video capsule endoscopy (WVCE) inside the human body, and in positioning in open areas where the multipath is negligible. In the multipath of the urban and indoor areas, WB ranging algorithms are preferable.

9.2.2 WB TOA Estimation Algorithms and Multipath

The most popular WB signal envelop for measurement of the TOA are DSSS signals[1] used in GPS, Zig-Bee, and 3G CDMA systems. Figure 9.1(b) shows the basic concept behind DSSS envelop signals, the envelop of the transmitted signal, $s(t)$, is a PN-sequence of length N with a chip duration of T_c. The autocorrelation function of this PN-sequence

$$R_{ss}(\tau) = p(\tau) = \frac{1}{T} \int_0^T s(t)s(t - \tau)dt \qquad (9.3a)$$

is a periodic sharp triangular pulse with period of $T = N \times T_c$. Therefore, the cross correlation between the received signal from direct path, $s(t - \tau_0)$, and the duplicate of the transmitted is:

$$R_{ss}(\tau - \tau_0) = p(\tau - \tau_0) = \frac{1}{T} \int_0^T s(t - \tau_0)s(t - \tau)dt, \qquad (9.3b)$$

which is the same periodic triangular pulse with a shifted peak at τ_0. Bandwidth of this system is $W = 1/T_c$.

To estimate the TOA of arrival, $\hat{\tau} = \tau_0$, we need to measure the shift in the peak of the received signal. One way to do that is to use sliding correlators[2] to generate the autocorrelation of the transmitted signal and the cross correlation of the transmitted and the received signal and measure the time difference between their peaks. The distance is then given by:

$$r = \hat{\tau}c. \qquad (9.3c)$$

This method for measurement of the TOA carries a distance ambiguity of $T \times c$ (close to 38 km for 1 MHz of bandwidth), which is a very reasonable ambiguity range for satellite navigation systems such as GPS. Some GPS receive the WB envelope of the DSSS for coarse estimation of the distance and the more precise NB carrier measurements for refining the results.

---

[1] For details, see Section 5.2.3.
[2] For details, see Section 5.5.1.

Similar to NB measurement of TOA, multipath also affects the WB measurements of TOA using DSSS, but in a much less harmful manner. In the absence of multipath, we receive the periodic signal with correlation function, $p(\tau - \tau_0)$. In multipath environment, correlation function becomes:

$$R_{ss}(\tau) = \sum_{i=0}^{L} \beta_i p(\tau - \tau_i)e^{-j\phi_i}. \tag{9.3d}$$

Figure 9.3(a) illustrates the expected received waveform for a TOA-based DSSS positioning system, and Figure 9.3(b) shows the received waveform in multipath. If the width of the pulses are narrower than the inter-arrival time of the multipath components, $2T_c < (\tau_i - \tau_{i-1})$, we can measure TOA of direct path, τ_0, without any effect from other paths. Therefore, a properly designed WB system with a bandwidth of $W < 2/(\tau_i - \tau_{i-1})$ can eliminate the effects of multipath on precision of TOA measurement. An additional benefit of such a wide-band TOA measurement system is that we can also measure the TOA, τ_i, amplitude, β_i, and phase, $\phi_i = \omega_c \times \tau_i$, of all other paths. Indeed, DSSS systems are commonly used in wireless communication systems to measure the channel multipath characteristics to enhance the quality of communication [Pah95].

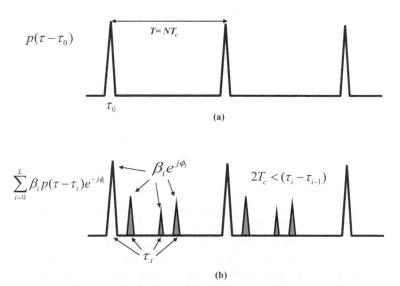

Figure 9.3 WB measurement of TOA in multipath, (a) the expected cross correlation of the received WB waveform, (b) the received WB waveform in multipath.

DSSS is used for TOA positioning in long-range GPS and cellular net-works as well as short-range positioning using ZigBee technology. There are other short-range positioning systems using UWB technology with direct pulse transmission. In these systems, the received signal is a peri-odic pulse similar to the autocorrelation function of the DSSS and we can use the peak of these pulses for calculation of TOA and consequently the distance. UWB systems have substantially larger bandwidths, mak-ing them a better choice for multipath rich indoor areas. In WB/UWB localization, bandwidth of the transmitted signal governs the width of the pulses used for TOA measurements, and the period of the transmit-ted signal governs the ambiguity range of the system. In NB measure-ment of the TOA, frequency of the sinusoid used for localization gov-erns precision and the ambiguity range. Smaller carrier frequencies pro-vide larger the ambiguity range as well as larger distance measurement errors.

9.3 TOA Estimation Algorithms for Multicarrier Signals

In the previous section, we showed that single frequency carrier phase posi-tioning techniques and algorithms are not a good choice for multipath rich environment and we introduced WB and UWB techniques and algorithms as the technology of choice for these environments. In this section, we focus on techniques and algorithms used for measurement of TOA using multicarrier measurements.

9.3.1 RLS Algorithm for Multicarrier Ranging in Multipath

In Figure 9.2, we demonstrated that NB measurement of TOA using a single carrier is only useful when the multipath is negligible. In this section, we explain how we can benefit from multi-carrier transmission and eliminate the effects of multipath. We begin by a simple example of a two-path channel, shown in Figure 9.4, to determine the number of carriers needed to elimi-nate the effects of a second path. Figure 9.4 shows the phasor diagram for calculation of the magnitude and phase of the received signal with a two-path channel, for two different frequencies. The normalized path amplitude[3] and arrival time of the paths, $(\tau_0, \tau_1, \beta_0, \beta_1)$, respectively, remain the same in both frequencies, because they are characteristics of the channel. However,

---

[3]Normalized to the receive power at 1 m at different frequencies.

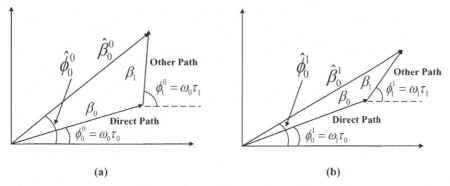

(a)   **(b)**

Figure 9.4 Phase and amplitude in a two-path channel at two different frequencies, (a) phasor diagram for the first frequency, (b) phasor diagram for the second frequency.

the phases of the paths are different,

$$
\begin{cases} \phi_0^1 = \omega_1\tau_0 \\ \phi_1^1 = \omega_1\tau_1 \end{cases}; \quad \begin{cases} \phi_0^2 = \omega_2\tau_0 \\ \phi_1^2 = \omega_2\tau_1 \end{cases}, \tag{9.4a}
$$

because the measurement frequencies are different. As a result, when we use different frequencies, (ω_1, ω_2), and measure the magnitude and phase of the received NB signal at these two different frequencies, $\left(\hat{\beta}_0^1, \hat{\phi}_0^1\right)$ and $\left(\hat{\beta}_0^2, \hat{\phi}_0^2\right)$, respectively, we have different results. This difference is caused by the effects of the multipath, and is the source of large DME in NB estimation of TOA, τ_0. In this section, we show that using multi-carrier transmission, we can estimate all channel parameters, $(\tau_0, \tau_1, \beta_0, \beta_1)$, and eliminate the effects of multipath.

In the two carrier transmission over a two-path channel, shown in Figure 9.4, we have:

$$
\begin{cases} \hat{\beta}_0^1 e^{j\hat{\phi}_0^1} = \beta_0 e^{j\omega_1\tau_0} + \beta_1 e^{j\omega_1\tau_1} \\ \hat{\beta}_0^2 e^{j\hat{\phi}_0^2} = \beta_0 e^{j\omega_2\tau_0} + \beta_1 e^{j\omega_2\tau_1} \end{cases}. \tag{9.4b}
$$

This is a two set of complex valued equations with four unknowns, $(\tau_0, \tau_1, \beta_0, \beta_1)$, from which we are interested in τ_0, to estimate the distance between the transmitter and the receiver, $d = \tau_0/c$. Expanding the complex valued, Equation (9.4a), into real and imaginary parts, we can form a set of four equations relating these four unknowns to one another:

$$\begin{cases} \hat{\beta}_0^1 \cos \hat{\phi}_0^1 = \beta_0 \cos \omega_1 \tau_0 + \beta_1 \cos \omega_1 \tau_1 \\ \hat{\beta}_0^1 \sin \hat{\phi}_0^1 = \beta_0 \sin \omega_1 \tau_0 + \beta_1 \sin \omega_1 \tau_1 \\ \hat{\beta}_0^2 \cos \hat{\phi}_0^2 = \beta_0 \cos \omega_2 \tau_0 + \beta_1 \cos \omega_2 \tau_1 \\ \hat{\beta}_0^2 \sin \hat{\phi}_0^2 = \beta_0 \sin \omega_2 \tau_0 + \beta_1 \sin \omega_2 \tau_1 \end{cases} \tag{9.4c}$$

Therefore, with two carriers, we will have four sets of equations with four unknowns, $(\tau_0, \tau_1, \beta_0, \beta_1)$. When we solve Equation (9.4b), we calculate the actual value of TOA, τ_0, eliminating the effects of the second path and the DME that it causes. For larger number of carriers, we will have redundancies and we can resort to LS formulation. The error function for the LS formulation for N-carriers is:

$$\min \{\varepsilon(\tau_0, \tau_1, \beta_0, \beta_1)\} = \min \left\{ \frac{1}{2} \sum_{k=1}^{2} \left| \hat{\beta}_0^k e^{j\hat{\phi}_0^1} - \beta_0 e^{j\omega_k \tau_0} - \beta_1 e^{j\omega_k \tau_1} \right|^2 \right\}.$$

We can calculate the Jacobian matrix and find an RLS solution for all four parameters. The discussion presented here demonstrated that using a multi-carrier system, we can eliminate the effects of the second path if we have at least two carriers. When we generalize the problem to L interfering paths, we need at least $N \geq L+1$ carriers, and Equation (9.4a) becomes a sequence of complex numbers:

$$Y(k) = \hat{\beta}_0^k e^{j\hat{\phi}_0^{ik}} = \sum_{l=0}^{L} \beta_l e^{j\omega_k \tau_l} ; \quad k = 0, 1, ...N - 1. \tag{9.5a}$$

Then, the error function of Equation (9.4c) changes to:

$$\min \{\varepsilon(\tau_i, \beta_i \; ; \; i = 0, 1, ...L)\} = \min \left\{ \frac{1}{N} \sum_{k=0}^{N-1} \left| \hat{\beta}_0^i e^{j\hat{\phi}_0^i} - \sum_{l=0}^{L} \beta_l e^{j\omega_i \tau_l} \right|^2 \right\},$$

$$\tag{9.5b}$$

and we can calculate that using the RLS algorithm when $N \geq L$. Therefore, we can eliminate the effect of multipath using multi-carrier transmission, if the number of carriers is at least the same as the number of multipath arrivals. This general solution indeed measures the channel multipath profile, $(\tau_i, \beta_i \; ; \; i = 0, 1, ...L)$, and uses that to eliminate the effects of multipath in estimation of the distance.

In practice, we can create a multi-carrier system by a simple system transmitting a carrier and measuring the magnitude and phase of the received signal at the receiver, while after each measurement, the frequency of the signal is changed. If we sweep the entire band in the coherence time of the channel we actually measure the band-limited discrete frequency response of the channel. The vector network analyzer described in Section 6.4.2 for frequency domain measurement of multipath characteristics of the radio channel, which is widely used for the analysis of the effects of multipath on TOA-based localization, operates based on this principle. For opportunistic localization using exiting wireless communication signals, one may use the amplitude and phase of carriers used for OFDM signaling or the amplitude and phase of the FHSS devices over all hops of the system. OFDM signals are used in Wi-Fi devices and 4G cellular networks. The FHSS was originally used in the legacy IEEE 802.11 and it is the modulation choice of Bluetooth and iBeacon technologies.

Another approach to solve this problem is to use the Fourier Transform techniques.

9.3.2 Peak Detection IFT Algorithm

The RLS algorithm is suitable for a small number of multipath arrivals and carrier frequencies. For a large number of carriers, RLS faces convergence problem and it is easier to use peak detection algorithm using inverse Fourier transform (IFT) to estimate the TOA, τ_0.

In a multi-carrier system, the complex valued sequence of the received multicarrier amplitudes and phases, $Y(k)$, given by Equation (9.5a), is a discrete frequency signal. The IFT of this sequence is a set of sinc shape pulses with the magnitude and phase of each path. The peak of the first sinc pulse occurs at τ_0. Therefore, if we use a peak detection algorithm on the IFT of the $Y(k)$, we can estimate the TOA. In Section 5.3.1, we briefly discussed this topic, and in this section, we provide the details.

Figure 9.5 shows the basic principles behind TOA measurement using peak detection algorithm on the IFT of the received multi-carrier signal without getting into account the effects of multipath and thermal noise. Figure 9.5(a) shows the transmitted signal in the frequency- and in the time-domain, assuming that multi-carrier system has N-carriers at frequencies:

$$f_k = f_0 + k\Delta f, \ 0 \le k < N. \tag{9.6a}$$

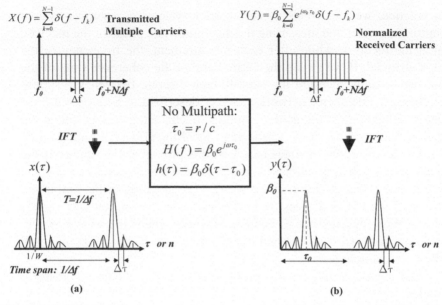

Figure 9.5   TOA arrival measurement using peak detection algorithm on the IFT of the received multi-carrier signal, (a) the transmitted signal, (b) the received signal.

The transmitted signal in frequency domain is a rectangular pulse sampled by impulses at the sampling rate of, Δf:

$$X(f) = Rec_W(f)|_{\Delta f} = \sum_{k=0}^{N-1} \delta(f - f_k) \qquad (9.6b)$$

and the bandwidth of the system is $W = N \times \Delta f$. The IFT of a rectangular pulse is a sinc pulse:

$$Rec_W(f) \Leftrightarrow W \, Sinc(W\tau),$$

and the sampling at the rate of Δf in frequency makes the signal in time-domain a periodic signal with period of $T = 1/\Delta f$ and a scaling factor of $1/W$. Therefore, the IFT of the transmitted signal is:

$$x(\tau) = \sum_{k=0}^{N-1} X(k) e^{j2\pi f_k \times \tau} = \sum_{n} Sinc\left[W(\tau - nT)\right], \qquad (9.6c)$$

where $\{X(k) = 1, \; k = 0, 1, ..N - 1\}$ is a sequence representing the discrete characteristics of the transmitted carriers in frequency. The period $T = 1/\Delta f$

is the span of time one can measure the TOA using a multi-carrier transmission system. We refer to that as the ambiguity time and the distance associated with this time is the distance ambiguity of the measurement system.

Characteristics of the channel in time and frequency are represented by:

$$\begin{cases} h(\tau) = \delta(\tau - \tau_0) \\ H(f) = e^{j\omega\tau_0} \end{cases}, \tag{9.7a}$$

where $\tau_0 = d/c$. As a result, the magnitude and phase of the received multicarrier sequence (Figure 9.5(b)) with normalized amplitude is $\{Y(k) = e^{j\omega_k \tau_0}; \ k = 0, 1, ... N - 1\}$, and the received signal in frequency is:

$$Y(f) = \beta_0 \sum_{k=0}^{N-1} e^{j\omega_k \tau_0} \delta(f - f_k) \tag{9.7b}$$

The IFT of this signal is again a periodic sinc pulse in time whose peak is shifted to τ_0:

$$y(\tau) = \frac{1}{N} \sum_{k=0}^{N-1} Y(k)e^{j2\pi\tau \times f_k} = \sum_{n} \text{Sinc}\left[W\left(\tau - \tau_0 - nT\right)\right]. \tag{9.7c}$$

If we use a peak detection algorithm on the received signal, we can estimate the magnitude and phase of the first path:

$$\begin{cases} \hat{\beta}_0 = y(0) \\ \hat{\tau}_0 = \tau_0 \end{cases}. \tag{9.7d}$$

To process the received time domain data digitally, in practice, we sample the time domain to obtain a complex digitized sequence of signal in time at the arbitrary sampling intervals of $\Delta\tau$ and M samples:

$$y(n) = y(\tau)|_{\tau=n\Delta\tau} = \frac{1}{N} \sum_{k=0}^{N-1} Y(k)e^{j2\pi n\Delta\tau \times f_k},$$

$$T_{start} \leq \tau < T_{start} + M\Delta\tau. \tag{9.8}$$

In the digital signal processing literature, the arbitrary sampling of the Fourier Transform is sometimes referred to as the chrip Z-transform (CZT)[4].

---

[4]If we take N-samples of Equation (9.7b) at $\Delta\tau = T_m/N$, we will have the traditional discrete Fourier transform (DFT), which is usually calculated by the fast Fourier transform (FFT) algorithms commonly available in the MATLAB or other scientific programming software packages.

9.3.3 Effects of Multipath on Peak Detection IFT Algorithm

As shown in Figure 9.5, peak detection using IFT of the results of multi-carrier transmission is a virtual pulse transmission technique (similar to DSSS) for measurement of the TOA. The received carriers magnitudes and phases after IFT form a sharp sinc pulse and the time for occurrence of the peak of the pulse provides the estimate for the TOA. Therefore, similar to other pulse transmission techniques, such as DSSS or UWB, it is very effective for measurement of TOA in multipath. The bandwidth of the system should be wide enough to resolve the multipath arrival and the ambiguity time should be large enough to accommodate the multipath spread of the channel. The system provides for estimate of magnitude and phase of all multipath components and it can be used for channel estimation purposes as well.

Figure 9.6 shows the time and frequency characteristics of TOA measurement in multipath using multi-carrier signal. In a multipath environment, signal arriving from each path forms a sinc pulse with amplitude and delay associated to that path. The multipath channel characteristics in time and in frequency are:

$$\begin{cases} h(\tau) = \sum_{l=0}^{L} \beta_l e^{i\phi_l} \delta(\tau - \tau_l) \\ H(f) = \sum_{l=0}^{L} \beta_l e^{i\phi_l} e^{j\omega\tau_l} \end{cases} \tag{9.9a}$$

The received sequence of magnitudes and phases of the multi-carrier system is then described by:

$$\left\{ Y(k) = \sum_{i=0}^{L} \beta_l e^{i\phi_l} e^{j\omega_k\tau_l} ; \quad k = 0, 1, \ldots N - 1 \right\} \tag{9.9b}$$

Using the sequence given by Equation (9.9b) in Equation (9.8) provides the samples of the received signal in the time domain. The lower part of Figure 9.6(b) shows a sample multipath arrival in time for the multi-carrier virtual pulse transmission system.

The multi-carrier virtual pulse transmission system associates a sinc pulse to each path. Sinc pulses have very strong side lobes, and they can be as high as 13 dB below the peak of the pulse. This allows more interference from paths close to the first path in the shape and the displacement of the peak of the direct path. We have sinc pulses because envelop of the transmitted carrier is a rectangular pulse. When we measure a discrete frequency response (Figure 9.7(a)), if we filter with a smother filter than the rectangular filter

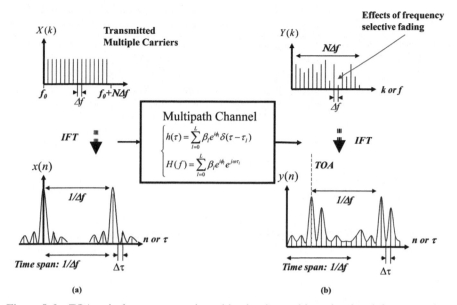

Figure 9.6 TOA arrival measurement in multipath using multi-carrier signal, frequency-time characteristics of, (a) the transmitted signal, (b) the received signal.

(Figure 9.7(b)) before taking the inverse Fourier transform, we can control the side lobes in time to reduce their effects on measurement of the TOA (Figure 9.7(c)). Triangular Bartlett, Hanning, and Hamming widow shapes are popular for control of the side lobes of the virtual received pulses. Figure 9.8 compares the time response of the rectangular filter with that of Bartlett, Hanning and Hamming windows. Hamming window in particular reduces the side lobes up to 44 dB below the peak and it is popular in the multipath profile measurement systems for wireless positioning. Result of channel measurements and modeling in Sections 6.4.2 and 6.5.1 use this filter for their TOA measurements. When we apply the filter in frequency response, Equation (9.8) for calculation of the time responses will change to:

$$y(n) = y(\tau)|_{\tau=n\Delta t} = \frac{1}{N} \sum_{k=0}^{N-1} W(k)Y(k)e^{j2\pi n\Delta\tau \times f_k},$$

$$T_{start} \leq \tau < T_{start} + M\Delta\tau, \tag{9.9c}$$

where $W(k)$ represents the samples of the frequency response characteristics of the filter.

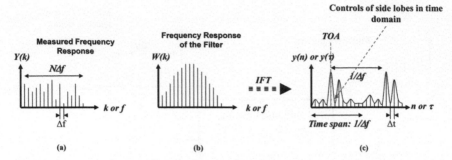

(a)   (b)   (c)

Figure 9.7   Effects of filtering the multi-carrier measurements on the measurement of TOA, (a) the measured channel frequency response using multiple carriers, (b) frequency response of the filter, (c) controlled side lobes with the fitter in the time domain.

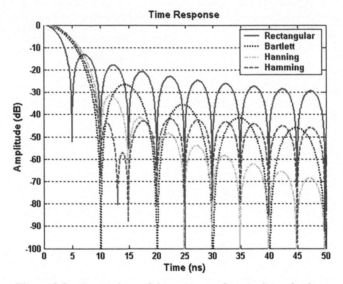

Figure 9.8   Comparison of time response for a variety of pulses.

9.3.4 Spectral Estimation Algorithms

Figure 9.9(a) shows the basic mathematical concept behind the TOA-based measurement using magnitude and phase of a multi-carrier system. We use a finite sequence of N-complex numbers in frequency domain to measure the TOA of the direct path in time domain. In multipath arrival, the paths arriving close to the direct path challenge accurate estimation of the TOA. As we explained in the last paragraph of the previous section, signal processing techniques such as filtering in frequency domain

(a) Multi-Carrier Estimation

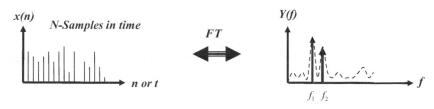

(b) Spectral Estimation

Figure 9.9 Similarity between, (a) measurement of TOA of a path in presence of neighboring paths using frequency domain data, (b) measurement of frequency of neighboring frequencies using time domain data.

(Figure 9.7) can help this process. In fact, measurement of TOA using multi-carrier transmission is a mathematical dual of the classical spectral estimation problem that contains a rich literature with a variety of algorithms that can be examined for TOA positioning. In spectral estimation techniques (Figure 9.9(b)), we have multiple frequencies arriving close to one another and using N-sample of the signal in the time domain we want to estimate these frequencies, so that we can differentiate them from one another. In Figure 9.9(a), we want to resolve multipath components, and in Figure 9.9(b), we want to resolve the frequency component. Using duality of time and frequency property of Fourier transform, we can examine all spectral estimation algorithms for multi-carrier measurements of TOA.

Spectral estimation techniques are traditionally divided into non-parametric and parametric estimation techniques [Hay06]. Non-parametric spectral estimation algorithms manipulate the data to refine the estimation process before taking the Fourier transform. Using filters, explained in the last paragraph of the previous section, is the simplest non-parametric spectral estimation method. Other non-parametric algorithms, such as periodogram or Welsh method, divide the data into segments and apply filters to each segment. Parametric algorithms model the signal and use the data to estimate the parameters of the model. Auto-regressive, moving-average, and auto-regressive-moving-average models are the most popular parametric models for the signal. In these methods, the measured sequence is assumed to be the output of a digital moving average or recursive digital filter, and with that assumption, we estimate the location of the poles and zeros of the filter. One of the earliest attempts in using parametric autoregressive modeling in spectral estimation for multipath parameter estimation using multi-carrier transmission is available in [How92]. A more advanced approach focused on the TOA is to use the so-called super-resolution algorithms to refine the time resolution of the measurements of multipath profiles.

9.3.5 Super-Resolution Algorithm

In the literature, the time-delay estimation problem has been studied with a variety of super-resolution techniques, such as minimum-norm [Pal91], root multiple signal classification (MUSIC) [Dum94], and total least-square estimation of signal parameters via rotational invariance techniques (TLS-ESPRIT) [Saa97]. While super-resolution techniques can increase time-domain resolution, it also increases complexity of system implementation. In this section, we introduce EV/FBCM (eigenvector forward backward correction matrix) frequency-domain super-resolution TOA estimation algorithm [Li03].

The impulse response of a radio channel in time domain and its Fourier transform are given by Equation (9.9a). In multi-carrier TOA estimation, we measure samples of frequency response of the channel, given by Equation (9.9b), and we intend to determine the TOA, magnitude, and phase of the first path in the multipath profile. This dual of this model is well known in spectral estimation literature and it is referred to as harmonic model approach [Man00]. Therefore, any spectral estimation techniques that are suitable for the harmonic model can be applied to the results of multi-carrier measurement to estimate the TOA. In [Li03], MUSIC algorithm is used as a spectral estimation technique to convert the frequency domain data into the time domain profile needed for determining the direct LOS path and TOA.

The discrete measurement of frequency domain data is achieved by sampling the channel frequency response

$$\left\{ Y(k) = H(f)|_{f=f_0+k\Delta f}; \ 0 \le k < N \right\} \tag{9.10a}$$

at N equally spaced frequencies. Considering additive white noise in the measurement, the sampled discrete frequency domain received signal using multi-carrier approach is given by

$$Z(k) = Y(k) + w(k) = \sum_{l=0}^{L} \beta_l e^{-j2\pi(f_0+k\Delta f)\tau_l} e^{j\phi_l} + w(k), \tag{9.10b}$$

where $k = 0,1,\ldots,$N-1 and $w(k)$ denotes the additive white Gaussian measurement noise with zero mean and variance $(\sigma_w)^2$. The signal model in vector form is:

$$\mathbf{Z} = \mathbf{VA} + \mathbf{W}, \tag{9.10c}$$

where

$$\begin{cases} \mathbf{Z} = \begin{bmatrix} z(0) & z(1) & ..... & z(L) \end{bmatrix}^T \\ \mathbf{W} = \begin{bmatrix} w(0) & w(1) & ..... & w(L) \end{bmatrix}^T \\ \mathbf{V} = \begin{bmatrix} \mathbf{v}(\tau_0) & v(\tau_1) & ..... & v(\tau_L) \end{bmatrix}^T \\ \mathbf{v}(\tau_0) = \begin{bmatrix} 1 & e^{-j2\pi\Delta f\tau_k} & ..... & e^{-j2\pi(L)\Delta f\tau_k} \end{bmatrix}^T \\ \mathbf{a} = \begin{bmatrix} \alpha_0' & \alpha_1' & ..... & \alpha_L' \end{bmatrix}^T \\ \alpha_k' = \alpha_k e^{-j2\pi f_0\tau_k} \end{cases} \tag{9.10d}$$

The MUSIC super-resolution algorithm is based on eigen-decomposition of the autocorrelation matrix of the signal model in Equation (9.10). The autocorrelation matrix is defined as:

$$\mathbf{R}_{xx} = E\left\{\mathbf{Z}\mathbf{Z}^H\right\} = \mathbf{V}\mathbf{A}\mathbf{V}^H + \sigma_w^2\mathbf{I}, \tag{9.11a}$$

where $\mathbf{A} = E\{\mathbf{aa}^H\}$ and superscript H is the Hermitian, conjugate transpose, of a matrix. Therefore, the $(L+1)$-dimensional subspace that contains the signal vector $\mathbf{Z}$ is split into two orthogonal subspaces, known as signal subspace and noise subspace, by the signal eigenvectors and noise eigenvectors, respectively. Since the vector, $\mathbf{v}(\tau_k)$; $k = 0, 1, .., L$, must lie in the signal subspace, we have:

$$\mathbf{P}_w\mathbf{v}(\tau_k) = 0, \tag{9.11b}$$

where $\mathbf{P}_w\mathbf{v}(\tau_k)$ is the projection matrix of the noise subspace. Thus, the multipath delays τ_k; $k = 0, 1, .., L$ can be determined by finding the delay values, at which the following MUSIC pseudo-spectrum achieves maximum value:

$$S_{MUSIC}(\tau) = \frac{1}{\|\mathbf{P}_w\mathbf{v}(\tau)\|^2} = \frac{1}{\sum\limits_{k=L_p}^{L-1} |\mathbf{q}_k\mathbf{v}(\tau)|^2} \qquad (9.11c)$$

where $\mathbf{q}_k$ are the noise eigenvectors. In practical implementation, when only one snapshot of length N is available, the data sequence is divided into M consecutive segments of length L and then the estimate of the correlation matrix is further improved using the *forward-backward correlation matrix* described in [Pal91]. In the analysis provided in [Li03], a slight variation on the MUSIC algorithm is used, which is known as the eigenvector (EV) method. In this approach, the pseudo-spectrum is defined as:

$$S_{EV/FBCM}(\tau) = \frac{1}{\sum\limits_{k=L_p}^{L-1} \frac{1}{\lambda_k} |q_w^H v(\tau)|^2}, \qquad (9.11d)$$

where λ_k; $k = 0, 1, .., L$ are the noise eigenvalues. Effectively, the pseudo-spectrum of each eigenvector is normalized by its corresponding eigenvalue. The performance of the EV method is less sensitive to inaccurate estimate of the parameter, L, which is highly desirable in practical implementation for TOA estimation.

Figure 9.10 shows a simplified block diagram of the super-resolution algorithm. The measured samples of frequency response of the channel measured by a multi-carrier system are used to form a pseudo-spectrum of the signal from its parametric model. The peak of the TOA of the first path in the pseudo-spectrum is the estimate of the TOA of direct path. Figure 9.11 represents two sample frequency domain measurements of the indoor radio channel with three different post processing techniques. The first post-processing algorithm uses the traditional IFT using Hanning window, and the second algorithm simulates a DSSS system by using a raised-cosine pulse with roll-off factor 0.25 rather than Hanning window[5]. The third algorithm is EV/FBCM given by Equation (9.11d). Figure 9.11(a) represents a case in which DLOS path is detectable, and Figure 9.11(b) represents an undetected direct path condition. In both cases,

---

[5]Raised cosine pulses are common in implementation of RF filters used in radio modems; for details, see Figure 6.26.

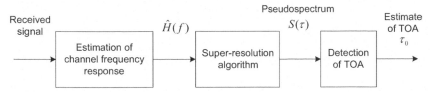

Figure 9.10 Simplified block diagram of the super resolution algorithm for measurement of TOA using finite sample of the frequency domain characteristics obtained from a multi-carrier system.

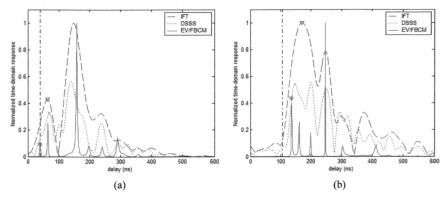

Figure 9.11 Effectiveness of the EV/FBCM super-resolution algorithm in, (a) DLOS path is detectable, (b) DLOS path is buried under the noise.

super-resolution algorithm resolves many more paths, allowing a more accurate estimate of the TOA of the direct path. The above discussion closely followed [Als07], in which more extensive results of measurements and post processing are available. More details of the EV/FBCM is available in [Li03]. Appendix 9.A provides MATLAB code for implementation of the super-resolution algorithm used for results presented in Figure 9.11.

9.4 DOA Estimation Using Multicarrier Signals

Direction of arrival (DOA) estimation techniques and algorithms are closely related to the TOA estimation techniques and algorithms. In Section 4.4, we used the CRLB of the TOA systems to derive the CRLB of DOA systems using antenna arrays. The DOA for positioning applications in a multipath environment is the angle of arrival of the direct path. If the direct path is blocked, we can position the transmitter through other paths (see

Section 9.5.1), and for that, the approach needs to estimate the angle of arrival of other paths. In the previous section, we presented a few algorithms using the multicarrier signals to estimate the TOA in multipath environment. In this section, we present two algorithms for DOA estimation in multipath using multicarrier signals.

If we consider the angle of arrival of different paths, the overall impulse response of the channel, given by Equation (9.9a), will be modified to:

$$h(\tau, \theta) = \sum_{i=0}^{L} \beta_i e^{j\phi_i} \delta(\tau - \tau_i) \delta(\theta - \theta_i), \tag{9.12}$$

where θ_i is the angle of arrival of the i-th path with $(\tau_i, \beta_i, \phi_i)$ the arrival delay, amplitude and phase of the path, respectively. Then, the measurement system has to be capable of associating arriving paths with the angle they arrive at. Obviously, omni-directional antennas cannot differentiate the angle of arrival of the paths and we need antenna system (see Section 4.4). These arrays could be linear (e.g., see Figure 9.12) or in any other geometric shape. The traditional method for measurement of DOA, used in radar and other applications, is to rotate an antenna arrays mechanically. In [Spe00], this approach is paired to multi-carrier frequency domain measurement system described in [How90] to measure the DOA of the multipath components. Another approach to measure the angle of arrival is to use a circular antenna array, which can select the channel electronically rather than mechanically. In this section, we describe the system and algorithms used in [Tin01], as examples, for measurements of DOA in extensive indoor multipath conditions using multicarrier signals.

Figure 9.12 shows the basic principles of the multicarrier frequency domain measurement system with an eight-element antenna used for measurements of angle of arrival in [Tin01]. The eight sequentially measured channel frequency responses are post processes with a parameter estimation algorithm to determine the magnitude, phase, and angle of arrival of different paths defined in Equation (9.12). The array antenna, shown in Figure 9.12(a), consists of eight nominally identical quarter-wave monopole elements, mounted at a constant radius and separated by one-third wavelength. Figure 9.12(b) shows basic post processing of the eight multi-carrier measurements of the channel frequency responses by the antenna elements. Measurements pass through a post processing algorithm to estimate the parameters of the channel given by Equation (9.12). Two algorithms, spatial filter periodogram (SFP)

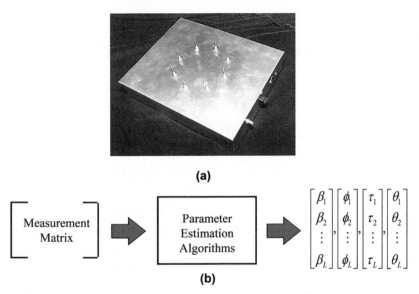

(a)

$$
\begin{bmatrix} \beta_1 \\ \beta_2 \\ \vdots \\ \beta_L \end{bmatrix}, \begin{bmatrix} \phi_1 \\ \phi_2 \\ \vdots \\ \phi_L \end{bmatrix}, \begin{bmatrix} \tau_1 \\ \tau_2 \\ \vdots \\ \tau_L \end{bmatrix}, \begin{bmatrix} \theta_1 \\ \theta_2 \\ \vdots \\ \theta_L \end{bmatrix}
$$

Measurement Matrix ⟹ Parameter Estimation Algorithms ⟹

(b)

Figure 9.12 (a) The 8-element antenna array used for angle of arrival measurements in [Tin00a,b], (b) post processing of the measurements of channel multipath profiles from different antennas.

and discrete maximum likelihood (DML), are used for the post processing of the measured data to estimate all multipath parameters including the angle of arrival.

This measurement system begins with a calibration procedure to generate a reference for relating the measured set of channel profiles to the angle of each antenna in the circular array and the direct path between the transmitter and the receiver. During the calibration process, the transmit antenna and receiving array are assembled in a small indoor anechoic range. The antennas are mounted atop turntables and separated by an arbitrary close distance of 3 m [Tin01]. In this configuration, the direct path with angle of arrival of zero is the dominant path. A series of eight-channel frequency domain measurements for the eight elements of the receiving antenna are conducted for each set of measurements used for calibration. Between each data set, the array is rotated for an arbitrary small angle of approximately $5.625 = 360/64$, degrees in azimuth, relative to its previous orientation. This measurement and rotation sequence is performed for one complete revolution of the array, which resulted in distinct data sets, given by:

$$
\mathbf{U}_m =
\begin{bmatrix}
H_{1,m}(0) & H_{2,m}(0) & \cdots & H_{8,m}(0) \\
H_{1,m}(1) & H_{2,m}(1) & & \\
\vdots & & \ddots & \\
H_{1,m}(101) & & & H_{8,m}(101)
\end{bmatrix}
$$

$$
=
\begin{bmatrix}
u_{1,1,m} & u_{1,2,m} & & u_{1,8,m} \\
u_{2,1,m} & u_{2,2,m} & & \\
& & & \\
u_{101,1,m} & & & u_{101,8,m}
\end{bmatrix},
\tag{9.13}
$$

where *m = 1, 2, ... , 64*. The indices are chosen such that *m = 1* corresponds to 0 degrees, *m = 2* corresponds to *5.625* degrees, and so forth. The *n-th* column of U_m represents the 101 samples of the frequency response between the transmit antenna and the *n-th* elements of the receiving array. The database of all 64 angles form a basic reference for the expected behavior of the channel for different angles between the transmitter and the receiver. This database is used during the post processing to detect angle and delay of arrival associated with individual paths in a measured set of channel frequency domain measurements. The aforementioned two post processing techniques, SFP and DML, described in the rest of this section use this training data.

9.4.1 Spatial Filter Periodogram Algorithm

In the SPF, rather than employing a conventional beam-forming algorithm, the collection of calibration matrices is used to design a least-squares (LS) spatial 2D filter that shapes the spectrum for optimal best angle of arrival estimation. We define this spatial filter by:

$$
\mathbf{W} =
\begin{bmatrix}
w_{11} & w_{12} & \cdots & w_{1M} \\
w_{21} & w_{22} & & \\
\vdots & & \ddots & \\
w_{81} & & & w_{8M}
\end{bmatrix},
\tag{9.14a}
$$

in which the mth column of Equation (9.14a) provides eight taps, which serve to steer the array in the desired direction, while minimizing the energy collected from all other directions. In other words, this filter is a transformation that applies to the eight measurements of the channel frequency responses received from eight-element antenna:

$$
\mathbf{V} = \begin{bmatrix} H_1(0) & H_1(1) & \cdots & H_1(N-1) \\ H_2(0) & H_2(1) & & \\ \vdots & & \ddots & \\ H_M(0) & & & H_M(N-1) \end{bmatrix}
$$

$$
= \begin{bmatrix} v_{11} & v_{12} & \cdots & v_{1N} \\ v_{21} & v_{22} & & \\ \vdots & & \ddots & \\ v_{M1} & & & v_{MN} \end{bmatrix} \tag{9.14b}
$$

and produces M-new channel frequency responses for M angle of arrivals. For the time being, let us defer the description of the method used for the design of this filter using calibration data to the end of this section and focus our attention on how to use this filter. Given an arbitrary measurement matrix, provided in Equation (9.14b), the space-time impulse response representing the behavior in delay and angle of arrival is estimated as:

$$
\hat{\mathbf{H}} = \mathbf{FXVW}, \tag{9.14c}
$$

where $\mathbf{F}$ is the inverse Fourier transform matrix and $\mathbf{X}$ serves as the standard smoothing window function, defined in Equation (9.9c) for 1D measurements, which minimizes the side lobes present in the equivalent time response.

Figure 9.13 represents two sampled 3D measurements of the channel impulse response in LOS and OLOS indoor areas. In the OLOS condition, the delay spread is roughly three times longer than the LOS, the arrivals are no longer tightly clustered about the source bearing, and the first arrival is weaker than many subsequent arrivals. The average spatial resolution of these measurements is approximately 40°, and the average time resolution is 7 ns.

The spatial filter is designed using the calibration data described in Equation (9.13). There are 64 channel frequency responses collected in the anechoic chamber, each containing eight frequency responses with 101 samples of the frequency domain signal. To determine the coefficients of the filter, we form the cost function:

$$
f(\mathbf{w}_m) = \sum_{j \neq m} \sum_{l=1}^{101} \left(\sum_{n=1}^{8} u_{l,n,j} w_{n,m} \right) \left(\sum_{n=1}^{8} u_{l,n,j}^* w_{n,m}^* \right), \tag{9.15a}
$$

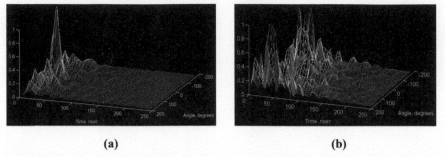

(a)   **(b)**

Figure 9.13   Measured 3D indoor channel impulse response using SFP filters and a frequency domain channel measurement system with 8-element circular antennas (a) LOS, (b) OLOS.

which is minimized, subject to the additional constraint:

$$c\left(\mathbf{w}_\mathrm{m}\right) = \sum_{l=1}^{101}\left(\sum_{n=1}^{8} u_{l,n,m} w_{n,m}\right)\left(\sum_{n=1}^{8} u_{l,n,m}^* w_{n,m}^*\right) - 1 = 0. \quad (9.15b)$$

The cost function serves to minimize the array's sensitivity to energy from all directions except the desired, while the constraint guarantees a solution with a fixed, constant gain in the desired direction. This approach is very similar to the well-known minimum-variance distortionless response method used in 2D adaptive filtering [Hay06], in which the cost function and constraint are satisfied jointly, using the method of complex Lagrange multipliers. The first step in this method is to combine the cost and constraint equations, defined in Equation (9.15), to form the *adjoint equation*:

$$\frac{\partial f\left(\mathbf{w}_\mathrm{m}\right)}{\partial \mathbf{w}_\mathrm{m}^*} + \lambda \frac{\partial c\left(\mathbf{w}_\mathrm{m}\right)}{\partial \mathbf{w}_\mathrm{m}^*} = 0. \quad (9.16a)$$

Performing the partial differentiation over each element of the complex-valued tap-weight vector given by Equation (9.15c) produces a system of equations of the form:

$$F_k\left(\mathbf{w}_m, \lambda\right) = \sum_{j \neq m}^{101} \sum_{l=1} u_{l,j,k} \sum_{n=1}^{8} u_{l,j,n}^* w_{n,m} + \lambda \sum_{l=1}^{101} u_{l,m,k}^* \sum_{n=1}^{8} u_{l,m,k}^* w_{n,m} = 0$$

$$(9.16b)$$

where $k = 1, 2, \ldots, 8$. Since the constraint equation must also be satisfied, we have

$$F_9\left(\mathbf{w}_m, \lambda\right) = \sum_{l=1}^{101}\left(\sum_{n=1}^{8} u_{l,n,m} w_{n,m}\right)\left(\sum_{n=1}^{8} u_{l,n,m}^* w_{n,m}^*\right) - 1 = 0. \quad (9.16c)$$

Taken Equations (9.16b) and (9.16c) together, a solution to the nine equations provides the optimal taps for the array processor, as well as the Lagrange multiplier. Although the eight equations for the array weights, described by Equation (9.16b), are linear in the independent variables, Equation (9.16c) for the ninth variable is quadratic. As a result, standard linear algebraic techniques cannot be used to compute the solution immediately and one may resort to RLS algorithm to solve the problem.

9.4.2 Discrete Maximum Likelihood Algorithm

While the SFP algorithm provides a simple method to estimate the space-time impulse response from a measurement matrix, it suffers from several limitations. The spatial resolution is strictly limited by the number and configuration of array elements, and in the design discussed in Section 9.4.1 stands at 40°. Likewise, the temporal resolution is given by the reciprocal of the sweep bandwidth, and is approximately 7 ns. Both of these limitations are independent of the signal-to-noise ratio of the data acquisition system. A third limitation is that the algorithm tends to introduce bias into the channel parameter estimates as a pair of closely spaced arrivals appears [Tin01].

These restrictions can be removed by developing an algorithm based on the maximum likelihood criteria. The basic principle behind ML estimation is very simple, and we consider a set of possible values for channel parameters and we reconstruct a channel frequency response for that set. Then, we find the error between the reconstructed response and the measured response. Finally, we do an exhaustive search to find the parameters of the channel that minimizes the error between the measurement and the reconstructed response. Since this is a very computationally extensive approach, the main obstacle is to find an efficient algorithm to search for optimal solution. In [Tin01], two simple serial and parallel algorithms are provided first, and since serial search is inaccurate and parallel search is computationally impractical, a new algorithm called recursive serial search is introduced.

The recursive serial algorithm uses the measured calibration matrices $\mathbf{U}_m$, m = 1, 2, ..., M, defined by Equation (9.13) as a fundamental basis functions. Given the overall channel impulse response defined by Equation (9.12), an arbitrary measurement $\mathbf{V}$ can be modeled as:

$$\hat{\mathbf{V}} = \sum_{l=1}^{L} \beta_l e^{j\phi_l} \mathbf{D}\left(\tau_l\right) \mathbf{U}_l \tag{9.17a}$$

where L is the number of discrete paths, $\beta_l e^{j\phi_l}$ is the complex weight of the l-th path, and $D(\tau_l)$ is the *101x101* diagonal time delay matrix defined as:

$$
\mathbf{D}(\tau) = \begin{bmatrix} e^{-j2\pi f_1 \tau} & 0 & \cdots & 0 \\ 0 & e^{-j2\pi f_2 \tau} & & \\ \vdots & & \ddots & \\ 0 & & & e^{-j2\pi f_N \tau} \end{bmatrix}, \tag{9.17b}
$$

where f_i is the *i-th* sample in the *101* set of a measured channel frequency response. The values of L and τ_λ, θ_λ, β_λ and ϕ_λ, where $l=\{1,2,...,L\}$ are selected to minimize:

$$
J_{DML} = \sum_{k=1}^{101} \sum_{n=1}^{8} |v_{kn} - \hat{v}_{kn}|^2. \tag{9.17c}
$$

Using the delay operator matrix $D(t)$ in Equation (9.17b), D-1 new versions of each of the calibration matrices are constructed, where the first is delayed τ_o s, the second $2\tau_o$ s, and so forth. This procedure yields the basis matrices C_k, where $k = 1, 2, ..., DxM$. The algorithm then begins with the assumption that a single path is to be found. A search is conducted over all values of k, until that basis is identified which minimizes J_{DML}. The index k_1 is retained in the path history, for use in subsequent searches. Once the index of the first path has been identified, the model order is incremented to consider a measurement composed of two discrete arrivals. The first arrival, found in the previous iteration, is assumed to be fixed at the index k_1. The second arrival is found by another search over the remaining $DxM-1$ values of k. Once found, the second arrival is assigned the index k_2.

   At each stage of iteration, the minimum value attained by the cost function is given as:

$$
\mathbf{J}_{DML_{min}} = m - \mathbf{p}^H \mathbf{R}^{-1} \mathbf{p}, \tag{9.18a}
$$

where $\mathbf{p}$ represents the cross correlation between each of the basis functions and the measurement matrix and $\mathbf{R}$ represents the correlation between the individual basis functions. Also,

$$
m = \sum_{k=1}^{101} \sum_{n=1}^{8} |v_{kn}|^2 \tag{9.18b}
$$

is the total energy contained in the measurement.

The process of incrementing the model order, followed by finding an additional path, is repeated until the desired number of paths have been identified. Since the sequential algorithm seeks to minimize J_{DML} at each stage, paths are identified in order of decreasing energy. This property provides a convenient means to exit the search, by monitoring the optimal value of J_{DML} achieved at each iteration. Once the residual drops below, say *10%* of *m*, we may assume that the model has accounted for most of the energy contained in the measurement, and the algorithm may be terminated. The recursive search requires $D \times M \times L$ evaluations of the cost function to identify L paths. However, in the recursive algorithm, these evaluations assume the order of the number of paths found. For example, searching for the *L-th* path requires evaluation of $D \times M$ systems of L linear equations. Although far more computation is required than the serial search, the algorithm is still quite practical. The *DML* algorithm is considered in more detail in [Tin01], where, using sample measurements in the anechoic chamber, the spatial and temporal resolutions are determined to be $2°$ and 1 ns, respectively.

Figure 9.14 shows the result of a typical simulation run for a case consisting of 20 paths. In Figure 9.14, the known path parameters are represented by an open circle, whereas the estimates using SFP or DML are represented by a cross-hatch. The SFP technique finds many of the paths exactly, but is seen to encounter difficulty for paths that are coincident in time or angle of arrival. By contrast, the DML algorithm identifies all paths exactly. This performance is to be expected so long as high measurement signal-to-noise ratio is maintained.

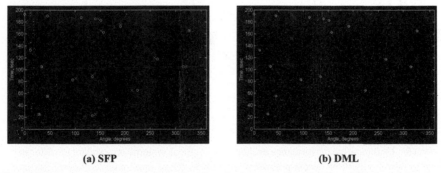

(a) SFP   (b) DML

Figure 9.14 Results of simulation of a 20-path channel and estimation of its location using, (a) SFP algorithm, (b) DML algorithm.

9.5 TOA Positioning in the Absence of Direct Path

Figure 9.15 shows a sample comparative performance evaluation of typical RSS- and TOA-based positioning algorithms on a square route in a corridor of an office building with three access points[6]. The route for a mobile device, location of APs and two large metallic objects blocking the direct paths, and the areas for which the direct paths are blocked are identified separately in both parts of the figure. Figure 9.15(a) shows the results of RSS-based localization using a maximum likelihood Kernel algorithm for a system with a bandwidth of 25 MHz. Figure 9.15(b) shows a similar result obtained for TOA-based localization using RLS algorithm for an UWB system with a bandwidth of 500 MHz. Performance of the RSS-based system appears moderately good but consistent over the entire route regardless of the propagation condition for the direct path. The TOA system, without using any signature database, performs better than the RSS system in the areas in which direct path is not blocked for any of the access points. When we have one blocked path, performance degrades noticeably, and with two paths blocked, we have extremely large errors. This example clearly displays sensitivity of the TOA to availability of the DP and the need for algorithms to mitigate the bipolar behavior of the TOA-based positioning systems in the absence of DP.

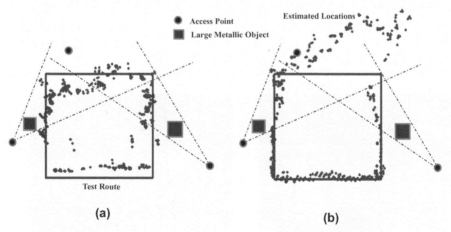

(a)   (b)

Figure 9.15   Comparative performance evaluation of typical systems on a square corridor in an office building with three access points using, (a) RSS based maximum likelihood Kernel algorithm with a bandwidth of 25 MHz, (b) using RLS algorithm and a TOA based UWB system with a bandwidth of 500 MHz.

---

[6]Physical map of this environment is shown in Figure 6.31.

For the practical implementation of algorithms to mitigate the effects of blockage of DP, we need the intelligence to discover the absence of DP. Detection of occurrence of these events is a challenging problem by itself. In WB and UWB measurements of the TOA, in addition to TOA of the first path, we can also measure the power of and TOA of individual multipath components. An empirical model relating the received power in multipath components of an UWB signal to occurrence of large ranging errors caused by undetected DP conditions is presented in [Als09]. In [Hei09], a neural network algorithm is used for detecting occurrence of undetectable DP conditions and refining the range estimate using several features of the received multipath profile. With the intelligence about the occurrence of undetected DP conditions, we need algorithms to mitigate their effects.

Diversity techniques such as time diversity, frequency diversity, and space diversity using multiple antennas are common basis for development of algorithms to mitigate the effects of multipath for communication systems. These traditional algorithms designed for radio communication are not effective in mitigating large ranging errors resulted from blockage of the DP. Two promising approaches to precise indoor localization in the absence of DP are localization exploiting non-direct paths and localization using extra RPs [Pah06]. The first approach measures the distance based on differential changes in TOA and DOA of any path. We refer to this approach as *ranging using multipath diversity*. The second approach uses redundancies in having extra antennas to bypass the effects of bad measurement from a blocked DP link with good measurements in links with DP. We refer to this approach as *localization using space diversity*. In opportunistic localization using existing wireless infrastructures, often we have more than three RPs to position a device. The redundancy in the minimum number of RPs provides for *space diversity* for localization of the target device. In ad-hoc wireless sensor networks, devices can communicate among themselves and with the RPs. Measurement of the distance among the devices provides additional redundant spatial information for positioning. The special diversity among the quality of these distances allow cooperative localization in the absence of DP. We discuss algorithms used for these three approaches in the remainder of this section.

9.5.1 Ranging Using Multipath Diversity

As a device moves the length and angle of arrival of all paths changes. Therefore, every path carries information on motion of the device. In this section, we present a technique for using multipath diversity to navigate a device, when the direct path is blocked by an object. Figure 9.16 shows a simple scenario to explain the basic concept behind using multipath diversity for TOA-based localization in a room. A reference point (RP) antenna is installed in the left top corner of a room with a large metallic object in the middle lower part of the room. A mobile device using TOA ranging moves along a straight route goes around metallic object when it approaches the object. A mobile moves towards the metallic object from top of a route until it turns around the object, we have an LOS connection between the RP and the mobile. In the middle of the lower part of the route around the object, the direct path is blocked and the mobile encounters an OLOS condition and the signal begins to arrive first from a first-order reflected path and then from a second-order reflected path. This is a simple two path scenario and we use it to explain the principles of ranging using multipath diversity.

Figure 9.17 illustrates the basic principle underlying the relationship between the TOA of the direct path and a path reflected from a wall, for a simple two-path scenario. This figure is focused on the portion of an arbitrary route behind a blocking object. We define the *x-coordinate* as the direction of movement of the mobile device. The length, representing the

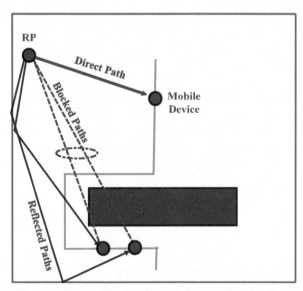

Figure 9.16   Basic concept of localization using multipath diversity. In locations where the direct path is blocked we use changes in TOA and DOA of a reflected path to track the location of the mobile device.

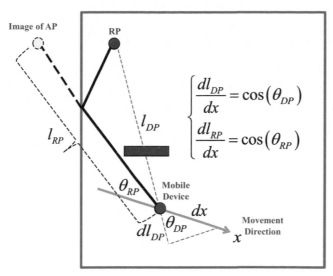

Figure 9.17   Basic geometric principles of navigation using a reflected path when the direct path is blocked.

TOA, of the direct path and the reflected path are represented by (l_{DP}, l_{RP}), respectively. The differential changes in the movements of the mobile device and differential change of the length of the direct path are related by:

$$dl_{DP} = dx \cos (\theta_{DP}), \tag{9.19a}$$

where θ_{DP} is the DOA of the direct path with respect to the direction of movement of the mobile. As shown in Figure 9.17, the same relationship between DOA of a path with respect to the direction of movement, differential changes of the length of a path, and differential changes of distance in direction of movement exists for the first-order path as well. A reflected path is indeed a direct path between the image of the RP with respect to the reflecting wall and the mobile. Therefore, knowing the angle of arrival of a reflected path with respect to direction of movement, we have:

$$dl_{RP} = dx \cos (\theta_{RP}). \tag{9.19b}$$

This relation can be extended to any number of reflections. For example, as shown in Figure 9.18, the second-order reflected path is a direct path between the image of the RP, and Equation (9.20b) still holds and we can calculate the differential changes in the amplitude of that path using the angle of arrival of that path with respect to the direction of movements. Therefore, in general, we

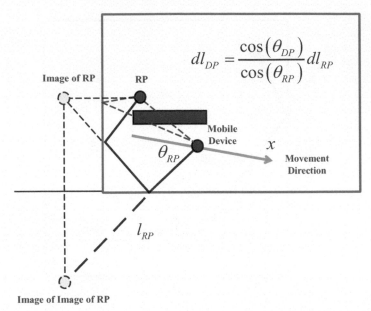

Figure 9.18   Relation between differential TOA and the differential time of arrival of a second order reflected path.

can calculate the differential changes in distance using differential changes of any path and angles of arrival of the path using:

$$dl_{DP} = \frac{\cos(\theta_{DP})}{\cos(\theta_{RP})} dl_{RP}. \tag{9.20a}$$

The same concept in terms of measurements of TOA is then given by:

$$d(TOA_{DP}) = \frac{\cos(\theta_{DP})}{\cos(\theta_{RP})} d(TOA_{RP}). \tag{9.20b}$$

In other words, knowing the DOA, θ_{RP}, between an arriving path and direction of movement and the angle, θ_{DP}, between the direction of movement and the direct path, we can estimate the differential changes in TOA of the direct path from changes in the TOA of a reflected path. If the receiver is capable of tracing angles of arrival of multiple paths, results of tracking using different paths can be combined for estimation of the TOA of the direct path with an enhanced performance. This technique is applicable to any localization and tracking system, from GPS to the radars operating in open spaces to indoor geolocation using WB or UWB signals. Application for indoor geolocation is more challenging because we have numerous paths with short life time [Pah06]. Emerging massive MIMO systems for wireless communications use multiple streaming, and for that, they measure the DOA, which is also useful for precise positioning.

9.5.2 RW-RLS Algorithm for Spatial Diversity

Figure 9.19 shows the positioning scenario of Figure 9.15(b) in a typical office, with three RPs and a mobile device moving around a square route, over the detailed layout of the building. The red line in the figure shows the location estimates using the estimated distance from the three RPs. Whenever the direct path is present from all three RPs, for example, in the lower and right hand routes, the ranging error is small and when we have at least one path blocked, performance degrades substantially. Therefore, if we have more than the minimum number of RP antennas, we can simply avoid links with blockage and achieve precise localization. Adding more antennas increases the special diversity of the received signals at the mobile device and that additional spatial diversity enables us to mitigate the effects of undetected DP conditions. To benefit from this type of special diversity for TOA positioning, the infrastructure of RPs needs to have redundancies and the positioning

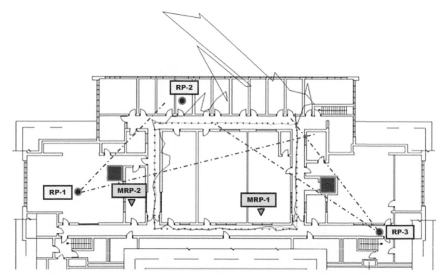

Figure 9.19 Demonstration of space diversity on the 3rd floor of the Atwater Kent Laboratory.

algorithm needs to have intelligence to identify the occurrence of a blockage in a link.

The existing wireless infrastructures in indoor and urban areas for opportunistic positioning often provide the needed redundancy to take advantage of special diversity in locating a target device. Figure 9.20 provides an abstract scenario in a grid deployment of infrastructure in an indoor area to explain this phenomenon. The target device in Figure 9.20 has four radio links with four access points (APs), and one of these links is blocked by a metallic object. However, three of the links have good quality of estimate of the TOA, and consequently, they can provide for good position estimate. In practice, we need to an algorithm to exploit this opportunity. The residual weighted RLS (RW-RLS) algorithm, explained in Section 7.4.2, is an example of a good algorithm for this scenario. This algorithm first uses RLS to estimate the location of the device using all possible three and four combinations of RPs and records the residual value of the location estimates with different position estimates of RPs. Then, it uses the weighted centroid over all estimated positioning from different combinations to find the final estimate of the position. The weight is determined from the inverse of the residual values. The intelligence of this algorithm is embedded in the assumption that if we have three good links, residual value of RLS is small and when we have blocked DP, the residual value becomes large. Therefore, smaller residues associate with better estimates of the location.

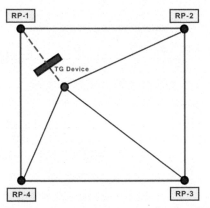

Figure 9.20   Operation scenario for residual weighted recursive least square (RW-RLS) algorithm for localization in the absence of DP.

9.5.3 Cooperative Localization for Spatial Diversity

In wireless ad-hoc and sensor networks, target devices are able to communicate with each other as well as the RP infrastructure of the wireless network. As a result, when we want to locate a mobile terminal in an ad-hoc or a sensor network, in addition to the distances from the respective fixed RPs, we can also use the location of other users as mobile reference points (MRP). We refer to this approach as cooperative localization since the localization is conducted through a cooperative method[7]. To demonstrate the effectiveness of cooperative localization to mitigate the bipolar behavior of the TOA-based ranging in the absence of DP, we return to our previous example in Figure 9.19. In this example, we had three fixed RPs and a mobile device moving along a square route. To change this to a cooperative localization scenario, consider we have two other mobile reference points, MRP-1 and MRP-2, which are located in good positions, where each of them has three direct path connections to the fixed RPs. As shown in the figure, when we use the three fixed RPs to estimate the location of MRP-1 and MRP-2, we have very good estimated locations for them. In this ad-hoc scenario, the target receiver moving along the loop route can also measure its distances from MRP-1 and MRP-2 with good quality. The green line in Figure 9.19 shows the estimate of location using RLS algorithm with the estimated locations of MRP-1 and MRP-2 and the actual location of AP-2. The drastic improvement in the precision of localization is caused by opportunistic utilization of the redundancy of the ad-hoc sensor networks. The positioning algorithm for this scenario needs the intelligence to differentiate paths with blockage from the other paths. The RW-RLS can be used for this scenario as well.

In the previous section, we used this algorithm to locate a target device when we have four RPs. In the cooperative localization example of Figure 9.19, we have three mobile devices and three RPs; therefore, we need to solve a six-dimensional matrix instead of a two-dimensional matrix. An increase in the dimension increases the convergence related aspects of the RLS algorithm. As the number of mobile devices increases RLS becomes less appealing and we may think of an iterative cooperative localization algorithm using spatial diversity, which locates mobile devices one by one. Our specific example is called CLOQ Cooperative LOcalization for Optimum Quality (CLOQ), and it was originally designed for first responder ad-hoc network operation scenarios [Als06][Pah11]. The algorithm uses characteristics of the

---

[7]For more details and derivation of CRLB for cooperative localization, see Section 8.2.5.

radio propagation for UWB signals to model the relation between the RSS and the quality of estimate of the TOA [Als08].

Figure 9.21 provides the application scenario and the basic concept for the CLOQ. The ad-hoc network consists of four fixed reference points in four sides of a building with access to the GPS and seven mobile terminals inside the building. Each mobile terminal can establish a link with some of the other mobiles and some of the fixed RPs. Associated with each measured TOA of the first path for each link, we can also measure other characteristics of the channel multipath profile such as amplitude and TOA of all multipath components. Using these parameters, we can calculate critical features of the received signal such as the power of the first path, total received power, and rms delay spread of the channel profile. Using the features, we can assign score to each link and refer to that as quality of the link (QoL), for example, we may give them a score of 1–3 according to the received signal strength of the first path [Pah11]. The first path and total received power with strong values may receive the score of one and the very weak first paths and total power may receive a score of 3, anything else may receive a score of 2. Then, we assign a Quality of Estimate (QoE) to each node. The QoE of the fixed nodes are zero because we know their exact location, to calculate the QoE of any other node we find the three links with lowest score for each and score

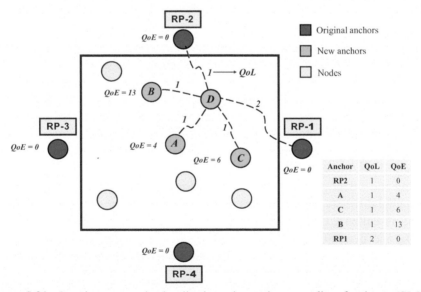

Figure 9.21   Iterative cooperative localization using optimum quality of estimate (CLOQ) using calculation of quality of link (QoL) and Quality of Estimate (QoE) scoring.

them by adding the value of 1 and 2 links linear and the 3 rated links by square of its index 3 × 3 = 9. This way, a node with one good link, one medium link, and one bad link (1, 2, 3) will have a QoE of (1 + 2 + 3 × 3 = 12). After assigning minimum QoE to all nodes, we begin the localization process.

In the localization process, first, we locate all nodes with at least three connections with the fixed reference points (nodes A, B, and C) in this case. For the localization of other nodes, for example D, we consider all of its links and the quality of estimate of the node that supports that link to create a table, shown in Figure 9.21. The table first sorts based on the QoL and whether the QoE is the same, based on the QoE of the connecting link to select the best anchors for localization (RP2 and nodes A and B). After that, D becomes an anchor and localization continues with other mobile nodes. With any new set of link measurement for a mobile node, this process continues and the network continually updating QoL, QoE, and the location estimate.

Appendix 9.A: MATLAB for Super-Resolution Algorithm (Prepared by Yunxing Ye)

```
clear;
clc;
close all;
    c=3e8;        %speed of light
    Dist_ft=[114 117 125.3 100 67.1 85.4 42.4 33.5 30.4 80.6
80.2 111 89.4 25 20.6]; %distance vector in feet
    dist=Dist_ft*0.3048;  %distance vector in meters
    error1=zeros(4,15);    %oringinalize the ranging error
matrix for the CZT method
    error2=zeros(4,15);    %oringinalize the ranging error
matrix for the MUSIC FCM method
    error3=zeros(4,15);    %oringinalize the ranging error
matrix for the EV/FCM method
    error4=zeros(4,15);    %%oringinalize the ranging error
matrix for the EV/FBCM method

% for q=1:4                %loop used for investigaint the
bandwidth effect
% disp(q)
% for iter=1:15               %loop used for different measurement
                        %points
        for q=1:1
        for iter= 3:3            % used to plot when looking at
the 3rd point when bandwidth is 80MHz
```

```
    pos_number = int2str(iter);      % string value of the
position number
    file_pre   = ['pos',pos_number]; % the prefix of the file
to open
    filename   = [file_pre,'.raw'];  % the complete file name
to open
    eval(['load ' filename]);            % load the
corresponding file
    meas_i     = eval(file_pre);     % the contents of the
file are now in meas_i

    f    = meas_i(:,1)';
    % first column of meas_i is the frequency in Hz
    mag = meas_i(:,2);
    % second column of mes_i is the magnitude in dB
    phs = meas_i(:,3);
    % third column of meas_i is the phase in degrees
    Fs = f(2)-f(1);                     % get the frequency
sampling interval
    Ts = 1/Fs;                              % corresponding
time sampling interval

    linmag  = 10.^(mag/20);                       % compute the
linear version of the frequency magnitude
    cmp_mag = linmag .* exp(j*phs*pi/180);
    % compute thecomplex frequency response

    % used for parse the frequency domain segment data
    if q==1
    cmp_mag=cmp_mag(1:40);    %20MHz Bandwidth
    f=f(1:40);
    mag=mag(1:40);
    phs=phs(1:40);
    Nf=length(f);
    elseif q==2
    cmp_mag=cmp_mag(1:80);    %40MHz Bandwidth
    f=f(1:80);
    mag=mag(1:80);
    phs=phs(1:80);
    Nf=length(f);
    elseif q==3
    cmp_mag=cmp_mag(1:160);    %80MHz Bandwidth
    f=f(1:160);
    mag=mag(1:160);
    phs=phs(1:160);
    Nf=length(f);
```

```
    elseif q==4
    cmp_mag=cmp_mag(1:240);    %120MHz Bandwidth
    f=f(1:240);
    mag=mag(1:240);
    phs=phs(1:240);
    Nf=length(f);
    end


    Nt   = 1601;                   % time response will have
1601 points
    t1   =   0 * 1e-9;             % the start time will be 0 ns
    t2   = 800 * 1e-9;             % the stop time will be 800 ns

    % Initialisation requirements for IFFT using ChirpZ
technqiue (czt)
    m    = Nt;
    w    = exp(-j*2*pi*(t2-t1)/(m*Ts));
    a    = exp(j*2*pi*t1/Ts);



    han = hanning(Nf);                      % compute the Hanning
window to put on frequency response
    cmp_han = cmp_mag .* han;               % frequency response
now windowed with the Hanning window

    % Computation of the time response and time axis using
Chirp Z
    tz_han  = (45/23)*conj((1/Nf)*czt(conj(cmp_han),m,w,a));
% complex time response
    tz       = ((0:length(tz_han)-1)'*(t2-t1)/length(tz_han))
+ t1;   % time values
    timz     = abs(tz_han); % magnitude of the time response

%~~~~~~~~~~~~~~~~~~~~~~~~~~~~~~~~~~~~~~~~~~~~~~~~~~~~~~~~~~~~~~
    % Extraction of the taps using a peak detection algorithm
    % The peak must be above a thresold to be valid
    dynamic_range_db = 30;                          % dynamic
range to consider in dB
    dynamic_range    = 10.^(dynamic_range_db/20);   % convert
to linear scale
    max_mag          = max(timz);                   % find
the maximum magnitude of the time response
    max_tim          = tz(find(timz == max_mag));   % find
the time at which it occurs
    threshold        = max_mag/dynamic_range;          % the
threshold to use in 30 dB below the maximum
```

```matlab
    % Algorithm that looks for a peak, checks if above
threshold
    % increments and stores the magnitude in taps_mag
    % and the time of the peak in taps_tim
    ntaps = 0;
    for k=2:Nt-1
        if timz(k)-timz(k-1)>0 & timz(k+1)-timz(k)<0
            if timz(k)>threshold
                ntaps              = ntaps + 1;
                taps_tim(ntaps) = tz(k);
                taps_mag(ntaps) = timz(k);
            end;
        end;
    end;


    %~~~~~~~~~~~~~~~~~~~~~~~~~~~~~~~~~~~~~~~~~~~~~~~~~~~~~~~
    % test the music algorithm with Forward correlation matrix
    [Ms,index1,t1]=imusic(cmp_mag,f,1600,1,1);%MUSIC method
with FCM correlation matrix
    [ee,index2,t2]=imusic(cmp_mag,f,1600,0,1);%EV method with
FCM correlation matrix
    [ff,index3,t3]=imusic(cmp_mag,f,1600,0,0);%EV method with
FBCM correlation matrix


    %~~~~~~~~~~~~~~~~~~~~~~~~~~~~~~~~~~~~~~~~~~~~~~~~~~~~~~~
    error1(q,iter)=abs(taps_tim(1)*c-dist(iter));    %calculate
the distance measurement error using CZT
    error2(q,iter)=abs(t1(index1(1))*c-dist(iter));  %calculate
the distance measurement error using MUSIC
    error3(q,iter)=abs(t2(index2(1))*c-dist(iter));  %calculate
the distance measurement error using EV/FCM
    error4(q,iter)=abs(t3(index3(1))*c-dist(iter));  %calculate
the distance measurement error using EV/FBCM
    end
    end

    figure(4)
    h1=plot(tz*1e9,timz/max(timz),'linewidth',1.5);
    hold on
    plot(taps_tim(1)*1e9,taps_mag(1)/max(timz),'b^','markersize'
      ,6);
    hold on
    h2=plot(t1*10^9,Ms/max(Ms),'g-.','linewidth',1.5);
    hold on
```

```
h=plot(t1(index1(1))*10^9,Ms(index1(1))/max(Ms),'ro');
set(h,'markersize',5);
hold on
h4=plot(t2*10^9,abs(ff)/max(abs(ff)),'k--','linewidth',2);
hold on
h5=plot(t2(index2(1))*10^9,abs(ff(index2(1)))/max(abs(ff)),
'ko');
set(h5,'markersize',5);
hold on
h3=plot([dist(3)/c*10^9,dist(3)/c*10^9],[0 1],'r-.');
legend([h1,h2,h4,h3],'CZT method','MUSIC with FCM','EV with
FBCM','actual distance');

clear taps_tim taps_mag                    % clear these
variables once fininshed

%%%% super resolution TOA estimation
%%%% Using Music algorithm
function [Ms,index,tau]=imusic(Zf,freq,Nt,flag1,flag2)
%Zf ==> the frequency domain response
%freq==> discrete frequency points
%Nt==> number of music spectrum in time domain
%flag1==> flag1==1 use normal music algorithm
% flag1==0 use EV method
%flag2==> flag2=1 use forward correlation matrix (FCM)
% flag2=0 use forward and backward correlation matrix
(FBCM)
%Output
%Ms==> time domain Music spectrum
%index==> time domain peaks found by the peak detection
algorithm
%tau==>time

%%%% define parameters %%%%
c=3*10^8; %speed of signal
j=sqrt(-1);

%%%% construct the correlation matrix
N=length(Zf);
fd=(freq(N)-freq(1))/(N-1); %frequency interval

%~~ codes used for contructing the correlation matrix~~
method refer to
%Xingrong's paper
L=floor(2/3*N);
M=N-L+1;
```

```
  X=zeros(L,L);
  x=zeros(L,1);
  for k=1:M
    x=Zf(k:1:k+L-1);
    X=x*x'+X;
  end
X=1/M*X;        %Forward correlation matrix
%%%%%%%%%%%%%%%%%%%%%%%%%%%%%%%%%%%%%%%%%%
if flag2==0
%%%using FBCM smoothing%%%%

J=fliplr(eye(L));
X=1/2*(X+J*conj(X)*J);    %Forward backward correlation matrix
else
end


%%%%%eigene decomposition
[V,D] = eig(X);
%%%%%%%%%%%%%%%%%%%%%%%%%%%%%%%

%%determine the number of Lp Using MDL method%%%
p=abs(sort(diag(D),'descend'));  %sort the eigenvalues in
descending order
o=abs(diag(D));                        %eigenvalues in ascending
order

%Using MDL criteria to determine the number of signals (the
dimension of %signal subspace )
for k=0:L-1
temp1{k+1,:}=p(k+1:L).^(1/(L-k));
temp2{k+1,:}=p(k+1:L);
tem1(k+1)=prod(temp1{k+1,:});
tem2(k+1)=sum(temp2{k+1,:});
temp3(k+1)=(tem1(k+1)/(1/(L-k))/tem2(k+1));
MDL1(k+1)=-log((temp3(k+1))^(M*(L-k)))+1/2*k*(2*L-k)*log(M);
% used for FCM
MDL2(k+1)=-log((temp3(k+1))^(M*(L-k)))+1/2*k*(L+k+1)*log(M);
% used for FBCM
end

if flag2==1
[y,index]=min(MDL1);
else
[y,index]=min(MDL2);
end
```

```
Lp=index-1;
%%%%%%%%%%%%%%%%%%%%%%%%%%%

%%%%%The eigenvectors corresponding to noise eigenvalues%%%%
   q=V(:,1:L-Lp);
%%%%%%%%%%%%%%%%%%%%%%%%%%%%%%%%%%%%%%%%%%%%%%%%%%%%%%%%%%%%

v=zeros(L,Nt);
tau=zeros(1,Nt);    %time

%~~calculate the music and EV spectrum
for ii=1:Nt
    tau(ii)=0+800*10^(-9)*(ii-1)/(Nt-1);   %time domain sample
points
    v(:,ii)=exp(-j*2*pi*fd*[0:1:L-1]'*tau(ii));
    Ms1(ii)=1./sum((abs(q'*v(:,ii)).^2));  %orin50 MUSIC
spectrum
    Ms2(ii)=1./sum(1./o(1:L-Lp).*abs((q'*v(:,ii)).^2));  % EV
method spectrum
end

if flag1==1
   Ms=Ms1;
else
    Ms=Ms2;
end

if flag2==1
   dynamic_range_db = 15;                          % dynamic
range to consider in dB used for MUSIC
else
   dynamic_range_db=35;                            % dynamic
range to consider in dB used for EV method
end
   dynamic_range     = 10.^(dynamic_range_db/20);   % convert
to linear scale
   max_mag           = max(Ms);                     % find the
maximum magnitude of the time response
   max_tim           = tau(find(Ms== max_mag));     % find the
time at which it occurs
   threshold         = max_mag/dynamic_range;       % the
threshold to use in ** dB below the maximum

   %~~peak detection algorithm
   peak_index=0;
   count=0;
```

```
for k=2:Nt-1
    if Ms(k)-Ms(k-1)>0 & Ms(k+1)-Ms(k)<0
        if Ms(k)>threshold
            if count == 0
                peak_index = k;
                count = 1;
                else
                peak_index = [peak_index, k];
            end;
          end;
        end;
    end;

index = peak_index;

return

%%%plot the mean and std deviation %%%%
x1=[19 39 79 119];
x2=[20 40 80 120];
x3=[21 41 81 121];
x4=[22 42 82 122];
y1=mean(error1');
y2=mean(error2');
y3=mean(error3');
y4=mean(error4');
std1=std(error1');
std2=std(error2');
std3=std(error3');
std4=std(error4');
figure(1)
h1=plot(x1,y1,'-rs');
set(h1,'markersize',5);
hold on
plot(x1,y1-std1,'r+',x1,y1+std1,'r+');
hold on
h2=plot(x2,y2,'-g^');
set(h2,'markersize',5);
hold on
plot(x2,y2-std2,'g+',x2,y2+std2,'g+');
hold on
h3=plot(x3,y3,'-bo');
set(h3,'markersize',5);
hold on
plot(x3,y3-std3,'b+',x3,y3+std3,'b+');
hold on
```

```
h4=plot(x4,y4,'-mo');
set(h4,'markersize',5);
hold on
plot(x3,y3-std3,'y+',x3,y3+std3,'y+');
hold on

for ii=1:4
plot([x1(ii) x1(ii)],[y1(ii)-std1(ii) y1(ii)+std1(ii)],'r');
hold on
plot([x2(ii) x2(ii)],[y2(ii)-std2(ii) y2(ii)+std2(ii)],'g');
hold on
plot([x3(ii) x3(ii)],[y3(ii)-std3(ii) y3(ii)+std3(ii)],'b');
hold on
plot([x3(ii) x3(ii)],[y3(ii)-std3(ii) y3(ii)+std3(ii)],'y');
end
legend([h1 h2 h3 h4],'CZT','music/fcm','EV/fcm','EV/FBCM');
xlabel('Bandwidth (MHz)');
ylabel('measurement error(m)');
title('comparison of different TOA estimation accuracy versus
bandwidth');
```

Assignments for Chapter Nine

Questions

1. Why algorithms used for RSS-based localization are more complex than algorithms used for TOA-based localization?
2. What are the similarities between spectral estimation and localization using the TOA that makes it easy to use similar algorithms in both applications?
3. Name four techniques to mitigate the effects of multipath on TOA-based localization.
4. What are the challenges for using phases of individual carriers in a multi-carrier system to eliminate the effects of multipath in indoor areas?
5. What is cooperative localization and how it counteracts the effects of multipath?
6. What are the challenges for localization using paths other than direct path when it is applied to indoor areas? Why these challenges do not exist for outdoor navigation using satellites as a reference point.

Problems

Problem 9.1:
Derive Equation (9.24d) using Figure 9.18.

Problem 9.2:
We want to use a DSSS system to measure distance between two points in a multipath environment using the TOA obtained from the peak of the autocorrelation function (ACF) of the received signal.

a) If the range of distances between the transmitter and the receiver is up to 100 m, what is the minimum time between the peaks of the ACF of the DSSS signal?

b) If we want a time resolution of $\Delta t = 1$ ns, what would be the bandwidth requirement for the system?

c) What is the necessary frequency resolution Δf to accommodate for the minimum period for pulse transmission? How many carriers we should measure within the bandwidth to accommodate the requirements?

Projects

Project 9.1:
For topology the mesh network shown in Figure P9.1, we know the location of 3 RPs and want to find the location of 2 unknown TGs.

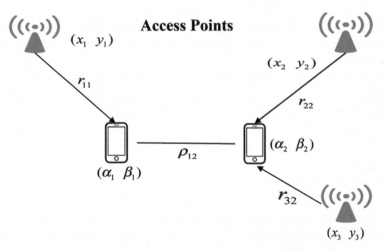

Figure P9.1   Topology of the mesh network used for Project 9.1.

a) If we have a noisy measurement of the distances between the nodes shown in the figure and the standard deviation of estimating each distance is σ, give the Fisher Matrix for the network.

b) Assuming the three RPs are located at $(0,0)$, $(10,0)$, and $(10,-5)$, the actual TGs are located at $(3, -2)$ and $(7, -2)$, variance of the DME is 8 m, and distances measured between the points are

    node UP-1 and AN-1 is 3 m
    node UP-2 and AN-2 is 4 m
    node UP-2 and AN-3 is 4 m
    node UP-1 and UP-2 is 3.5 m

Calculate the CRLB for the system.

c) Use the RLS algorithm to determine the locations of the terminals and calculate the DME for each location. Compare the errors from the algorithm with the CRLB for each point

10

Sensor Fusion for Hybrid Localization

10.1 Introduction

Location sensors measure certain characteristics related to the location of a target (TG) object, such as a device, person, robot, or drone, to position that object on a map. The map could be a universal map of the world, such as a Google map, a local map, such as layout of a building, or sketch of movement of a tiny robot on a CAD Scan or X-ray pictures inside of a human body. The sensors could either provide information on the distance from a reference point (RP) to help finding the *absolute location* of the TG on the map, or they provide the velocity and direction of movement for finding the displacement or *relative location* of the TG. Since absolute RF localization is the most popular and the most complex element of positioning systems, we studied them in Chapters 2–9 of this book. Our discussions on RF-based positioning in these chapters were focused on sensing the RSS, TOA, and DOA for positioning in indoor and urban areas. We discussed the behavior of these sensors, we introduced popular systems using them, we derived performance bounds for them, and we introduced applied positioning algorithms for them. This chapter provides an overview of the behavior of sensors for finding the velocity and direction of movements for relative localization and presents hybrid positioning algorithms for integration of absolute and relative location information.

In the remainder of this section, we introduce the concept of universal and local coordinates for integration of absolute and relative localization. Then, we classify sensors in a typical smart phone into mechanical, electromagnetic, and environmental sensors. In the following three Sections (10.2–10.4), we provide details of these classes of sensors and discuss their effectiveness in positioning. In Section 10.5, we discuss geometric method for calculation of speed and direction of movements using RF sign posts and the similarities of pictures taken by a camera. Section 11.6 is devoted to the review of classical methods for estimation of speed using Doppler spectrum of an RF signal. In Section 10.7, we provide examples of using Kalman and Particle filters for fusion of results of absolute and relative positioning.

10.1.1 Coordinate System for Integration

The fundamental difference between the absolute and relative localization is that the absolute localization needs an infrastructure of RPs to position the TG, but in relative localization position of the TG is relative to its previous location. As a result, the way that positioning error is manifested in an experience using absolute and relative localizations is different. Figure 10.1 shows two examples of the results of localization using absolute and relative positioning systems on a square route in a typical office building. Figure 10.1(a) shows the results of absolute Wi-Fi localization. The estimated locations (dots) are randomly distributed around the path of movement (solid line) of the object. Figure 10.1(b) shows the results of localization using an Inertia Motion Unit (IMU) device measuring the speed and direction of

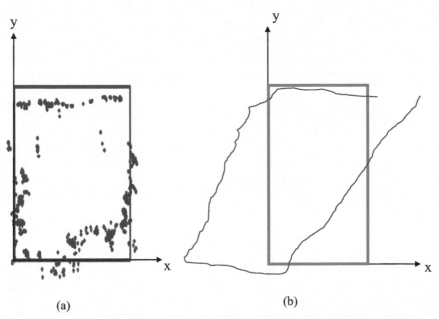

(a)   (b)

Figure 10.1   Results of localization with two classes of location sensing in a square route in an office building, (a) absolute localization using RSS of Wi-Fi access points, (b) relative localization using an IMU device.

the movement. The estimated path of movement follows a straight line but, because of errors in angle and speed of movements, the location estimate gradually drifts from the actual path. Hybrid algorithms combine the results of the two class of sensor to achieve a more reliable navigation. These algorithms use the absolute localization to reduce the drift of IMU and use the IMU result to smooth the location estimates of the absolute location.

In order to integrate the results of absolute and relative location, we need to define coordinate systems. In absolute localization, the position of a TG is represented by a vector form the origin of the coordinate of the fixed frame coordinate to the position of a device. In relative localization, the velocity and direction are vectors at the position of the TG in the moving coordinates of the TG platform. Figure 10.2 shows an example of the Cartesian coordinate systems for absolute localization in the Earth frame and relative localization in a smart phone platform. Figure 10.2(a) shows the fixed Earth 3D coordinate system, an acceleration or velocity vector in X-axis defines a vector tangential to the ground at the TG's current location pointing to the East. The vector in Y-axis is tangential to the ground at the TG's current location and pointing towards magnetic north. A vector in Z-axis points towards the sky and perpendicular to the ground at the location of the TG. Figure 10.2(b) shows a typical local coordinates of a smart phone: the x-axis is horizontal and points to the right side of the device, the y-axis is vertical and points up to the top of the device, and the z-axis points toward the outside of the

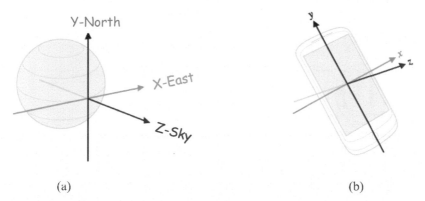

(a)   (b)

Figure 10.2   Cartesian coordinate system for absolute and relative localization, (a) the Earth frame fixed coordinate, (b) a local moving coordinate system of a smart phone for relative localization.

screen face. This is a moving coordinate system with respect to the world fixed coordinate and it changes as the user moves the screen position. The mechanical location sensors are installed in the device and they measure the velocity and direction in the device coordinates. We need to transfer these coordinates to the universal coordinates to integrate them with the results of absolute localization.

The most popular 3D representation of the Earth coordinates is a polar coordinate identifying a location on the surface of the Earth by three angles: longitude, latitude, and altitude, (θ, ϕ, ρ), shown in Figure 10.3(a). Longitude expresses angular deviation from east to west on the surface of the Earth in degrees, and latitude is the angular deviation from north to south in degrees. Elevation shows the distance from the center of the Earth and it is usually given with respect of the sea level height. In the aerial vehicle applications, elevation is referred to as altitude.

Ground navigation systems for outdoor and indoor areas track movements of the vehicles or robots on 2D maps such as layout of a floor of a building or 2D map of streets of an urban area in area for which altitude does not change substantially. Even in multi-floor 3D applications we use a 2D map with the floor number (a representative of altitude). Cartesian coordinates provide a better tool for visual tracking for most of popular applications in indoor and urban areas. Therefore, we usually convert longitude, latitude, and altitude to

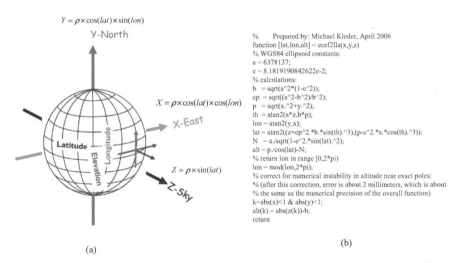

(a)   (b)

Figure 10.3 (a) Earth 3D polar coordinate system with elevation, longitude, and latitude, (b) MATLAB code for conversion of (lat, lon, alt) coordinates to Cartesian coordinates.

a Cartesian form using:

$$\begin{cases} X = \rho \times \cos(lat) \times \cos(lon) \\ Y = \rho \times \cos(lat) \times \sin(lon), \\ Z = \rho \times \sin(lat) \end{cases} \tag{10.1}$$

where ρ is the radius of the Earth. Figure 10.3(b) shows the MATLAB code to map the polar coordinates to 3D Cartesian coordinates using the above equations. The LLA2ECEF in MATLAB converts latitude, longitude, and altitude to earth-centered, earth-fixed (ECEF) Cartesian coordinates [Kle06].

10.1.2 Classification of Relative Location Sensors

Traditionally, more expensive relative localization systems are used for positioning in the GPS-denied environments such as inside the tunnels, jungles, indoors, and under the water, to navigate vehicles. With the emergence of inexpensive micro-electro-mechanical systems (MEMS) chips in smart devices, such as smart phones, inexpensive robots, and autonomous vehicles, other applications of the relative localization sensors for activity and motion detection have emerged in electronic gaming and health industry. Information provided by these sensors enables a device to enhance its positioning in areas in which RF positioning is challenged.

As an example, a typical smart phone carries RF-based GPS, Wi-Fi, Cell Phone, and Bluetooth chip sets, which support a variety of opportunistic absolute RF localization capabilities for different environments and precision requirement. In addition, they carry a number of other sensors. A typical smart phone carries the traditional relative positioning sensors such as gyroscope, magnetometer, accelerometer, and barometer as well as sensors for ambient light, temperature, and humidity, force of gravity levels, proximity, and step counters. These sensors can be beneficial for positioning applications to enhance precision of RF localization. We can use the data collected by these sensors for 3D localization, calculation of speed and direction of movements, and specifying the environment of operation. The data from all sensors is available as a resource to a positioning engineer to design different algorithms to achieve the precision requirements in the environment of operation.

Other platforms carry different sets of sensors. A ground robotic platform usually carries a camera similar to a smart phone, but the camera is always on because robots do not have battery restrictions of a smart phone. As a result, robots generally use the camera images and the so-called Simultaneous

Localization and Mapping (SLAM) algorithms [Bai06] to map the environment and localize the position of the robot on the map. Robots may also carry RF-based localization tools such as Wi-Fi, Bluetooth, RFID readers, and optical proximity check devices. The speed of the wheel of the ground robot provides an odometer and the robot may carry an accelerometer, gyroscope, magnetometer, or many other sensors useful for localization. Depending on the environment and precision needed for the application, a navigation engineer may use any or a set of theses sensors. For example, a vacuum cleaner robot may use SLAM with one or multiple cameras to create a map of the environment to be cleaned and the location of the robot in the map to provide an intelligent cleaning service with comprehensive coverage of the area to be cleaned. A robot operating in a warehouse for automated movement of the material on the warehouse floor may use an RFID reader and install RFID enabled floor tiles to support sub-meter accuracies in warehouse floors.

Table 10.1 provides a summary of typical sensors used in smart devices. To study these sensors, we categorize them into four groups: mechanical, electromagnetic, environmental, and geometric sensors. Mechanical sensors include accelerometer, gyroscope, step-counter, and odometer providing information on speed and direction of a device. The electromagnetic sensors are: gravity sensor and magnetometer, using principles of electromagnetics to provide for direction and relative distance. The environmental sensors measure relative distance from an obstacle using ambient temperature, air pressure (barometer), ambient light, and relative humidity. These sensors are used to sense closeness to crashing with an obstacle and changes in the environment of operation as a mobile passes through a non-homogeneous environment. Geometric sensors are sign-post counter and video cameras, which use RF signal from known RPs or similarities between images taken by a camera installed in a moving object to calculate the speed and direction of movements. In the following four sections, we provide an overview of these sensor.

10.2 Mechanical Location Sensor

As we discussed earlier, IMU devices are perhaps the most popular relative localization systems traditionally used in aircrafts, satellites, unmanned aerial vehicles (UAV), and missiles to support the control system to maintain stability of the motion. Another application of these devices is for navigation in GPS-denied environments such as inside the jungles, tunnels, water, or

Table 10.1 Typical sensors in an Android device

Sensor Name	Description	Type	Common Uses
ACCELEROMETER	Measures the acceleration force in m/s^2 that is applied to a device on all three physical axes (x, y, and z), including the force of gravity.	Mechanical	Motion detection (shake, tilt, etc.).
GYROSCOPE	Measures a device's rate of rotation in rad/s around each of the three physical axes (x, y, and z).	Mechanical	Rotation detection (spin, turn, etc.).
STEP_COUNTERS	Counts the number of steps taken by the person carrying the device.	Mechanical	Measuring travelled distance.
ODOMETER	Counts the number of turns of the wheel of a robot or a car.	Mechanical	Measuring travelled distance.
MAGNETOMETER	Measures ambient geomagnetic field strength along the x, y, and z axes in micro-Tesla (uT).	Electromagnetic	Creating a compass.
GRAVITY	Measures the force of gravity in m/s^2 that is applied to a device on all three physical axes (x, y, z).	Electromagnetic	Motion detection (shake, tilt, etc.).
AMBIENT_TEMPRATURE	Measures ambient room temperature in degree Celsius.	Environmental	Detecting environment changes
BAROMETER	Measures the atmospheric pressure in hPa (millibar).	Environmental	Elevation for floor numbering
AMBIENT_LIGHT	Measures ambient light level in SI lux.	Environmental	Detecting environment changes

Table 10.1   Continued

RELATIVE_HUMIDITY	Measures relative ambient air humidity in percent.	Environmental	Detecting environment changes
PROXIMITY	Measures distance from the sensor to the closest visible surface measured in centimeters.	Environmental	Mapping the environment
SING-POST_COUNTER	Counts the number of RF sign-post read by a device.	Geometric	Measuring speed and direction.
CAMERA	Measures the velocity and direction of movement of the device in all three physical axes (x, y, and z).	Geometric	Simultaneous localization and mapping (SLAM).

indoor areas. Later on, IMU devices have penetrated consumer device market in smart phone, robot, PC mouse, drone, interactive electronic game controllers, and even inside iBeacon sensors. IMU devices can be manufactured with a wide range of sizes and prices going from less than a dollar for a penny size MEMS chip to over hundred thousand dollars for complete navigation back pack unit.

IMU devices carry accelerometer, gyroscope, and sometimes magnetometer in one unit and use them for relative localization. The accelerometer and gyroscope are mechanical devices measuring the speed and direction of movement of the device. Magnetometer is an electromagnetic device used for measurement of angular rotations of the device against the gravity of the earth [V-Gyr09]. We discuss accelerometer and gyroscope in this section and magnetometer in Section 10.3.

Another popular mechanical sensor is the odometer used in devices moving on wheels such as vehicles and land robots. Odometers measure the number of rotations of the wheel in time to measure the speed of motion on the ground. Smart devices use step counter for the same purpose, to measure the speed of movement of the device on the ground. Step counters count the steps from changes to rapid changes of the gyroscope readings during the landing of the step of the person carrying the device.

10.2.1 Accelerometer

Accelerometer is a electro-mechanical sensor measuring the acceleration, the differential rate of variations of velocity, in meters per square second (m/s^2). This measurement is in the device's Cartesian coordinates. Figure 10.4 shows the basic concept behind the mechanical parts of an IMU device carrying a 3D accelerometer and a 3D gyroscope mounted on the same plate. Figure 10.4(a) shows the basic concept behind the mechanics of the accelerometer. A weight is hanging between two springs in a box and the displacement of the weight due to external forces moving the box is measured to calculate the acceleration in the direction of the movement. When we connect the fixed and moving parts to plates of a capacitance in a simple circuit, the movement changes the distance between the plates causing changes in the capacitance. Changes in the capacitance change the current of the circuit that we can measure and relate that to the displacement. The left side of Figure 10.4(c) shows how we can measure 3D acceleration of a plate carrying three accelerometer boxes placed in orthogonal directions.

Conceptually, an accelerometer measures the forces applied to the sensor in different directions and using Newton's second equation calculates the acceleration by dividing the force by mass of the sensor. When an IMU device rests in a plate on a table in parallel to the earth surface, the accelerometer of the device should read a magnitude of $g = 9.81 \; m/s^2$, the gravity of the earth. When the IMU device is in free-fall and therefore accelerating towards the ground at 9.81 m/s^2, its accelerometer should read a magnitude of 0 m/s^2.

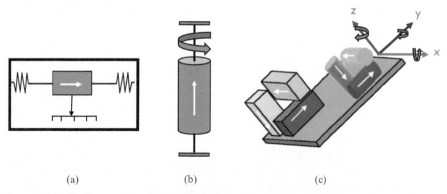

(a)   (b)   (c)

Figure 10.4 Basic concept of an IMU device with, (a) an accelerometer, (b) a gyroscope, with (c) the 3D implementation of the two devices on one plate.

In order to measure the real acceleration of the device, also referred to as **linear-acceleration**, the contribution of the force of gravity must be eliminated. Applying a high-pass filter to the results of accelerometer readings eliminates the slow changes of the readings caused by gravity and detects the fast changes introduced by movements of the device.

10.2.2 Gyroscope

Traditional gyroscope is a mechanical sensor measuring the differential rotation momentum around the local x, y, and z axes of the device's Cartesian coordinates in radians/second. MEMS devices implement electro-mechanical versions of a gyroscope. Figure 10.4(b) shows the basic concept behind the mechanical part of a gyroscope. A cylindrical mass is hanging around a central shaft like a wheel, as we change the direction, wheel tits and we measure the angular displacement of the wheel. The coordinate system is the same as the one used for the acceleration sensor. The right side of Figure 10.4(c) shows how we can measure the rotation in 3D around each of the three axes of the device by placing three gyroscopes in orthogonal directions. The output of the gyroscope is integrated over time to calculate a rotation describing the change of angles over time.

Rotation is positive in the counter-clockwise direction. That is, an observer looking from some positive location on the x, y, or z axis at a device positioned on the origin would report positive rotation if the device appeared to be rotating counter clockwise. If we refer to angular speed around the x-axis as ψ, angular speed around the y-axis as θ, and angular speed around the z-axis as ϕ (Figure 10.5(a)), Euler's transformation maps the accelerometer readings, $a(t)$, from the device coordinates to the universal coordinate using these angles:

$$\begin{bmatrix} a_X(t) \\ a_Y(t) \\ a_Z(t) \end{bmatrix} = \begin{bmatrix} \cos\theta\cos\psi & \sin\phi\sin\theta\cos\psi - \cos\phi\sin\psi & \cos\phi\sin\theta\cos\psi + \sin\phi\sin\psi \\ \cos\theta\sin\psi & \sin\phi\sin\theta\sin\psi + \cos\phi\cos\psi & \cos\phi\sin\theta\sin\psi - \sin\phi\cos\psi \\ -\sin\theta & \sin\phi\cos\theta & \cos\phi\cos\theta \end{bmatrix} \begin{bmatrix} a_x(t) \\ a_y(t) \\ a_z(t) \end{bmatrix}$$

$$(10.2)$$

Figure 10.5(b) shows a typical IMU device with both the device and the Earth-fix coordinates. To find the location from the above equation, we need to take a double integral. The first integration provides the velocity in each direction, and the second integral provides the distance traveled by the device in each direction.

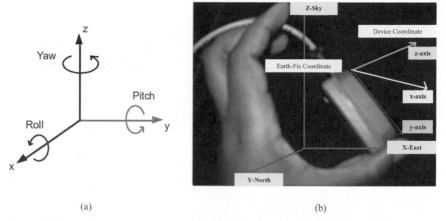

(a)   (b)

Figure 10.5   (a) Definition of angular rotations Roll, Pitch, and Yaw. Direction of rotation is counter clockwise if we look from the origin of the coordinate, (b) A typical IMU device with local device coordinates and universal earth-fix coordinates.

Example 10.1: (Motion prediction for a typical IMU unit)

Figure 10.6(a) shows a simple laboratory experiment moving the IMU device shown in Figure 10.5(b) forward and then turning to the right. This IMU device is a relatively expensive device purchased in mid-2000 for approximately $3500. Figures 10.6(b,c) show the three-dimensional readings of the gyroscope and accelerometer of the device. The gyroscope reads the pitch, yaw, and roll in radian/second. The only rotation with this scenario of motion is around the z-axis for the yaw angle and pitch and roll remain at zero. Acceleration only has reading in x and y coordinates because we have no vertical motion. These values changes up and down. Figures 10.7(a,b) show the results of first and second integral of the results for acceleration after applying Equation (10.2) of Euler's equation combining the results of accelerometer and the gyroscope. The two figures represent the speed and travelled distance recorded by the laptop. The linear speed that is the square root of the sum of squares of the speeds on the x and y axes should remain relatively constant representing the speed of movement. Figure 10.7(c) shows the trajectory of the movement of the device as predicted from the results of IMU sensors readings. Drift from the actual path of movement demonstrates the typical behavior of relative positioning systems.

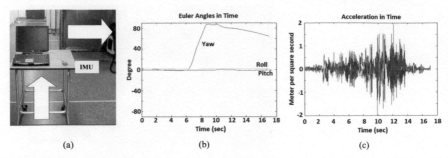

Figure 10.6   (a) Scenario of movement of an IMU device resting on a table and controlled by a laptop, (b) readings of angular movements of the gyroscope, (c) readings of the accelerometer.

10.2.3 Odometer and Step Counter Sensor

Odometers measure the traveled distance by counting the number of rotations of the wheel of a vehicle. Traditionally, odometers were mechanical devices, and in the modern time, the electro-mechanical odometers are widely used in cars and in land robots. Using the value of the distance in time, we can calculate the speed of a robot or a car very accurately. Using the angle of steering wheel to provide the direction of movements, we can form a relative localization system in GPS denied areas such as cars inside the tunnels or robots operating in indoor areas. Less expensive commercial robots may not have accurate measures of steering wheel, but the odometer results are reliable. In cases like that, one may use the gyroscope to find the direction [Ye12].

Step counters are the counter part of the odometers for handheld devices. Rather than counting the number of rotations of the wheel, step counter counts the number of steps taken by a pedestrian carrying a smart device. The number of steps is an integer and it resets to zero on a system reboot. The timestamp of the event is set to the time when the last step for that event was taken. The obvious application of this sensor is in fitness tracking applications, but it is also useful for indoor geolocation using smart devices.

Smart devices implement the step counter using sharp variations of the acceleration during the walk. Figure 10.8 shows hundred samples of the normalized measurements of magnitude of an accelerometer:

$$a(t) = \sqrt{a_x(t)^2 + a_y(t)^2 + a_z(t)^2} - g, \qquad (10.3)$$

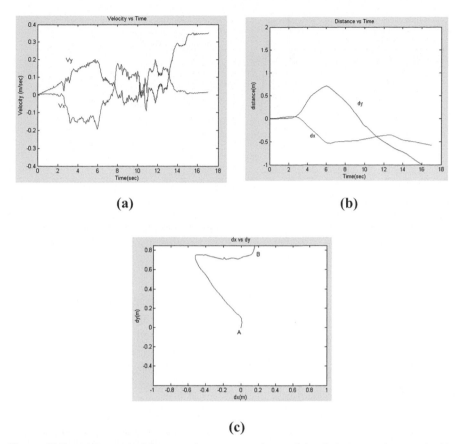

Figure 10.7 (a) Integral of the rotated movement after applying Euler's equation, (b) double integral representing motions in x and y axes in time, and (c) trajectory of movement of the device.

obtained from an Android phone carried by a walking person. In each step, first, the person increases accretion of the body torso using the landing foot. Then, the person reduces the speed by applying negative acceleration to land on the other foot. As a result, variation of the magnitude of the accelerometer measurements follows a zigzag pattern among the steps. We can detect the number of steps by counting the zigzags and that can be done either by peak detection or by detecting the zero crossing. With the peak detection algorithm (Figure 10.8(a)), we detect all peaks above a threshold and we select the highest in a window of time associated to the time for one step. For zero

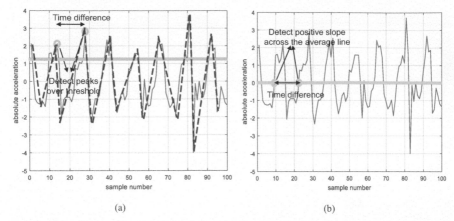

(a)   (b)

Figure 10.8   Principles of two algorithms for implementation of a step-counter using the results of accelerometer of an Android phone, (a) peak detection algorithm, (b) zero crossing algorithm.

crossing algorithm (Figure 10.8(b)), we find the intersect between the line representing the average acceleration and rising (or falling) parts of the curve.

Using the step counter and the time stamps associated with each step, the average speed and the average step size can be determined. To position the location of the person in a 2D map, we also need to measure the direction. Direction can be measured with the results of a gyroscope or a compass.

Example 10.2: (Positioning using step counter and gyroscope)

Figure 10.9 shows the results of an example relative localization system using the step counter and gyroscope of an Android phone at the square corridor route of the third floor of the Atwater Kent Laboratory, WPI. The schematic of the route with respect to the layout of the exterior walls of this scenario is shown in Figure 10.9(b). A user carrying the Android phone enters the floor from top right corner of the route and walks straight and then around the central corridors before turning back to the original location. User holds the smartphone in front of the torso in parallel to the ground. In this posture, the z-axis of the smartphone is towards the ceiling and the only rotation of the device is around this axis. Figure 10.9(a) shows the results of gyroscope readings in the three dimensions. There are four jumps in rotations around the z-axis (Azimuth or yaw) associated with four 90° turns in the route of movement of the device. The red line in Figure 10.9(b) illustrates the estimated path of movement when the results of step counts are complemented with the gyroscope readings. The typical increasing pattern of

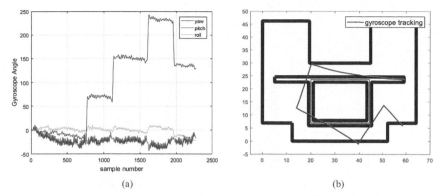

Figure 10.9 Using gyroscope and step counter to localize a person in an office building, (a) results of 3D gyroscope measurements, (b) results of positioning using step counter and gyroscope.

the drift of the location estimate away from the actual path of movement is observed. This type of error pattern is caused from accumulation of errors in estimating the speed and direction of movement for relative positioning.

10.3 Magnetic Location Sensors

In the previous section, we introduced mechanical and electro-mechanical sensors to show how they can be beneficial for calculation of velocity and direction of movement. In this section, we introduce two magnetic sensors, the gravity sensor and magnetometer. Gravity sensor is implemented using the results of the accelerometer, and magnetometer is an independent sensor measuring the ambient electromagnetic field. An important application of the magnetometer is implementation of the electric compass for smart devices. In the next two sub-sections, we first introduce the gravity sensor and then we describe magnetometer and how we can use it to implement an electronic compass.

10.3.1 Gravity Sensor

In Section 10.2.1, we mentioned that if we want to eliminate the effects of gravity from the accelerometer, we pass the readings through a high-pass filter to eliminate the slow varying gravity effects from the relative acceleration of the device. If rather than high-pass filtering, we low-pass filter the measurements from an accelerometer, we measure the force of Gravity in

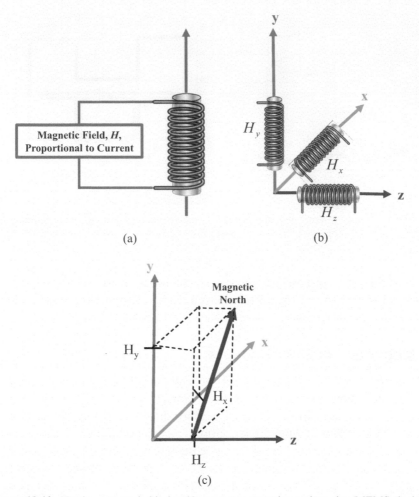

Figure 10.10   Basic concept behind a 3D magnetometer in modern day MEMS devices, (a) the induced current in a coil is used to measure the magnetic field, (b) implementation of a 3D magnetometer on device coordinate, (c) plot of the magnitude of pointing to the magnetic North pole.

the device coordinate. The magnitude of this vector is 9.81 m/s², and its 3D (x, y, z) values reflect the effects of gravity on each of the three axis. These are values that we should subtract from the acceleration vector components to determine the linear acceleration. The value of the gravity vector in device coordinate is used for turning the smartphone and pads screens to be in parallel to the eye of the person reading the screen. Considering device coordinates shown in Figure 10.3(b), if the gravity force along y-axis is the highest component, the device is in up position and a screen with a smaller top is used. When the x-axis is the highest device in the side position, the screen with long top is used. When device is resting on the floor, z-axis is the largest value and the screen can be turned off. Results of gyroscope and magnetometer measurements are often used to complement the result of gravity force measurement obtained from the accelerometer. When the device is at rest, the output of the gravity sensor should be identical to that of the accelerometer.

10.3.2 Magnetometer and Electronic Compass

Magnetometer is a magnetic field sensor measuring the ambient magnetic field along the sensor device in local x, y, and z coordinates in micro-Tesla (mT). Modern MEMS magnetometers measure the current induced in a coil due to changes in the magnetic field in the surrounding environment. Figure 10.10 shows the basic concept behind a modern magnetometer in a MEMS. Figure 10.10(a) shows how the electric current induced in a coil is utilized to measure the magnitude field in the surrounding environment. The intensity of the induced current in the coil is used as a measure for the strength of the magnetic field, H. Figure 10.10(b) shows how the 3D magnetometer can be constructed from three coils mounted in orthogonal directions. Since magnetometers measure the strength and direction of magnetic field, $(H_x \quad H_y \quad H_z)$, they are also utilized for measurement of the direction in localization and tracking. As shown in Figure 10.10(c), the direction of vector, $(H_x \quad H_y)$, in any location on the earth shows the direction to the magnetic North pole. The traditional device measuring the direction using the Earth's magnetic field is a compass. Therefore, the MEMS magnetometers in smartphones are utilized to implement an electronic compasses app that we describe in more details in the remainder of this section.

Compass is one of the oldest mechanical magnetometers historically used as an important navigation tool. If we lay a compass in parallel to the earth surface, the heading of the compass magnet points to the magnetic North. As a result, by rotating the compass to point towards a specific direction, we can measure the angle of that direction with respect to the magnetic North. This angle provides us with direction for relative navigation. Implementation of a compass in a smart phone with a magnetometer is very simple, and we need a graphical user interface (GUI) to display the direction of the $(H_x \quad H_y)$ on the screen of the device. The H_x and H_y measurements of the magnetometer define the direction of compass heading in degrees using:

$$\theta = \begin{cases} 90 - \arctan\left(\frac{H_x}{H_y}\right) \times \frac{180}{\pi}; & H_y > 0 \\ 270 - \arctan\left(\frac{H_x}{H_y}\right) \times \frac{180}{\pi}; & H_y < 0 \\ 180°; & H_y = 0, \ H_x < 0 \\ 0°; & H_y = 0, \ H_x > 0 \end{cases} \qquad (10.4)$$

Magnitude of $\begin{pmatrix} H_x & H_y \end{pmatrix}$ vector on the screen always points to the magnetic North, regardless of the position of the screen. Therefore, to design the GUI, we need to map these readings on the device screen coordinates using Euler's rotation equations. Figure 10.11 shows a snapshot of an electric compass app on iPhone. The letter N always points to the magnetic North pole regardless of the changes in orientation of the device.

The traditional 2D map of an urban area or the layout of a floor of a building are based on the geographical Earth coordinates pointing to the geographical North and South poles. To use a compass with a traditional map, we need to understand the difference between magnetic and geographical

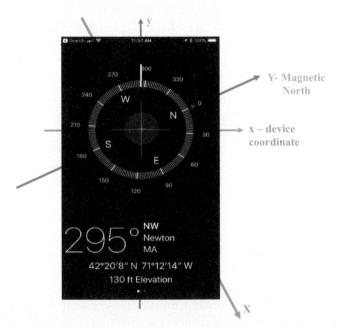

Figure 10.11 Electronic compass app on an iPhone with device and magnetic Earth coordinates, compass heading shows the magnetic North direction.

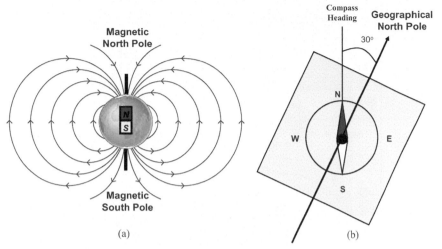

Figure 10.12 Basic concept behind a compass, (a) Magnetic field of the Earth and North-South Poles, (b) declination angle is the angle between geographical and magnetic north in a location.

North poles. A compass is designed based on the Earth's electromagnetic field because, as shown in Figure 10.12(a), Earth acts as a large dipole magnet with magnetic field lines originating from the magnetic South pole, located near the geographic South pole, terminated in magnetic North pole, and located near the geographic North pole. A compass has a small magnet in the center, aligning with the magnetic North pole in all locations on the surface of the Earth. The angle between the compass heading (the magnetic North) and the geographic North is called the magnetic declination angle, as shown in Figure 10.12(b). The value of the decantation angle changes in different locations on the Earth surface and it changes slightly as in time with the changes in the location of the planet with respect to universe. Therefore, calculation of declination angle in different locations of the earth is relatively complex. The National Oceanic and Atmospheric Administration (NOAA) has a public web calculator for calculation of magnetic declination angle around the Globe.

Example 10.3: (Finding declination angle in a location)

Figure 10.13 shows a snapshot of the NOAA web calculator at https://www.ngdc.noaa.gov/geomag-web/#declination for calculation of the declination angle of the Newton, MA, with the zip code of 024591451. We read $-14°35'W$ on July 3, 2018 at a location with 130 ft elevation and

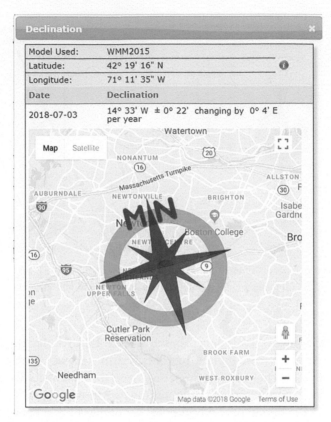

Figure 10.13   Snapshot of the NOAA web calculator for calculation of the declination angle of the Newton, MA, with the zip code of 024591451.

longitude, latitude of (42°19′16″N, 71°, 11′35″W). Results indicate that the accuracy of values are within 22″ 33′W, and the estimated value declines 4′E each year.

In the east coast of the United States, the inclination angle can go as low as −22° (22°W), and on the west coast, it can be as high as 22° (22°E).

10.4 Environmental Location Sensors

Environmental location sensors measure changes in the environment. Modern positioning and navigation systems are designed for ubiquitous mobile and personal operations in different environments. The sensors and algorithms used for localization and the map used to visualize the results of positioning and navigation are different in different environments. Therefore, we need to sense changes in the environment to automate selection of sensor, algorithm, and the map for navigation. In particular, in urban and indoor applications, we want to know whether we are operating outdoors or indoors. If we are indoors, we want to know in which specific floor of the building, or if we are in the stairs or in an elevator. Here, we provide a short description of the environmental sensors and their potential application for intelligent environment recognition.

10.4.1 Barometer

Barometer is an air pressure sensor measuring the relative atmospheric pressure in different heights in hector-Pascal (hPa) or millibar. Barometer is used to estimate elevation changes and consequently floor number of operation inside a multi-floor building. The relation between air pressure, p, and height, h, is given by [Yin15]:

$$h \approx -\frac{R \times T_0}{g \times M} \times \ln(\frac{p}{p_0}). \tag{10.5}$$

The parameters of this equation are defined in Table 10.2. Using Equation (10.5) and the parameters in Table 10.2, we can calculate altitude from air pressure.

Figure 10.14 shows the results of a smartphone barometer measurement in a typical three-story office building. Figure 10.14(a) is produced from the results obtained in an elevator elevating three floors of the building, and Figure 10.14(b) shows the results for climbing one floor of the building using stairs. Both figures demonstrate gradual change of the barometer reading during climbing up and down among the floors regardless of the speed and the pattern of motion. The range of variation of measurements in the elevator experiment for climbing three floors are approximately three times larger than that of the measurements obtained from climbing one floor on the stairs. These observations demonstrate that barometer can be utilized to measure the change in altitude inside the buildings. Figure 10.15 shows results of

Table 10.2 Parameters used for calculation of relation between height, h, and pressure, p

Parameter	Description	Value
p_0	Standard atmospheric pressure	101325 Pa
R	Universal gas constant	8.31447 J/(mol*K)
T_0	Sea level standard temperature	288.15 K
g	Gravitation acceleration	9.81 m/s^2
M	Molar mass of dry air	0.0289644 kg/mol

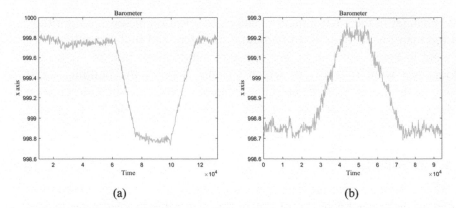

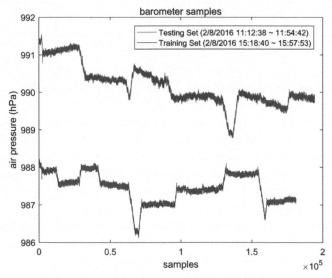

Figure 10.14 Barometer measurement using smart phones in a typical three-story office building, (a) lifting three floors of a building in an elevator, (b) climbing one floor of the building through stairs.

Figure 10.15 Barometer measurements of random walk in a four-story office building in different times of the same day. The red line is taken in late morning and the blue line is taken in mid-afternoon.

barometer measurements from a smartphone when a user carries the device in four floors of a typical office building in two different times of the same day. The pattern of measurement in different floors are clearly different allowing us to distinguish the floor of operation. However, the results of barometer are corrupted by additive Gaussian noise, bias, and it is sensitive to the time of measurement. Methods to mitigate these effects to have a better estimate of the floor number in an office building are discussed in [Yin15].

10.4.2 Ambient Light, Temperature, and Humidity Sensors

As shown in Table 10.1, smart devices carry a few other sensors useful for monitoring changes in the environment. AMBIENT LIGHT SENSORS measure the ambient light level in the unit of lux. Smartphones use the ambient light sensor measurements to adjust the background color of the display to the changes in the illumination of the environment from light to dark. AMBIENT TEMPERATURE SENSORS measure the ambient environment temperature in Celsius degrees, and RELATIVE AMBIENT HUMIDITY SENSORS measure the relative ambient air. These features of the environment enable intelligence to differentiate the environments to select an appropriate localization algorithm or map of the environment.

10.4.3 Proximity Sensors and LiDAR

Proximity sensors in intelligent personal and mobile systems measure the distance of the sensor to the closest visible surface. Proximity sensors in smartphone use the infrared (IR) technology, a light-emitting diode (LED) transmits an optical signal, and a photo-sensitive diode measures the received reflected light to estimate the distance. Some proximity sensors only support a binary "near" or "far" measurement. In this case, the sensor should report its maximum range value in the far state and a lesser value in the near state. Typical application of this sensor is to detect when the screen is close to ear, when we make a call, turn off the screen light to save the battery. Other technologies such as ultrasound are considered to replace infrared. Smart cars, commercial drones, and robots use light detection and ranging (LiDAR) technology for more accurate measurement of the distance of obstacles close to the sensor. LiDAR uses laser pulses that focus the direction of the light as it is compared with the diffused IR, which propagates in all directions. LiDAR has a longer range of operation and a more accurate estimate of distance. The focused beam allows LiDAR to map the proximity obstacle by measuring the distances at different angles of emission. This features allows implementation of more complicated applications such as automatic parking of a vehicle or landing of a drone or adjustment of speed according to the distance for automatic driving.

10.5 Geometric Methods for Relative Positioning

Geometric methods can be used for measurement of speed and direction of motion for any sensor capable of differentiating a specific sign. This can be a Wi-Fi device measuring power from an AP in a known location or a camera with intelligence to spot the location of a feature of a photo. In this section, we begin by reviewing the geometric relation between the speed and direction of movement using RF reference points and we explain how this concept is used in WPS systems to calculate the speed and direction of movement. Then, we extend that technique for calculation of speed and direction of movement in a device equipped with a camera, using consecutive images of a video, followed by a brief description of simultaneous localization and mapping (SLAM) method widely used in robotics navigation and mapping of the environment. Finally, we describe how video pictures are used for SLAM inside the human body.

10.5.1 RF Geometric Methods

As shown in Figure 10.16, in mobile localization and tracking systems using WPS or CPS, the absolute location of a mobile is measured in discrete time instances presented by $\begin{pmatrix} x_i & y_i & t_i \end{pmatrix}$. Therefore, we can use a geometers approach to calculate the velocity and direction of movement using estimated absolute location in time using:

$$\begin{cases} v_i = \frac{\sqrt{(x_i-x_{i-1})^2+(y_i-y_{i-1})^2}}{t_i-t_{i-1}} \\ \theta_i = \tan^{-1}\left(\frac{y_i-y_{i-1}}{x_i-x_{i-1}}\right) \end{cases} . \qquad (10.6a)$$

To improve the quality of estimate, we may average the speed and direction over several samples or find an LS solution by fitting a line to the estimated values in time. The speed of location estimate in typical WPS or CPS systems is around 1 s, and during a few seconds, speed and direction of movement of the device remain constant. Figure 10.17 shows the best fit line for the sequence of readings of the locations, $\begin{pmatrix} x_i & y_i & t_i \end{pmatrix}$, as[1]:

$$y = a\,x + b. \qquad (10.6b)$$

---

[1]Detailed derivation of the LS solution to find a best fit line is given in Example 2.1 for empirical path-loss modeling.

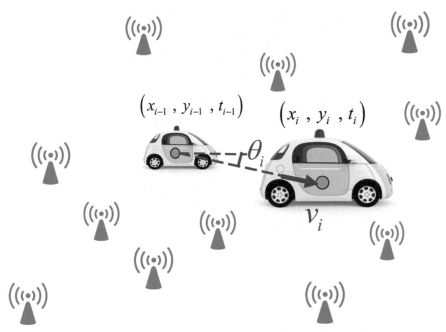

Figure 10.16 Geometric calculation of velocity and direction of movement using consecutive location estimate.

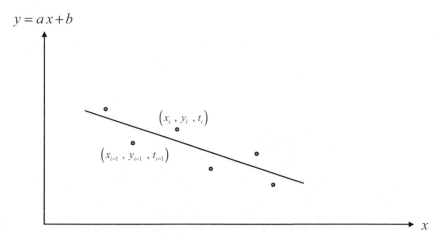

Figure 10.17 LS solution for the calculation of speed and direction of motion.

Using the best line fit LS solution parameters, Equation (10.6a) for calculation of speed and direction of movements will be modified to:

$$\begin{cases} v = \sqrt{dx^2 + dy^2} = \sqrt{1 + a^2} \\ \theta_i = \tan^{-1}\left(\frac{dy}{dx}\right) = \tan^{-1}(a) \end{cases} \tag{10.6c}$$

Application of this approach for CPS is available at [Hel97], and a modified version of this approach that includes the speed estimation using Doppler spectrum for WPS is patented at [Ali09]. As we explain in Chapter 12, using Doppler spectrum is the classical method for calculation of speed of a moving device using RF signals. Doppler spectrum is the Fourier transform of the variations of the RSS, and it provides a very accurate estimation of speed of movement and it is commonly used in radars as well as in GPS systems. However, in WPS, we do not have large variations of the signal power because access points are inside the building and when we read them in the streets, we often read very low values of the RSS. Calculation of Doppler spectrum in these situations is not as reliable as what we can obtain from RSS of the GPS signal. In [Ali09], an adaptive filter is used for fusion of the estimates from the Doppler as well as geometric techniques to achieve a more reliable estimate of the velocity and direction of movement for WPS.

10.5.2 Geometric Methods Using Video Cameras

Camera images of a flat surface have a one-to-one geometric correspondence with their 2D images taken by a camera. Figure 10.18 shows this fundamental

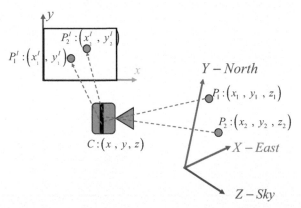

Figure 10.18 Fundamental geometric relation between the physical location of objects in 3D Earth coordinates and the 2D coordinates of their photo pictures on screen of a device.

geometric relation between the physical location of objects in 3D Earth coordinates and the 2D coordinates of their photo pictures on the screen of a device. Two points on the surface of the earth are mapped to two points on the 2D photo. If we know the location and direction of the focal point, C: (x, y, z), of the camera, we can use geometry to find the 3D Earth coordinate location of a point, P_i: (x_i, y_i, z_i), from its image on the 2D camera picture of the point, $P_i^I (x_i^I, y_i^I)$. Therefore, cameras can be used as RPs for localization, and similarities between consecutive pictures of a video camera can be used for calculation of speed and direction of movement. As shown in Figure 10.19, if we have two consecutive images of a moving object and we can identify a fixed point on the object, for example, the center of the image of the object, we can calculate the velocity and angle of movement of the object using Equation (10.6a). The method to implement this system is to use a boundary detection algorithm to identify boundaries of the moving object and then find the centroid of the boundaries. Boundary detection is a classic field in image processing and a variety of algorithms exists in this field.

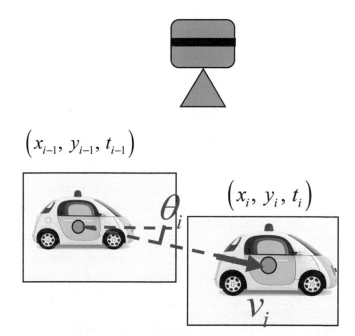

Figure 10.19 Geometry of direction of movement with respect to location of a mobile and multiple reference points used for RF localization.

10.5.3 Simultaneous Localization and Mapping (SLAM)

Similarities between consecutive pictures of a video camera installed on a moving object, such as a robot, can also provide for relative localization. Robots commonly use this technique for simultaneous localization and mapping (SLAM), where the video images and other robot sensor are used to generate a map of the environment and location of the robot on that map [Bai06]. A common approach to SLAM is to use feature detection algorithms to identify certain feature points in a pictures and then use the displacement of the features in consecutive images to determine the distance of fixed objects from the robot. Feature detection algorithm look into the similarities among neighboring pixels of an image to define boundaries of a feature point. The boundary is detected when the change in color similarities changes drastically among neighboring pixels. Figure 10.20 [Est08] shows the detected features of two consecutive pictures of a video camera of a robot in an office environment. Odometer and IMU devices on board of the robots provide a more precise measure of the speed and direction of movement of the robot. Using odometer and IMU of the robot, we can project the motion path of the device, and using location of fixed feature points of the camera, we can reconstruct the map of the area. Figure 10.20(b) shows a sample result of SLAM in an office building [Est08]. As the mobile robots moves along the blue reconstructed route, the building map is reconstructed from the detected location of fixed feature points.

10.5.4 Camera for Relative Positioning Inside the Body

In traditional indoor robotic applications, the video camera feature detection is used as a part of the SLAM algorithm to map the building, while the route of the robot is estimated from the results of odometer and IMU of the robot. The rich literature available for the feature detection algorithms is also beneficial for estimation of speed and velocity for other localization applications, where the moving platform does not have an odometer or an IMU device. One recent application emerging in this area is in localization of wireless video capsule endoscope (WVCE) inside the human GI-tract [Mi14, Bao15]. In these works, feature detection is used for measurement and modeling of the speed and direction of movements of the WVCE inside the small intestine of the human body.

Movement of the WVCE inside the small intestine can be modeled as the movement of camera inside a tube. The geometric model for this movement is shown in Figure 10.21. Figure 10.21(a) illustrates the relation between

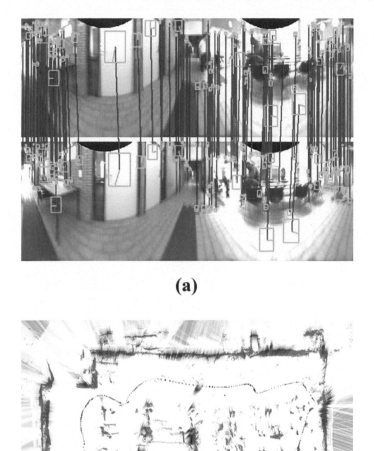

(a)

(b)

Figure 10.20   (a) Comparison of the feature points of two consecutive images of the robot camera, (b) results of SLAM in an office building.

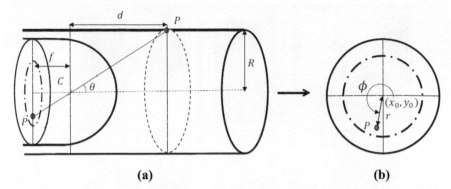

Figure 10.21   Geometric model for movement of a camera inside a tube, (a) relation between location of a point on the surface of the tube and its image on the picture, (b) cylindrical image representing the angle of the image of the point on the tube.

location of a point on the surface of the tube, the focal point of the camera, C, and the image on the point taken by the camera. The line connecting the location P to its image has an angle θ, with the direction of motion of the camera. The relation between the surface distance of the point and the plane of the focal point, d, is $\theta = \tan(R/d)$. Knowing f, the distance of the focal point from the location of the image and the distance of the image from its center, r, we can also calculate $\theta = \tan(r/f)$, and from that, we calculate d. Figure 10.21(b) shows the cylindrical image showing the angle of the image of the point on the tube, φ. This geometry can be used for calculation of speed and direction of movement in two consecutive images.

Figure 10.22(a) shows geometric model for calculation of speed from consecutive images of a camera moving along a tube. When the camera travels a distance of d inside the tube, the focal point of the camera moves from location C to location C', and we have:

$$\begin{cases} \frac{R}{D} = \tan \theta_1 \\ \frac{R}{D-d} = \tan \theta_2 \end{cases} \Rightarrow d = R \left(\frac{1}{\tan \theta_2} - \frac{1}{\tan \theta_1} \right). \tag{10.7a}$$

Therefore, by calculating angles θ_1 and θ_2 from the two consecutive images, we can calculate the traveled distance between the two consecutive images. If the time between taking the two images is Δt, the velocity of the moving camera inside the tube is:

$$v = \frac{d}{\Delta t} = \frac{R}{\Delta t} \left(\frac{1}{\tan \theta_2} - \frac{1}{\tan \theta_1} \right). \tag{10.7b}$$

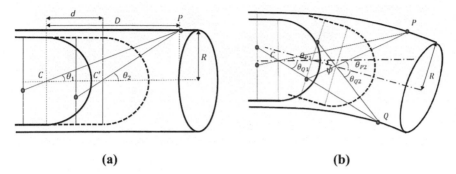

(a) **(b)**

Figure 10.22 Geometric model for consecutive images used for calculation of, (a) speed of the camera, (b) angle of the tube [Bao15].

Figure 10.22(b) shows how we can calculate the tilt in the angle of the camera inside the tube. We have two points on the consecutive images, P and Q, which are of the same distance from the initial position of the camera C. After the camera moves to C' and tilted with angle φ towards Q, the angular depths of the two points change with different amount because of the additional tilt and:

$$\varphi = (\theta_{Q2} - \theta_{Q1}) - (\theta_{P2} - \theta_{P1}).$$ (10.7c)

Using the tilt angle, we can simply determine the pitch and yaw angles in the image plane from $\alpha = \varphi . \cos\phi$ and $\beta = \varphi . \sin\phi$. For calculation of roll angle γ, we can use unrolled image of the signal, which is explained in [Bao15]. Knowing the pitch, yaw, and role angles on the image plane, we know the direction of movement.

In practice, a feature point detection algorithm is applied to the consecutive images to track the transformations such as rotation, translation, and scaling between frames to reflect the motion of the camera. There is a rich literature in image processing that addresses these algorithms and they are beyond the scope of this book. In [Bao15], the Affine Scale Invariant Feature Transform (ASIFT) described in [Moe09] is used for feature point detection to determine the directional angles of movement of a wireless video capsule endoscope inside the small intestine. Figure 10.23 shows the results of feature detection for calculation of velocity and direction of the WVCE [Bao15]. Figure 10.23(a) shows the detected features in two consecutive images, and Figure 10.23(b) shows the motion vectors by linking corresponding feature points. In Figure 10.23(a), the green "O" represents the coordinates of detected feature points in the reference image frame, and red "△" represents the coordinates of

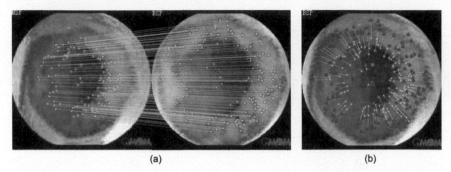

(a)   (b)

Figure 10.23   Feature detection for calculation of velocity and direction of the WVCE, (a) detected features in two consecutive images, (b) motion vectors by linking corresponding feature points.

matched featured point on the next image frame. Figure 10.23(b) shows the motion vector connecting these points when we link the pairs on the same frame representing the displacements between the frames. The magnitudes and distribution of these motion vectors reflect the motions of the WVCE during that time interval. To determine the real direction of movement of the capsule, we need to map these local motions to a universal coordinate.

10.6 RF Signals and Relative Positioning

Traditionally, RSS of RF signals have been used in Doppler radars and for GPS signals to measure the speed. Antenna arrays use TOA of the RF signals to measure the direction of arrival of the signal. In recent years, using RSS signal in time and in frequency with intelligent algorithms has attracted attention in new emerging fields for big data such as gesture [Pu13] and motion detection [Fu11, Gen16].

Principles of using RF signal for DOA estimation was explained in Section 4.4, here we describe measurement of speed using Doppler spectrum.

10.6.1 Doppler Spectrum and Speed of Motion

Principles of understanding of Doppler spectrum begin with understanding of Doppler shift in physics. Figure 10.24 shows the basic principle of Doppler shift; a mobile device transmits a signal with a carrier frequency of f_c. This signal with normalized amplitude is represented by:

$$x(t) = \cos 2\pi f_c t. \tag{10.8a}$$

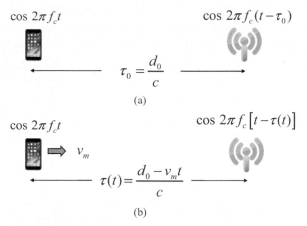

$$\cos 2\pi f_c t \qquad\qquad \cos 2\pi f_c (t - \tau_0)$$

$$\tau_0 = \frac{d_0}{c}$$

(a)

$$\cos 2\pi f_c t \qquad\qquad \cos 2\pi f_c \left[t - \tau(t) \right]$$

$$v_m$$

$$\tau(t) = \frac{d_0 - v_m t}{c}$$

(b)

Figure 10.24 Basic concept of Doppler shift, (a) a fixed device with respect to an AP, (b) a mobile device moving towards the AP.

If the device is fixed in a location (Figure 10.24(a)), the received signal with normalized amplitude in an access point at the distance of d_0 is then given by:

$$y(t) = \cos 2\pi f_c(t - \tau_0) = \cos(2\pi f_c t - \phi_0), \tag{10.8b}$$

where $\tau_0 = d_0/c$ is the delay of arrival of the signal, and $\phi_0 = \omega_c \tau_0$ is the phase of the received carrier. If the device begins to move towards the access point with a velocity of v_m (Figure 10.24(b)), the received signal becomes:

$$y(t) = \cos 2\pi f_c[t - \tau(t)] \tag{10.9a}$$

where

$$\tau(t) = \frac{d_0 - v_m t}{c} = \tau_0 - \frac{v_m}{c} t. \tag{10.9b}$$

Substituting (10.9b) and (10.9a), we have:

$$y(t) = \cos\left[2\pi \left(f_c + f_D\right) t - \phi_0\right], \tag{10.10a}$$

where

$$f_D = \frac{v_m}{c} f_c = \frac{v_m}{\lambda}. \tag{10.10b}$$

Therefore, the transmitted frequency, f_c, of a mobile moving with the velocity of v_m towards a receiver is changing to $f_c + f_D$, when it is measured at the receiver. This shift in frequency caused by motion is referred to as Doppler shift.

The Doppler shift shown in Figure 10.24 is caused by differential changes of the length of the direct path between the transmitter and the receiver. If a mobile device is moving in a different direction (Figure 10.25(a)), the measured Doppler shift associates with the projection of the velocity on the direct path. In a multipath environment, each path has its own Doppler shift, and the received signal is given by:

$$y(t) = \sum_{i=1}^{L} \beta_i \cos\left[2\pi \left(f_c + f_{D-i}\right) t - \phi_i\right], \qquad (10.10c)$$

where β_i and $\phi_i = \omega_c \tau_i$ are the magnitude and phase of the multipath components and $f_{D-i} = v_{m-i}/\lambda$ is the Doppler shift of that calculated from projection of velocity in the direction of departure of that path.

For multipath arrivals, as shown in Figure 10.25(b), the Doppler shift associates with differential changes in the distance of the path. To calculate this Doppler, we need to map the velocity to the direction of that path, $v = v_m \cos\theta$, where θ is the angle between the direction of the mobile and the direction of the path. The largest possible value of Doppler shift associates

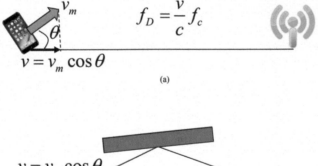

(a)

(b)

Figure 10.25   (a) Velocity used for calculation of Doppler shift associated with direct path is the velocity of movement of the device in direction of the direct path, (b) multipath components cause Doppler shift with the projected velocity on direction of departure for the path.

with $\theta = 0$, which is the path that is in the direction of movement of the device.

The complex envelope of the received signal in multipath at the frequency of ω_c, given by Equation (10.10c), is a baseband signal given by:

$$D(t) = \sum_{i=1}^{L} \beta_i \cos(2\pi f_{D-i} t - \phi_i).$$
(10.11a)

Doppler spectrum is the Fourier transform of this signal:

$$D(\lambda) = \frac{1}{2} \sum_{i=1}^{L} \beta_i e^{-j\tau_i} \left[\delta(f + f_{D-i}) + \delta(f - f_{D-i}) \right].$$
(10.11b)

A simple method to measure the Doppler spectrum is to use the vector network analyzer (VNA) described in Figure 6.12. If we set the system for NB measurement using a single carrier frequency, the VNA transmits a single carrier and provides samples of envelope of the received signal in time. The Fourier transform of these samples provides the Doppler Spectrum. Figure 10.26 provides sample measurements of envelop of the received signal and its Fourier transform, $D(\lambda)$, used for motion detection in body area networking application in [Fu11]. Figure 10.26(a) shows the basic measurements in an indoor LOS multipath condition when the transmitter and the receiver are fixed and no one is moving close by the experiment. The transmitted

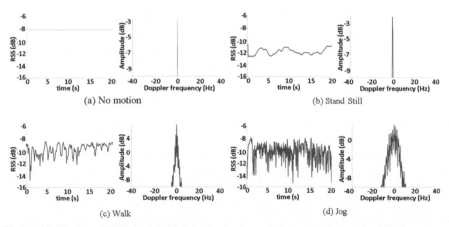

(a) No motion

(b) Stand Still

(c) Walk

(d) Jog

Figure 10.26 The RSS in an AP from a body mounted sensor in indoor multipath, next to its Doppler spectrum, for different human motions, (a) fixed transmitter, (b) the person stands, (c) walks, and (d) jogs.

signal by the network analyzer is a carrier frequency at 2.25 GHz, and the received signal envelope is a straight line representing the magnitude of the received sinusoid. Since there is no motion in the environment, all multipath arrivals have fixed amplitude and phase and the resulting signal is a sinusoid with fixed amplitude. The Fourier transform of this signal is an impulse, and there is no Doppler spread. Figure 10.26(b) shows the experiment when a person is standing still. Small variations of the signal are caused by breathing and heart beats of the person, and the Doppler spectrum has a very small spread. As the person begins to walk (Figure 10.26(c)), we observe around 10 dB of variations in the signal amplitude and a Doppler spread of around 5 Hz in frequency domain. As the person begins to jog (Figure 10.26(d)), fluctuation of the received signal envelop becomes faster and the Doppler spread approximately doubles.

10.6.2 Speed Measurement Techniques for Doppler Spectrum

As we explained mathematically by Equation (10.11), the maximum value of the Doppler shift for a transmitted sinusoid can be used for calculation of the speed of the mobile, because its Doppler shift value is associated with the path in the direction of motion. In practice, however, the sinusoid used for measurement has a finite duration in time, and we always suffer from thermal measurement noise at the receiver. Finite duration of measurement time changes impulses in the spectrum into sinc pulses with infinite side lobes in frequency domain. Adding noise on top of the side lobes, it would be difficult to detect the end of the spectrum to find what the highest value of the Doppler shift is. We can reduce the side lobes in the frequency domain using widows in time domain, but we cannot eliminate them and the noise is always there. We need an algorithm to estimate the maximum value of the Doppler shift, which we want to use to calculate the speed of the device.

The simplest algorithm to solve this problem is to define a threshold under the peak of the Doppler spectrum and use the maximum value of a Doppler shift before the threshold to estimate the speed. We can use empirical measurements to calibrate the threshold for the application environment of choice. This approach is independent of the shape of the Doppler spectrum.

Another approach is to calculate the rms Doppler spread of the empirical data:

$$f_{D-rms} = \sqrt{\frac{\int \lambda^2 D(\lambda) d\lambda}{\int D(\lambda) d\lambda}}, \qquad (10.12a)$$

and relate that to the speed of the mobile. We may also consider finding a best fitting curve to the noisy Doppler spectrum as a model, for example:

$$D(\lambda) = \frac{a}{b + \lambda^2}. \tag{10.12b}$$

Then, using LS or other optimization algorithms, calculate the parameters of the model and relate them to the speed. This approach smoothens the shape of the Doppler spectrum and provides an analytical model for Doppler spectrum, which can be used for a variety of general applications in wireless access and localization.

We can also calculate the speed from the rate of fluctuation of the envelope of the received signal caused by multipath fading. It is well known that in Rayleigh fading channels, the threshold crossing rate, $N(\rho)$, and average duration of fade, $\tau(\rho)$, are related to the rms Doppler spread f_{D-rms}. Figure 10.27 [Pah05] shows the definition of the fade rate and the duration of the fade as well as equations relating them.

Defining the normalized crossing threshold as $\rho = A/A_{rms}$, in which A and A_{rms} are the threshold level and rms amplitude of the received signal, respectively, these relations are given by:

$$f_{D-rms} = \begin{cases} \dfrac{N(\rho)}{\sqrt{2\pi}\rho e^{-\rho^2}} \\ \dfrac{e^{\rho^2} - 1}{\sqrt{2\pi}\rho\tau(\rho)} \end{cases}. \tag{10.12c}$$

The advantage of using Equation (10.12c) is that we can implement them with simple hardware, which was very important before the emergence of the digital technology.

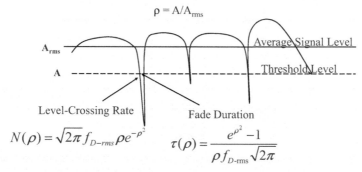

$$N(\rho) = \sqrt{2\pi} f_{D-rms} \rho e^{-\rho^2} \qquad \tau(\rho) = \frac{e^{\rho^2} - 1}{\rho f_{D-rms} \sqrt{2\pi}}$$

Figure 10.27 Definition of level crossing rate and fade duration statistics and their relation to rms Doppler spectrum.

Another method to determine the speed is to use the correlation function of the samples of the received complex envelope to calculate the coherence time of the channel using the normalized correlation function in time. For N received sample of envelop of the received signal:

$$y(n) = y(t)|_{t=nT_s}; \quad n = 1, \ldots, N,$$

the normalized auto correlation function is given by:

$$R_{yy}(m) = \frac{\sum\limits_{n=1}^{N-m} \{y(n) - m_y\} \{y^*(n+m) - m_y^*\}}{|r(n)| \times |r(n+m)|}, \quad (10.13a)$$

where:

$$m_y = \frac{1}{N}\sum_{n=1}^{N} y(n) \quad \text{and} \quad r(n) = \sqrt{\frac{1}{N}\sum_{n=1}^{N}[y(n) - m_y]^2} \quad (10.13b)$$

Figure 10.28 [Fu11] shows the normalized autocorrelation function of the time domain data for calculation of coherence time for the empirical data presented in Figure 10.26. The value of the plot at intersect with 50% line is used as a measure of coherence time [Pah95]. The inverse of the coherence time is a measure for the effective Doppler shift of the channel, which can be used for calculation of speed of a mobile.

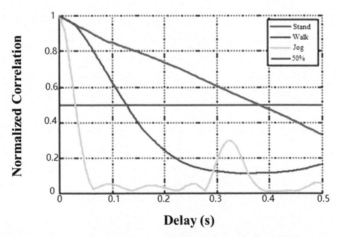

Figure 10.28 Normalized autocorrelation function of the time domain data for calculation of coherence time for the empirical data presented in Figure 10.26. The value of the plot at the intersect with 50% line is used as a measure of coherence time.

10.7 Hybrid Algorithms for Sensor Fusion

We use the term hybrid localization algorithms to algorithms that integrate relative and absolute localizations. These algorithms have been existing since the early days of positioning and navigation and control. The most popular of them are Kalman Filtering and Particle Filtering, used in airplanes, cars, ships, and robot. The Kalman filter is restricted to linear processes and Gaussian distribution noise, while the particle filter can deal nonlinearity of the process and non-Gaussian noise distribution. In cases where abrupt sensor noise is rarely observed, both filters work fairly well. However, when sensor noise exhibits jerky error, Kalman filter results in location estimation with hopping, while particle filter still produces robust localization [Ko12]. In the remainder of this section, we provide examples of application of these two filtering methods for indoor and in-body localizations. We begin with particle filtering example, because it is more intuitive and easier to understand.

10.7.1 Particle Filter for 2D Indoor Positioning of a Robot

Particle filtering is a recursive Bayesian estimation method similar to probabilistic grid algorithm (Section 8.5), which can integrate the results of noisy absolute localization with the results of relative localization with drift in estimated location. We introduce the algorithm using a simple example for robot localization inside a typical office building [Ye10] based on the particle filter described in [Gus02]. In this example, we have a robot equipped with Wi-Fi, a high-quality IMU device[2], and the robot odometer measuring the speed from the wheels of the robot. Figure 10.29 shows the general block diagram of this hybrid localization system. Measurements of the Wi-Fi device is compared with the signature database using a Wi-Fi localization algorithm (Kernel algorithm described in Section 9.4.3) to produce the absolute estimate of location of the robot, $[x_k \; y_k]^T$. In the relative localization segment of the system, the gyroscope of an IMU device provides for the angle of motion, θ_k, and the odometer of the robot measures the speed, v_k. Since gyroscope provides the differential value of the angular motion, the angle used for calculation of motion is calculated as:

$$\theta_k = \theta_{k+1} + \Delta\theta_k. \tag{10.14a}$$

---

[2]This is the IMU device shown in Figure 10.5(b), a typical output of the gyroscope of this device.

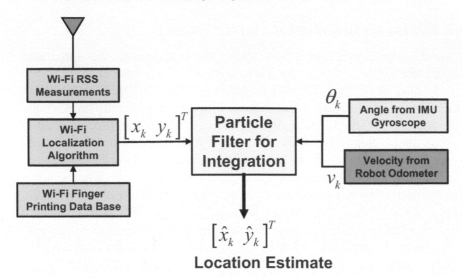

Figure 10.29 System architecture of a multi-sensor hybrid localization system integrating Wi-Fi localization with the speed of Odometer and direction of movement of the gyroscope in a high quality IMU device.

Results of Wi-Fi localization would be similar to Figure 10.1(a), and the results of relative localization using odometer and gyroscope would be similar to Figure 10.1(b). The particle filter combines the results of absolute and relative localization to provide a smoother and more accurate estimate of the locations, $[\hat{x}_k \ \hat{y}_k]^T$. The algorithm postulates the problem as observation of two parameters in Gaussian measurement noise:

$$\mathbf{O}(k) = [x_k \ y_k]^T = [\hat{x}_k \ \hat{y}_k]^T + \mathbf{N}, \qquad (10.14b)$$

where $\mathbf{N}$ is the complex measurement noise of the absolute localization with the variance of σ^2. Algorithm is not very sensitive to the accuracy of this variance, and we can either use the CRLB to calculate it or we can collect a few training samples to measure it.

Our example particle filter begins with an initiation steps and continues with the normal steps. Figure 10.30 shows the initiation steps, upon the arrival of the first Wi-Fi estimate of the location, $(x_1 \ y_1)$, the algorithm generates a large number of random locations for the particles around Wi-Fi estimated location:

$$[x_i(1) \ y_i(1)]^T \ ; \ i = 1, ...N. \qquad (10.15a)$$

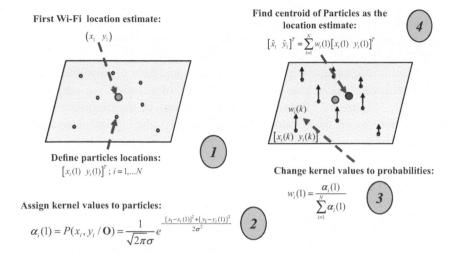

Figure 10.30 Initial steps after arrival of first Wi-Fi location: (1) Define random particles around Wi-Fi location estimate, (2) assign a kernel value to each particle, (3) change kernel values to probability for each particle, (4) calculate the weighted centroid of particles as the estimate of the location of the robot.

In the second step, the algorithm calculates a kernel value for by assigning a probability to each of these particles according to their distance from the Wi-Fi estimate location using the Gaussian measurement noise assumption of Equation (10.8b):

$$\alpha_i(1) = P(x_i, y_i/\mathbf{O}) = \frac{1}{\sqrt{2\pi}\sigma} e^{-\frac{1}{2\sigma^2}\{[x_1-x_i(1)]^2+[y_1-y_i(1)]^2\}} \qquad (10.15b)$$

As the third step in initialization (Figure 10.30), we use the kernel probability values to define a discrete probability density function for the particles:

$$w_i(1) = \frac{\alpha_i(1)}{\sum_{i=1}^{N} \alpha_i(1)}. \qquad (10.15c)$$

This is the probability of *i-th* particle be the location of the robot. Using these probabilities and location of the particles, the algorithm uses the weighted centroid algorithm on all particles to have the first estimate of the location:

$$[\hat{x}_1 \ \hat{y}_1]^T = \sum_{i=1}^{N} w_i(1) \, [x_i(1) \ y_i(1)]^T. \qquad (10.15d)$$

This new estimate of location, $(\hat{x}_1 \quad \hat{y}_1)$, is obtained from the particles, and if we have adequate number of particles, it should be close to the original Wi-Fi location estimate, $(x_1 \quad y_1)$.

After initiation stage, the particle filter uses samples of absolute Wi-Fi location, estimate, $(x_k \quad y_k)$, as well as estimates of velocity, v_k, and angle of motion, θ_k. Figure 10.31 shows the regular steps of the algorithm upon arrival of a new set of these values. In the first step, the algorithm uses the past location of the particles with the speed and direction of motion to find the new location of the particles from:

$$\begin{bmatrix} x_i(k) \\ y_i(k) \end{bmatrix} = \begin{bmatrix} x_i(k-1) \\ y_i(k-1) \end{bmatrix} + v_k \begin{bmatrix} \cos(\theta_k) \\ \sin(\theta_k) \end{bmatrix}. \qquad (10.16a)$$

In the second step, the algorithm calculates a probability for each of the newly located particles using the newly arrived estimate of the location from Wi-Fi localization, $\mathbf{O}$: $(x_k \quad y_k)$:

$$\alpha_i(k) = P(x_i, y_i/\mathbf{O}) = \frac{1}{\sqrt{2\pi}\sigma} e^{-\frac{1}{2\sigma^2}[x_k-x_i(k)]^2 + [y_k-y_i(k)]^2}. \qquad (10.16b)$$

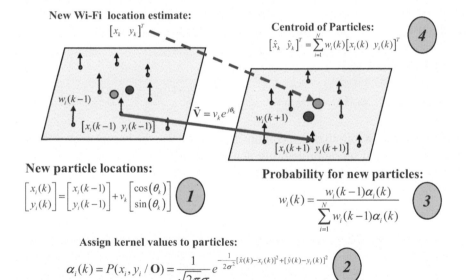

Figure 10.31 Normal steps after arrival of k-th Wi-Fi location estimate, (1) Find the new particle locations using velocity and direction, (2) assign a kernel value to each particle, (3) change kernel values to probability for each particle, (4) calculate the weighted centroid of particles as the estimate of the location of the robot.

To link the estimate to the past history, we calculate the new probability distribution function of the particles, based on $w_i(k - 1)\alpha_i(k)$. Then, the third step to calculate the probability distribution function of the new particle locations becomes:

$$w_i(k) = \frac{w_i(k - 1)\alpha_i(k)}{\sum_{i=1}^{N} w_i(k - 1)\alpha_i(k)}. \qquad (10.16c)$$

The final location estimate using the particle locations and their probabilities is given by:

$$[\hat{x}_k \ \hat{y}_k]^T = \sum_{i=1}^{N} w_i(k) \left[x_i(k) \ y_i(k)\right]^T. \qquad (10.16d)$$

Figure 10.32 shows a test route in a typical office building, at the third floor of the Atwater Kent Laboratory at the Worcester Polytechnic Institute, for performance evaluation of the particle filter that we described in this section, combining Wi-Fi localization with the results of odometer and gyroscope of a robot [Ye10]. Figure 10.32(a) shows the motion route, initial Wi-Fi estimated location, and location of initial particles at the beginning of the route. Figure 10.32(b) shows the estimated locations with particle filter and the location of particles and Wi-Fi localization after completion of the route[3]. Figure 10.33 shows the cumulative distribution function of localization error of K nearest neighbor (K-NN), Wi-Fi localization (Section 9.4.1), kernel

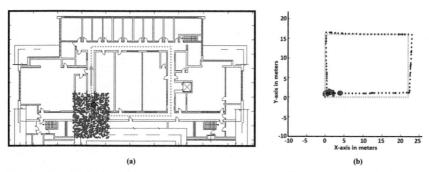

(a)   (b)

Figure 10.32 (a) A typical office building, a square motion route (green), initial Wi-Fi estimated location (red) and initial particles (blue) at the beginning of the route, (b) estimated locations with particle filter (black) and the location of particles and Wi-Fi localization after completion of the route.

---

[3]The video animation of operation of this particle filter is available at YouTube: https://www.youtube.com/watch?v=VtU_lmJcC9A&feature=youtu.be

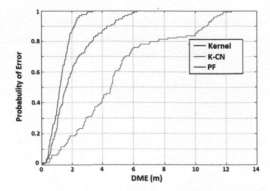

Figure 10.33   Cumulative distribution function of localization error of K closest neighbor (K-CN) Wi-Fi localization, Kernel Wi-Fi localization, and hybrid localization using Particle Filtering (PF).

Wi-Fi localization (Section 9.4.3), and hybrid localization using particle filtering described in this section. Particle filter has improved 90 percentile localization error of the kernel algorithm from close to 4 to 2 m.

10.7.2 Kalman Filter for Indoor 2D Positioning of a Robot

Kalman Filtering was originally introduced for control of complex dynamic systems such as continuous manufacturing process, and navigation of aerial, naval, and land vehicles but soon its applications extended to estimation theory in many diversified applications [Gre01]. Kalman filtering was originally designed for control problem applications when we want to estimate the states of a linear stochastic process, $\mathbf{X_k}$, with a state transition matrix $\mathbf{A_k}$, and a control signal, $\mathbf{u_k}$, relating to the states by matrix $\mathbf{B_k}$. The observations, or measurements, $\mathbf{O_k}$, are related to the state vector with another matrix   . Both state transition and observations are made in zero mean white Gaussian noise with covariance matrixes of $\mathbf{N(0,Q)}$ and $\mathbf{N(0,R)}$, respectively:

$$\begin{cases} \mathbf{X_k} = \mathbf{A_k X_{k-1}} + \mathbf{B_k u_k} + \mathbf{w_{k-1}}; & \mathbf{N(0, Q)} \\ \mathbf{Z_k} = \quad \mathbf{X_k} + \mathbf{v_k}; & \mathbf{N(0, R)} \end{cases} \qquad (10.17a)$$

The a priori estimate of the state error, $\hat{\mathbf{e}}_k^-$, and its covariance matrix, $\mathbf{P}_k^-$, are defined as:

$$\begin{cases} \hat{\mathbf{e}}_k^- = \mathbf{X_k} - \hat{\mathbf{X}}_k^- \\ \mathbf{P}_k^- = E[\hat{\mathbf{e}}_k^- \, \hat{\mathbf{e}}_k^{-\mathbf{T}}] = \mathbf{E}\left[(\mathbf{X_k} - \hat{\mathbf{X}}_k^-)(\mathbf{X_k} - \hat{\mathbf{X}}_k^-)^\mathbf{T}\right] \end{cases}, \qquad (10.17b)$$

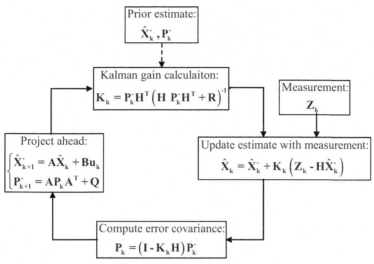

Figure 10.34 A flowchart for principles of Kalman filter operation. Kalman filter begins with an initial estimate and a guess for error covariance, using measurements of the desired parameters the algorithm updates the estimated parameters iteratively.

where $\hat{X}_k^-$ is a priori estimate of the states of the linear stochastic process. Kalman gain matrix, K_k, minimizes the a posteriori covariance matrix, P_k, and can be calculated iteratively from:

$$
\begin{cases}
\hat{X}_k = \hat{X}_k^- + K_k \left(Z_k - \hat{X}_k^- \right) \\
P_k = E\left[e_k\, e_k^T \right] = E\left[\left(X_k - \hat{X}_k \right) \left(X_k - \hat{X}_k \right)^T \right] = (I - K_k)\, P_k^-
\end{cases}
$$

$$
K_k = P_k^{-\,T} \left(P_k^{-\,T} + R \right)^{-1} \tag{10.17c}
$$

$$
\begin{cases}
\hat{X}_{k+1}^- = A\hat{X}_k + Bu_k \\
P_{k+1}^- = AP_k A^T + Q
\end{cases}
$$

Figure 10.34 shows a flowchart for principles of Kalman filter operation. Kalman filter begins with an initial estimate of parameters plus a guess for error covariance. Using measurements of the desired parameters, the algorithm updates the estimated parameters iteratively. The estimate of the parameters and the variance of the estimate is recursively updated and improved. When parameters change in time, the Kalman filter tracks these changes. In localization and navigation industry Kalman filtering is commonly used for fusion of multi-sensor for relative and absolute localization.

Here, we provide two examples of these applications, one for 2D localization of a robot[4], and another for localization of WVCE inside the small intestine of the human body.

To apply Kalman filter to any application, the process begins by formulation of the problem in terms of state transition and observation matrix of Equation (10.17a). A group of undergraduate students [Shi10] at WPI used the extended Kalman filtering (EKF) as a part of their senior project to localize a robot in an indoor environment. Similar to Section 10.6.1 [Ye10], they used Wi-Fi localization for absolute location estimate and odometer and gyroscope for the velocity and direction of movements. In extended Kalman filtering, state transition and observation model do not need to be a linear function and they can be any differentiable function:

$$\begin{cases} \mathbf{X_k} = f(\mathbf{X_{k-1}}, \mathbf{u_k}) + \mathbf{w_{k-1}}; \ \mathbf{N}(\mathbf{0}, \mathbf{Q}) \\ \mathbf{Z_k} = h(\mathbf{X_k}) + \mathbf{v_k}; \ \mathbf{N}(\mathbf{0}, \mathbf{R}) \end{cases} \qquad (10.17d)$$

To calculate the transition using covariance instead of the function, the Jacobian matrix of the function is used. To formulate this problem in the format of (10.11d) in [Shi10], the state matrix is defined as:

$$\mathbf{Z_k} = \mathbf{X_k} = \begin{bmatrix} x_k & y_k & \theta_k \end{bmatrix}^T \qquad (10.18a)$$

and the state transition function as:

$$f(\mathbf{X_k}) = \begin{bmatrix} f_x \\ f_y \\ f_\theta \end{bmatrix} = \begin{bmatrix} x_k \\ y_k \\ \theta_k \end{bmatrix} = \begin{bmatrix} x_{k-1} + \Delta D_k \cos \theta_k \\ y_{k-1} - \Delta D_k \sin \theta_k \\ \theta_{k-1} + \Delta \theta_k \end{bmatrix}. \qquad (10.18b)$$

The state transition matrix used for calculation of the Kalman filter iterations is the Jacobian of this matrix:

$$\mathbf{k} = \begin{bmatrix} \frac{\partial f_x}{\partial x_k} & \frac{\partial f_x}{\partial y_k} & \frac{\partial f_x}{\partial \theta_k} \\ \frac{\partial f_y}{\partial x_k} & \frac{\partial f_y}{\partial y_k} & \frac{\partial f_y}{\partial \theta_k} \\ \frac{\partial f_\theta}{\partial x_k} & \frac{\partial f_\theta}{\partial y_k} & \frac{\partial f_\theta}{\partial \theta_k} \end{bmatrix} = \begin{bmatrix} 1 & 0 & \Delta D_k \sin \theta_k \\ 0 & 1 & \Delta D_k \cos \theta_k \\ 0 & 0 & 1 \end{bmatrix}. \qquad (10.18c)$$

Figure 10.35 shows the cumulative distribution function of localization error of three different approaches on the path of Figure 10.32(a). These three results belong to localization using robot odometer and gyroscope, Wi-Fi localization, and hybrid localization using extended Kalman filtering (EKF)

---

[4]The same problem was solved in Section 10.6.1 by using a particle filter.

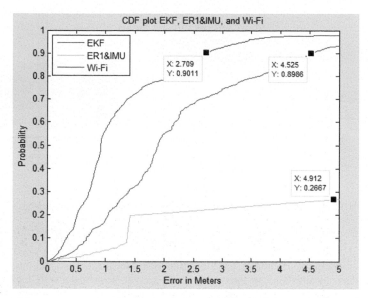

Figure 10.35 Cumulative distribution function of localization error of localization using ER1 robot odometer, Wi-Fi localization, and hybrid localization using extended Kalman filtering (EKF).

to integrate the results of two approaches [Shi10]. The EKF improves the result of 90% error for Wi-Fi localization from mid–4 m to mid–2 m. As compared with the results of Particle filtering of Figure 10.2, we observe that hybrid localization, regardless of the type of filter increase the accuracy, smoothens the result of localization and makes them appear in sequence, as it is expected for a mobile.

10.7.3 Kalman Filter for 3D In-Body Localization

In the last two sub-sections, we gave two examples of using hybrid algorithms, particle filtering, and extended Kalman filtering, for 2D positioning a robot inside an office building. In both examples, the hybrid algorithm was combining results of absolute Wi-Fi localization with relative localization using odometer and gyroscope of the robot. In this section, we provide an example for using Kalman filter to integrate the absolute RSS-based localization and results of estimation of speed and direction of movements from the video camera of a wireless video capsule endoscope (WVCE).

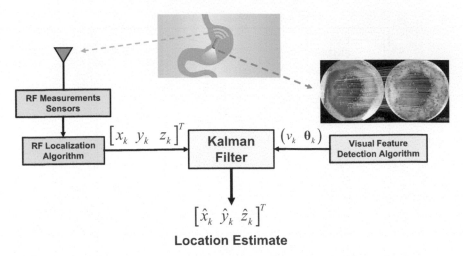

Figure 10.36   System architecture of a multi-sensor hybrid localization system integrating RSS-based absolute localization with the speed and direction of movement a WVCE obtained from consecutive images taken by the camera inside the capsule.

Figure 10.36 shows the general system architecture of the WVCE positioning system. A WVCE swallowed by a patient takes pictures from inside the GI tract using wireless communication RF signals. A feature detection algorithm processes these video images to find the speed, v_k, and directional angles of movement of the WVCE, $\mathbf{R}_k = [\alpha_k \ \beta_k \ \gamma_k]^T$. Additional body-mounted RF sensors detect the features of the received signal on the surface of the body to find the absolute location of the micro-robot, $(x_k \ \ y_k \ \ z_k)$. A Kalman filter integrates the results of RSS-based absolute localization with the speed and direction of movement of the WVCE to estimate the refined 3D position of the micro-robot, $(\hat{x}_k \ \ \hat{y}_k \ \ \hat{z}_k)$.

The feature detection algorithm used for producing the results of this section was explained in Section 10.5.4. Figure 10.28 shows the results of this algorithm on actual pictures taken inside the small intestine of human body. The magnitudes and distribution of these motion vectors reflect the motions of the WVCE during that time interval. To determine the real direction of movement of the capsule, we need to map these local motions to a universal coordinate. Figure 10.37 explains the relation between the universal and local coordinates. The world coordinate is represented as *(X Y Z)*, and WVCE coordinate as *(X' Y ' Z')*. The moving direction of the capsule is given by a norm vector $(n_x \ n_y \ n_z)^T$ in the universal coordinate. After the WVCE rotated with pitch, yaw, and roll angles of $(\alpha_k \ \beta_k \ \gamma_k)^T$, the new directions

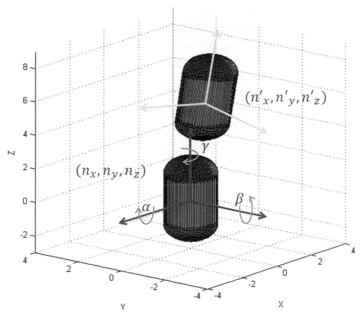

Figure 10.37   Modeling the direction of movement of the WVCE.

of the WVCE become $(n'_x \ n'_y \ n'_z)^T$. Defining **R** as the rotation matrix, the relation between the old and new directions of the capsule is given by:

$$\begin{bmatrix} n'_x \\ n'_y \\ n'_z \end{bmatrix} = \mathbf{R} \begin{bmatrix} n_x \\ n_y \\ n_z \end{bmatrix}. \tag{10.19a}$$

If we assume the WVCE coordinate system was initially aligned with the universal coordinate system with its focal axis pointed to the Z axis, then, the initial value of **R** is a 3×3 identical matrix. As the capsule moves away from the original position, **R** updates after each frame by:

$$\mathbf{R} = \mathbf{R} \times \mathbf{R_t} \times \mathbf{R}^{-1}, \tag{10.19b}$$

where $\mathbf{R_t}$ is Euler's transformation matrix:

$$\mathbf{R_t} = \begin{bmatrix} \cos\alpha\cos\gamma & \sin\alpha\sin\beta\cos\gamma - \cos\alpha\sin\gamma & \cos\alpha\sin\beta\cos\gamma + \sin\alpha\sin\gamma \\ \cos\beta\sin\gamma & \sin\alpha\sin\beta\sin\gamma + \cos\alpha\cos\gamma & \cos\alpha\sin\beta\sin\gamma - \sin\alpha\cos\gamma \\ -\sin\beta & \sin\alpha\cos\beta & \cos\alpha\cos\beta \end{bmatrix}. \tag{10.19c}$$

The rotation matrix $\mathbf{R}$ in (10.19b) is the 3D version of the rotation angle θ, which we used for 2D rotations in extended Kalman filtering of Equation (10.18). In 2D formulation, our state matrix had three elements, two for the location and one for the angle. In 3D formulation of localization problem, our state vector will have six elements, three for location and three for the direction. For the state transition matrix to formulate the Kalman filter for hybrid localization using RSS-based absolute and video-based relative localization, we have:

$$\begin{bmatrix} x(k) \\ y(k) \\ z(k) \\ n_x(k) \\ n_y(k) \\ n_z(k) \end{bmatrix} = \begin{bmatrix} 1 & 0 & 0 & v_k\Delta t & 0 & 0 \\ 0 & 1 & 0 & 0 & v_k\Delta t & 0 \\ 0 & 0 & 1 & 0 & 0 & v_k\Delta t \\ 0 & 0 & 0 & & & \\ 0 & 0 & 0 & & \mathbf{R_k} & \\ 0 & 0 & 0 & & & \end{bmatrix} \begin{bmatrix} x(k-1) \\ y(k-1) \\ z(k-1) \\ n_x(k-1) \\ n_y(k-1) \\ n_z(k-1) \end{bmatrix}.$$

$$(10.19c)$$

Figure 10.38 shows the modeling of small intestine for performance evaluation of localization algorithms presented in [Bao15]. Figure 10.38(a) shows the location of small intestine inside the GI-tract of the body. Small intestine is a curly organ with a length of 69 m condensed in small area in the middle of the belly. To emulate the small intestine for the motion of the WVCE, we can use a tube shape (Figure 10.38(b)). When WVCE moves along the small intestine, the center of the tube (Figure 10.38(c)) shows the route of the WVCE in travelling through the small intestine. To emulate the environment, pictures from inside of the tube are used with a speed of motion model derived from empirical data using the actual videos from inside the human body. Using the formulation of Kalman filtering described by Equation (10.19), the performance of the localization using only speed and direction of movements obtained from the video images and performance of RSS-based localization are compared with that obtained when we combine all of them with the Kalman Filter.

Figure 10.39 illustrates the comparative performance evaluation of localization using video images, RSS-base localization, and hybrid localization using Kalman filtering. The NIST model introduced in Section 2.2.1 is used for the emulation of RF propagation inside the human body. Figure 10.39(a) shows the performance along the route of motion for WVCE, and Figure 10.39(b) shows the same results when the route is opened straight. As expected, using motion from measurements of speed and direction from the videos accumulate the error as the WVCE moves along the route from

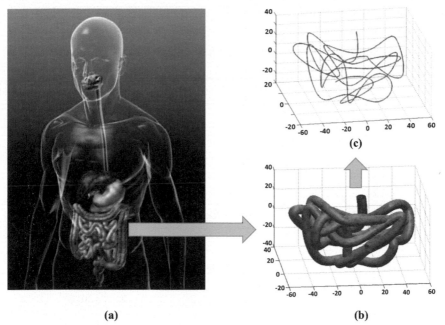

(a)   **(b)**

Figure 10.38 Modeling of the small intestine for performance evaluation of localization algorithms: (a) location of the small intestine in the body, (b) a tube emulating the 3D shape of the small intestine, and (c) the center route of the tube taken by the WVCE [Bao15].

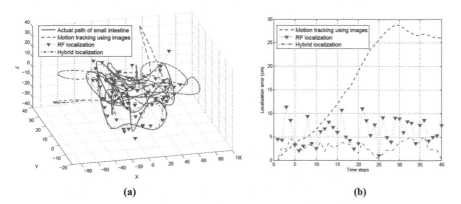

(a)   **(b)**

Figure 10.39 Comparative performance evaluation of localization using video images, RSS-based localization, and hybrid localization using Kalman filtering: (a) performance along the route of motion for WVCE and (b) error along the route when it is opened straight [Bao15].

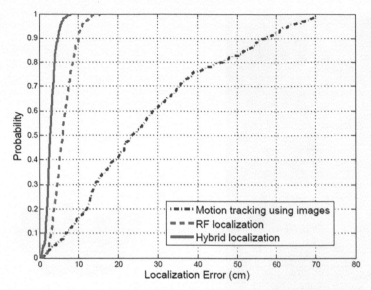

Figure 10.40   Cumulative distribution function of localization error inside the small intestine using image-based motion tracking, RSS-based localization, and hybrid localization using Kalman filtering [Bao15].

very small to relatively large values, RSS-based localization provides random but consistent errors along the way, and hybrid localization smoothens and reduces the error. Figure 10.40 shows the cumulative distribution function of localization error inside the small intestine using image-based motion tracking, RSS-based localization, and hybrid localization using Kalman filtering. The 90% error of the RSS-based localization drops from 10 cm to 5 cm, when we use hybrid localization.

The interesting conclusion, which one can draw from all three examples we presented in Section 10.6, is that the hybrid localization reduces the error to approximately a half of the traditional RF localization algorithms, regardless of the type of hybrid algorithm (Kalman, extended Kalman, or particle filters) or the method used for measuring the speed and direction of movements (mechanical or image-based).

Appendix 10.A: MATLAB Codes

The following MATLAB code implements these operations:

```
w = rotationW(i); % values of ROTATION VECTOR SENSOR
x = rotationX(i);
y = rotationY(i);
z = rotationZ(i);
```

```
transMat(1,1) = w^2+x^2-y^2-z^2; % Calculation of rotation matrix
transMat(2,1) = 2*(x*y+w*z);
transMat(3,1) = 2*(x*z-w*y);
transMat(1,2) = 2*(x*y-w*z);
transMat(2,2) = w^2-x^2+y^2-z^2;
transMat(3,2) = 2*(y*z+w*x);
transMat(1,3) = 2*(x*z+w*y);
transMat(2,3) = 2*(y*z-w*x);
transMat(3,3) = w^2-x^2-y^2+z^2;
```

```
magneticAll(:,i) = [Ex(i),Ey(i),Ez(i)]; % values of MAGNETIC FIELD
SENSOR
magneticAllWorld(:,i) = transMat\magneticAll(:,i);
```

```
compass = atan2(magneticAllWorld(1,:),magneticAllWorld(2,:));
```

Assignments for Chapter Ten

Questions

1. What is the difference between absolute and relative localization and which features of a mobile device are typically measured in each class?
2. Why do we need fusion algorithms for hybrid localization?
3. What are the differences between Kalman and particle filters?
4. What are the typical sensors in a smartphone that can help localizations?
5. What are the differences among localization sensors in an autonomous ground moving vehicle and a car with a driver?
6. How can one measure the speed and direction of movement of a car from reading the number of Wi-Fi access points as the car is driving?
7. How a video camera on a mobile robot can be used to calculate the speed and direction of movement of the robot?
8. How do you compare particle filtering with Kalman filtering?

Appendix A: Review of Classical Estimation Theory for Positioning

In this appendix, we review the classical parameter estimation theory used in the analysis of positioning systems. We begin with estimation of a single and multiple observations of a parameter in zero mean additive Gaussian noise and we extend that to observations of function of a parameter in noise. Finally, we generalize that to multi-parameter estimation in additive Gaussian noise. These techniques are reviewed for the classical Maximum Likelihood estimation and Minimum Mean Square Error estimation and it is further extended to the derivation of the Cramer–Rao Lower Bound (CRLB) for calculation of variance of estimation. The application of single parameter estimation is in ranging, where we want to estimate the distance, and multi-parameter estimation is used for positioning, where we need to estimate the coordinates of a target.

A.1 Maximum Likelihood Estimation

In TOA-based localization, we measure the TOA, τ, and then, we use $d = c \times \tau$ to measure the distance. Since this measurement involves error, our observation is distorted with noise and we model it as: $O = d + \eta$, where η, is an independent Gaussian random variable with a variance of σ^2. In classical estimation theory, this problem is referred to as observation of a single parameter in noise. In RSS-based localization, we measure RSS as a function of distance. In this case, our observation in noise is given by: $O = P_r = P_o - 10\alpha \log d + \eta = g(d) + \eta$. In classical estimation theory, this problem is referred to as observation of function of a parameter in noise. In this section, we derive the Maximum Likelihood (ML) estimates of observation of a parameter and function of a parameter in additive white Gaussian noise.

419

A.1.1 ML Estimate of Single Observation of a Parameter

In its simplest form, if we observe an unknown parameter, α, in a zero mean Gaussian noise, η, with a variance of σ^2, our measured observation is represented by:

$$O = \alpha + \eta \qquad (A.1a)$$

To relate it to range measurement, assume that we have a device that measures the distance using TOA in noise. A single measurement or observation using this device is a sample of a Gaussian random variable with the same variance but a mean of α:

$$f(O \mid \alpha) = \frac{1}{\sqrt{2\pi}\sigma} e^{-\frac{(O-\alpha)^2}{2\sigma^2}}, \qquad (A.1b)$$

where $f(O \mid \alpha)$ is the probability density function of the observation given that we know the parameter α. Since we are interested in probability density function of the parameter given the observation, we can use Bayes' theorem:

$$f(\alpha \mid O) = \frac{f(O \mid \alpha)P(O)}{P(\alpha)} \approx \frac{1}{\sqrt{2\pi}\sigma} e^{-\frac{(O-\alpha)^2}{2\sigma^2}}, \qquad (A.1c)$$

where $P(O)$ is the probability of observation and $P(\alpha) = 1$ is the probability of the parameter having a value. Since $P(O)$ is a fixed number, $f(\alpha \mid O)$ becomes proportional to $f(O \mid \alpha)$ and:

$$f(\alpha \mid O) = P(O)f(O \mid \alpha) \approx \frac{1}{\sqrt{2\pi}\sigma} e^{-\frac{(O-\alpha)^2}{2\sigma^2}} \qquad (A.1d)$$

This PDF is referred to as the likelihood function of the estimation process. For different values of α, we have different likelihoods to observe what we have measured. The value of α, which maximizes the likelihood function, is obtained when the likelihood function is maximized and that happens when the exponent of the function is maximized. Therefore, we define the log likelihood function as:

$$\Lambda(O \mid \alpha) = \ln\left[f(O \mid \alpha)\right] = \ln\left[\frac{1}{\sqrt{2\pi}\sigma}\right] - \left[\frac{(O-\alpha)^2}{2\sigma^2}\right]$$

$$\Lambda(\alpha \mid O) = \ln\left[f(\alpha \mid O)\right] = \ln\left[\frac{1}{\sqrt{2\pi}\sigma}\right] - \left[\frac{(O-\alpha)^2}{2\sigma^2}\right] \qquad (A.1e)$$

The value of α, which maximizes the log likelihood, also maximizes the likelihood function. This optimum value is referred to as the Maximum Likelihood (ML) estimate of the parameter. To obtain this value, we need to take the derivative of the log likelihood function. Therefore, for observation of a single parameter in noise, we have:

$$\begin{cases} \frac{\partial}{\partial x}\{\Lambda(O\ \alpha)\} = -\frac{2(-1)(O-\alpha)}{2\sigma^2} = 0 \\ \hat{\alpha}_{ML} = O \end{cases} \quad ,$$

$$\begin{cases} \frac{\partial}{\partial \alpha}\{\Lambda(\alpha\ O)\} = -\frac{2(-1)(O-\alpha)}{2\sigma^2} = 0 \\ \hat{\alpha}_{ML} = O \end{cases} \quad , \tag{A.1f}$$

where $\hat{\alpha}_{ML}$ is the ML estimate of the parameter. It maximizes the likelihood function, and it is equal to the value that we measure or observe. This simply means that if we have a distance of 5 m, but due to Gaussian noise involved in the measurement process using TOA, we measure it as 6 m, and our ML estimate is 6 m. Because of distortion in the measurement process caused by the noise, we have 1 m of distance measurement error (DME).

A.1.2 ML Estimate of Function of a Parameter

Now that we have the general formulation for calculation of the ML estimate, let us assume that we observe function of a parameter in Gaussian noise with variance of σ^2:

$$O = g(\alpha) + \eta \tag{A.2a}$$

To relate this to a practical positioning situation, assume that we measure the noisy value of the RSS from a reference point (RP) and we want to measure the distance using the path-loss model relating the observation to the parameter of interest. Then, the likelihood and log likelihood functions become:

$$\begin{cases} f(\alpha\ O) = \frac{1}{\sqrt{2\pi}\sigma}e^{-\frac{[O-g(\alpha)]^2}{2\sigma^2}} \\ \Lambda(\alpha\ O) = \ln\left[\frac{1}{\sqrt{2\pi}\sigma}\right] - \frac{[O-g(\alpha)]^2}{2\sigma^2} \end{cases} \quad , \tag{A.2b}$$

and we can calculate the ML estimate of the parameter:

$$\begin{cases} \frac{\partial}{\partial \alpha}\{\Lambda(\alpha\ O)\} = -\frac{2(-1)g'(\alpha)[O-g(\alpha)]}{2\sigma_i^2} = 0 \\ g(\hat{\alpha}_{ML}) = O \Rightarrow \hat{\alpha}_{ML} = g^{-1}(O) \end{cases} \quad . \tag{A.2c}$$

Therefore, if we measure or observe a function of a parameter in Gaussian noise, the ML estimate of the parameter is the inverse of that function of the observation. For example, if we measure the RSS, since $O = P_r = P_0 - 10\alpha \log d + \eta$, the ML estimate of the distance is given by:

$$\hat{d}_{ML} = 10^{-\frac{O-P_0}{10\alpha}} = 10^{-\frac{P_r-P_0}{10\alpha}}.$$

Results of these ML estimates are obvious and it may seem that this formulation only explains an obvious matter in complicated terminologies. However, as problem becomes more complex, we will begin to appreciate the benefits of this formulation.

A.1.3 ML Estimate of N-Observations of a Parameter

In some TOA ranging problems, sometimes we have multiple independent observations or measurements of the range; here, we derive the ML estimate for these situations. We begin with the simple case of N-observations of a single parameter in independent zero mean Gaussian noises with variances of σ_i^2:

$$\begin{cases} O_i = \alpha + \eta_i \ ; \ i = 1, 2, ...N \\ \mathbf{O} = \alpha + \boldsymbol{\eta} \end{cases} , \tag{A.3a}$$

where:

$$\begin{cases} \mathbf{O} = \begin{bmatrix} O_1 & O_2 & . & . & O_N \end{bmatrix}^T \\ \boldsymbol{\eta} = \begin{bmatrix} \eta_1(\sigma_1) & \eta_2(\sigma_2) & . & . & \eta_N(\sigma_N) \end{bmatrix}^T \end{cases}$$

The likelihood function for this case is the joint probability density function of the N observations, given by:

$$f(\mathbf{O} \ \alpha) = \prod_{i=1}^{N} \frac{1}{\sqrt{(2\pi)^N} \sigma_i} e^{-\frac{(O_i-\alpha)^2}{2\sigma_i^2}}$$

$$= \frac{1}{\sqrt{(2\pi)^N} \prod_{i=1}^{N} \sigma_i} e^{-\sum_{i=1}^{N} \frac{(O_i-\alpha)^2}{2\sigma_i^2}},$$

$$= \prod_{i=1}^{N} \frac{1}{\sqrt{(2\pi)^N} \sigma_i} e^{-\frac{[\mathbf{O}-\alpha\mathbf{I}_N]^T \boldsymbol{\Sigma}^{-1} [\mathbf{O}-\alpha\mathbf{I}_N]}{2}} \tag{10.1}$$

$$f(\alpha \mid \mathbf{O}) = \prod_{i=1}^{N} \frac{1}{\sqrt{(2\pi)}\sigma_i} e^{-\frac{(O_i - \alpha)^2}{2\sigma_i^2}}$$

$$= \frac{1}{\sqrt{(2\pi)^N} \prod_{i=1}^{N} \sigma_i} e^{-\sum_{i=1}^{N} \frac{(O_i - \alpha)^2}{2\sigma_i^2}}$$

$$= \prod_{i=1}^{N} \frac{1}{\sqrt{(2\pi)}\sigma_i} e^{-\frac{[\mathbf{O} - \alpha\mathbf{I}_N]^T \Sigma^{-1}[\mathbf{O} - \alpha\mathbf{I}_N]}{2}} \qquad \text{(A.3b)}$$

where $\Sigma = E\left\{\eta^T \eta\right\}$ is the covariance matrix of the noise and $\mathbf{I}_N = \begin{bmatrix} 1 & 1 & \dots & 1 \end{bmatrix}^T$.

Then, the log likelihood function is:

$$\Lambda(\mathbf{O} \mid \alpha) = \ln\left[f(\mathbf{O} \mid \alpha)\right]$$

$$= \ln\left[\prod_{i=1}^{N} \frac{1}{\sqrt{(2\pi)^N}\sigma_i}\right] - \sum_{i=1}^{N} \frac{(O_i - \alpha)^2}{2\sigma_i^2}$$

$$= \ln\left[\prod_{i=1}^{N} \frac{1}{\sqrt{(2\pi)^N}\sigma_i}\right] - \frac{[\mathbf{O} - \alpha\mathbf{I}_N]^T \Sigma^{-1}[\mathbf{O} - \alpha\mathbf{I}_N]}{2}$$

$$\Lambda(\alpha \mid \mathbf{O}) = \ln\left[f(\alpha \mid \mathbf{O})\right]$$

$$= \ln\left[\prod_{i=1}^{N} \frac{1}{\sqrt{(2\pi)}\sigma_i}\right] - \sum_{i=1}^{N} \frac{(O_i - \alpha)^2}{2\sigma_i^2}$$

$$= \ln\left[\prod_{i=1}^{N} \frac{1}{\sqrt{(2\pi)}\sigma_i}\right] - \frac{[\mathbf{O} - \alpha\mathbf{I}_N]^T \Sigma^{-1}[\mathbf{O} - \alpha\mathbf{I}_N]}{2} \qquad \text{(A.3c)}$$

The ML estimate is obtained when the derivative of log likelihood function is minimized:

$$\frac{\partial}{\partial \alpha}\left\{\Lambda(\mathbf{O} \mid \alpha)\right\} = -\sum_{i=1}^{N} \frac{2(-1)(O_i - \alpha)}{2\sigma_i^2}$$

$$= \frac{\mathbf{I}_N^T \Sigma^{-1}[\mathbf{O} - \alpha\mathbf{I}_N] - [\mathbf{O} - \alpha\mathbf{I}_N]^T \Sigma^{-1}\mathbf{I}_N}{2} = 0,$$

$$\frac{\partial}{\partial \alpha} \{\Lambda(\alpha \ \mathbf{O})\} = - \sum_{i=1}^{N} \frac{2(-1)(O_i - \alpha)}{2\sigma_i^2}$$

$$= \frac{\mathbf{I}_N^T \Sigma^{-1} [\mathbf{O} - \alpha \mathbf{I}_N] + [\mathbf{O} - \alpha \mathbf{I}_N]^T \Sigma^{-1} \mathbf{I}_N}{2} = 0$$

which results in:

$$\hat{\alpha}_{ML} = \frac{\sum_{i=1}^{N} O_i \ \sigma_i^2}{\sum_{i=1}^{N} 1 \ \sigma_i^2} = \mathbf{I}_N^T \Sigma^{-1} \mathbf{O} \left[\mathbf{I}_N^T \Sigma^{-1} \mathbf{I}_N \right]^{-1}. \qquad (A.3d)$$

This means that the ML estimate of multiple observations in independent noise with different variance is the average of the observations scaled with the inverse of the variance of their noises. If the noise for all measurements are identical, $\sigma_i^2 = \sigma^2$, the ML estimate becomes:

$$\hat{\alpha}_{ML} = \frac{1}{N} \sum_{i=1}^{N} O_i = \frac{1}{N} \mathbf{I}_N^T \mathbf{O}, \qquad (A.3e)$$

which is the average of all observations or measurements. In practice, we can normalize the observations so that the variance of the noise for all observations is $\sigma_i^2 = \sigma^2$, by multiplying the observation with $\sigma^2 \ \sigma_i^2$. Then, the derivations for Equation (A.3d) become redundant and the ML estimate is always calculated from Equation (A.3e). In the information theory literature, they refer to this problem as multiple observations in independent, identical Gaussian noise. In the remainder of our discussion for multiple observations, we follow this assumption for our observations.

Another interesting observation from multi-observation of a single parameter in identical independent noise is that the log likelihood function that we want to minimize becomes:

$$\Lambda(\mathbf{O} \ \alpha) = \ln [f(\mathbf{O} \ \alpha)] = \ln \left[\prod_{i=1}^{N} \frac{1}{\sqrt{(2\pi)^N} \sigma} \right] - \frac{1}{2\sigma^2} \sum_{i=1}^{N} (O_i - \alpha)^2$$

$$\Lambda(\alpha \ \mathbf{O}) = \ln [f(\alpha \ \mathbf{O})] = \ln \left[\prod_{i=1}^{N} \frac{1}{\sqrt{(2\pi)} \sigma} \right] - \frac{1}{2\sigma^2} \sum_{i=1}^{N} (O_i - \alpha)^2$$

The value of α that maximizes the likelihood function is the same as the value of α that minimizes the error function:

$$\varepsilon(\alpha) = \sum_{i=1}^{N}(O_i - \alpha)^2$$

because the first term in $\Lambda(O \ \alpha)$ is a constant and it does not play a role in finding the minimum by taking a derivative. Therefore, the ML is indeed the same as LS estimate of multiple measurement of a parameter obtained by solving N-equations for one unknown:

$$\mathbf{O}^T - \alpha \mathbf{I}_N^T = 0.$$

A.1.4 ML Estimate of N-Observation of Function of a Parameter

In many RSS ranging problems, sometimes we have multiple observations or measurements of the RSS function of the range; here, we derive the ML estimate for these situations. We begin with the simple case of N-observations of a function of a single parameter in independent zero mean identical independent Gaussian noises with variances of σ:

$$\begin{cases} O_i = g_i(\alpha) + \eta_i; \quad i = 1, 2, ...N \\ \mathbf{O} = \mathbf{G}(\alpha) + \boldsymbol{\eta} \end{cases} , \tag{A.4a}$$

where:

$$\begin{cases} \mathbf{O} = \begin{bmatrix} O_1 & O_2 & . & . & O_N \end{bmatrix}^T \\ \mathbf{G} = \begin{bmatrix} g_1(\alpha) & g_2(\alpha) & . & . & g_N(\alpha) \end{bmatrix}^T \\ \boldsymbol{\eta} = \begin{bmatrix} \eta_1 & \eta_2 & . & . & \eta_N \end{bmatrix}^T \\ E\left(\boldsymbol{\eta}^T \boldsymbol{\eta}\right) = \sigma^2 \mathbf{I} \end{cases} .$$

The likelihood function for this case is the joint probability density function of the N observations, given by:

$$f(\mathbf{O} \ \alpha) = \prod_{i=1}^{N} \frac{1}{\sqrt{(2\pi)}\sigma} e^{-\frac{[O_i - g_i(\alpha)]^2}{2\sigma^2}} = \frac{1}{\sqrt{(2\pi)^N}\sigma^N} e^{-\frac{1}{2\sigma^2}\sum_{i=1}^{N}[O_i - g_i(\alpha)]^2}$$

$$= \frac{1}{\sqrt{(2\pi)^N}\sigma^N} e^{-\frac{[\mathbf{O} - \mathbf{G}(\alpha)]^T[\mathbf{O} - \mathbf{G}(\alpha)]}{2\sigma^2}}. \tag{A.4b}$$

$$f(\alpha \mid \mathbf{O}) = \prod_{i=1}^{N} \frac{1}{\sqrt{(2\pi)}\sigma} e^{-\frac{[O_i - g_i(\alpha)]^2}{2\sigma^2}} = \frac{1}{\sqrt{(2\pi)^N}\sigma^N} e^{-\frac{1}{2\sigma^2}\sum_{i=1}^{N}[O_i - g_i(\alpha)]^2}$$

$$= \frac{1}{\sqrt{(2\pi)^N}\sigma^N} e^{-\frac{[\mathbf{O} - \mathbf{G}(\alpha)]^T[\mathbf{O} - \mathbf{G}(\alpha)]}{2\sigma^2}}$$

Then, the log likelihood function is:

$$\Lambda(\mathbf{O} \mid \alpha) = \ln[f(\mathbf{O} \mid \alpha)]$$

$$= \ln\left[\frac{1}{\sqrt{(2\pi)^N}\sigma^N}\right] - \frac{1}{2\sigma^2}\sum_{i=1}^{N}[O_i - g_i(\alpha)]^2$$

$$= \ln\left[\frac{1}{\sqrt{(2\pi)^N}\sigma^N}\right] - \frac{1}{2\sigma^2}[\mathbf{O} - \mathbf{G}(\alpha)]^T[\mathbf{O} - \mathbf{G}(\alpha)] \quad \text{(A.4c)}$$

$$\Lambda(\alpha \mid \mathbf{O}) = \ln[f(\alpha \mid \mathbf{O})]$$

$$= \ln\left[\frac{1}{\sqrt{(2\pi)^N}\sigma^N}\right] - \frac{1}{2\sigma^2}\sum_{i=1}^{N}[O_i - g_i(\alpha)]^2$$

$$= \ln\left[\frac{1}{\sqrt{(2\pi)^N}\sigma^N}\right] - \frac{1}{2\sigma^2}[\mathbf{O} - \mathbf{G}(\alpha)]^T[\mathbf{O} - \mathbf{G}(\alpha)]$$

The ML estimate is obtained when:

$$\frac{\partial}{\partial\alpha}\{\Lambda(\mathbf{O} \mid \alpha)\} = \frac{1}{2\sigma^2}\sum_{i=1}^{N} 2g_i'(\alpha)[O_i - g_i(\alpha)]$$

$$= \frac{1}{2\sigma^2}\left\{\mathbf{G}'^T(\alpha)[\mathbf{O} - \mathbf{G}(\alpha)] + [\mathbf{O} - \mathbf{G}(\alpha)]^T\mathbf{G}'(\alpha)\right\} = 0,$$

$$\frac{\partial}{\partial\alpha}\{\Lambda(\alpha \mid \mathbf{O})\} = \frac{1}{2\sigma^2}\sum_{i=1}^{N} 2g_i'(\alpha)[O_i - g_i(\alpha)]$$

$$= \frac{1}{2\sigma^2}\left\{\mathbf{G}'^T(\alpha)[\mathbf{O} - \mathbf{G}(\alpha)] + [\mathbf{O} - \mathbf{G}(\alpha)]^T\mathbf{G}'(\alpha)\right\}$$

$$= \frac{1}{\sigma^2}\left\{[\mathbf{O} - \mathbf{G}(\alpha)]^T\mathbf{G}'(\alpha)\right\} = 0$$

which is the LS solution to solving N-equations with one unknown: $\mathbf{G}(\alpha) = \mathbf{O}$, with

$$\varepsilon(\alpha) = \sum_{i=1}^{N} [O_i - g_i(\alpha)]^2 .$$

To find the general solution, we can plot $\sum_{i=1}^{N} g_i'(\alpha) [O_i - g_i(\alpha)]$ as a function of α and find the intersect with horizontal axis. For the special case where $g_i(\alpha) = g(\alpha)$, we have:

$$\begin{cases} \frac{\partial}{\partial \alpha} \{\Lambda(\mathbf{O} \ \alpha)\} = \frac{g'(\alpha)}{\sigma^2} \sum_{i=1}^{N} [O_i - g(\alpha)] = 0 \\ g(\hat{\alpha}_{ML}) = \frac{1}{N} \sum_{i=1}^{N} O_i \\ \hat{\alpha}_{ML} = g^{-1} \left(\frac{1}{N} \sum_{i=1}^{N} O_i \right) \end{cases} \qquad , \qquad \text{(A.4d)}$$

$$\begin{cases} \frac{\partial}{\partial \alpha} \{\Lambda(\alpha \ \mathbf{O})\} = \frac{g'(\alpha)}{\sigma^2} \sum_{i=1}^{N} [O_i - g(\alpha)] = 0 \\ g(\hat{\alpha}_{ML}) = \frac{1}{N} \sum_{i=1}^{N} O_i \\ \hat{\alpha}_{ML} = g^{-1} \left(\frac{1}{N} \sum_{i=1}^{N} O_i \right) \end{cases}$$

which means we take the average of observations and we insert that in the inverse of the function.

A.2 Minimum Mean Square Error Estimation

Another popular classical method in estimation theory is the Minimum Mean Square Error (MMSE) estimation. The MMSE estimation provides the same estimated value as ML estimation except in exceptional cases. Similar to the last section, we proceed to introduce this approach using examples relevant to single parameter estimation for ranging.

A.2.1 MMSE Estimation of a Parameter

Formulation of observation of a single parameter in noise and the distribution function of the observation given the value of the parameter is given in

Equations (A.1a) and (A.1b). The MMSE of a parameter is the value of the parameter, which minimizes the square of the expected error between the parameter and the observation given by:

$$\varepsilon(\alpha) = E\left[(O - \alpha)^2\right], \tag{A.5a}$$

where expectation, E, designates the mean or average operator. To determine the MMSE estimate, we need to take the derivative of the error function with respect to the parameter and set that to zero:

$$\frac{\partial}{\partial \alpha}\left[\varepsilon(\alpha)\right] = \frac{\partial}{\partial \alpha}\left\{E\left[(O - \alpha)^2\right]\right\}$$

$$= \frac{\partial}{\partial \alpha}\left[E(O^2) - 2\alpha E(O) + \alpha^2\right] = -2E(O) + 2\alpha = 0 \tag{A.5b}$$

To understand the details of this derivation, the reader should notice that expectation and derivative are both linear operators and we can exchange their orders arbitrarily. Also, observation is a stationary process and its first- and second-order statistics are fixed values; therefore, in taking derivative, we consider them as constant numbers. Therefore, the derivative of the first term in the middle is zero and the derivative of the second term is *2E(O) = 2O*, because expectation of a number is the same as the number itself. Solution to the Equation (A.5b) provides the MMSE estimation of the parameter:

$$\hat{\alpha}_{MMSE} = E(O) = O. \tag{A.5c}$$

Since we have only one observation, its expected value is the same as the measured or observed value. Therefore, ML and MMSE estimation of a parameter in zero mean Gaussian noise are the same as our common sense, which says if we measure a parameter one time, the estimation is the same as the measured value.

A.2.2 MMSE Estimation of Function of a Parameter

Similar to derivations for ML estimation, moving from a parameter to a function of a parameter is very simple. Following notations in Equations (A.2a) and (A.2b) for formulation and distribution function, here we have to minimize the MSE:

$$\varepsilon(\alpha) = E\left\{[O - g(\alpha)]^2\right\}, \tag{A.6a}$$

which leads to solution of the following linear equation:

$$
\frac{\partial}{\partial \alpha} E\left[(O - g(\alpha))^2\right] = \frac{\partial}{\partial \alpha}\left[E(O^2) - 2g(\alpha)E(O) + g^2(\alpha)\right]
$$
$$
= -2E(O)g'(\alpha) + 2g(\alpha)g'(\alpha)
$$
$$
= -2E(O) + 2g(\alpha) = 0 \qquad \text{(A.6b)}
$$

Therefore, for the MMSE, we have:

$$
\begin{cases}
g(\hat{\alpha}_{MMSE}) = E\left[O\right] = O \\
\hat{\alpha}_{MMSE} = g^{-1}(O)
\end{cases} . \qquad \text{(A.6c)}
$$

This is the same as ML estimate of observation of a function of a parameter in noise given by Equation (A.2c).

A.2.3 MMSE Estimation of N-Observations of a Parameter

The MMSE estimation is a statistical formulation of the LS problem. For N-observation of a single parameter in noise, the error function will be:

$$
\varepsilon(\alpha) = E\left[\sum_{i=1}^{N}(O_i - \alpha)^2\right]. \qquad \text{(A.7a)}
$$

Taking derivative to minimize, we have:

$$
\frac{\partial}{\partial \alpha}\left[\varepsilon(\alpha)\right] = \frac{\partial}{\partial \alpha}\left\{E\left[\sum_{i=1}^{N}(O_i - \alpha)^2\right]\right\}
$$
$$
= \sum_{i=1}^{N} \frac{\partial}{\partial \alpha}\left[E(O_i^2) - 2\alpha E(O_i) + \alpha^2\right]
$$
$$
= \sum_{i=1}^{N}\left[-2E(O_i) + 2\alpha\right]
$$
$$
= -2\sum_{i=1}^{N} O_i + 2N\alpha = 0. \qquad \text{(A.7b)}
$$

Therefore:

$$\alpha_{MMSE} = \frac{1}{N} \sum_{i=1}^{N} O_i \qquad (A.7c)$$

This is again the same as Equation (A.6c) regardless of the variance of the estimated noise because it does not scale the observations with the inverse of their variance of noise given by Equation (A.6b). The ML estimate always provides the best estimate by scaling the observation with their level of noise.

Similar derivations for N-observations of function of a parameter in additive noise shows that:

$$\hat{\alpha}_{MMSE} = \hat{\alpha}_{ML} = g^{-1} \left[\frac{1}{N} \sum_{i=1}^{N} O_i \right]. \qquad (A.8)$$

A.2.4 MMSE Estimation of N-Observations of Function of a Parameter

For N-observation of a single parameter in noise, the error function will be:

$$\varepsilon(\alpha) = E \left\{ \sum_{i=1}^{N} [O_i - g(\alpha)]^2 \right\}. \qquad (A.9a)$$

Taking derivative to minimize, we have:

$$\begin{aligned}
\frac{\partial}{\partial \alpha} [\varepsilon(\alpha)] &= \frac{\partial}{\partial \alpha} \left[E \left\{ \sum_{i=1}^{N} [O_i - g(\alpha)]^2 \right\} \right] \\
&= \sum_{i=1}^{N} \frac{\partial}{\partial \alpha} \left[E(O_i^2) - 2g(\alpha)E(O_i) + g^2(\alpha) \right] \\
&= \sum_{i=1}^{N} \left[-2E(O_i)g'(\alpha) + 2g'(\alpha)g(\alpha) \right] \\
&= -2g'(\alpha) \left\{ \sum_{i=1}^{N} [O_i - Ng(\alpha)] \right\} = 0. \qquad (A.7b)
\end{aligned}$$

Therefore:

$$\alpha_{MMSE} = g^{-1} \left(\frac{1}{N} \sum_{i=1}^{N} O_i \right) \qquad (A.7c)$$

This is again the same as Equation (A.4c) for ML estimate when the noise for all observations is identical.

A.3 Performance Analysis Using CRLB

In classical estimation theory terminology, the *smallest* variance of the estimate of a parameter based on noisy Gaussian observations is the CRLB, and it can be calculated by inversing the Fisher Information Matrix (FIM) [Van04, Kay13, Poo13]:

$$CRLB = Var\left[\hat{\alpha}(O) - \alpha\right] \geq F^{-1}, \tag{A.8a}$$

The FIM matrix is given by:

$$\mathbf{F} = E\left[\frac{\partial \Lambda(\mathbf{O}\ \alpha)}{\partial \alpha}\right]^2 = -E\left[\frac{\partial^2 \Lambda(\mathbf{O}\ \alpha)}{\partial \alpha^2}\right], \tag{A.8b}$$

where $\Lambda(\mathbf{O}\ \alpha)$ is the log likelihood function in ML estimation process. This is a very useful mathematical tool for comparative performance evaluation of different positioning techniques. In this section, we provide examples of derivation of the CRLB for observation of a parameter, function of a parameter, multiple observation of a parameter, and multiple observations of function of a parameter in noise.

A.3.1 CRLB of Single Observation of a Parameter

For a single observation of a parameter that is corrupted by zero mean Gaussian noise with variance σ^2, Equation (A.1a), the log likelihood function is given by Equation (A.1e). Therefore, using Equation (A.8b), the FIM is:

$$\mathbf{F} = -E\left[\frac{\partial^2 \Lambda(O\ \alpha)}{\partial \alpha^2}\right] = E\left[\frac{\partial^2}{\partial \alpha^2}\left\{\ln\left[\frac{1}{\sqrt{2\pi}\sigma}\right] - \left[\frac{(O-\alpha)^2}{2\sigma^2}\right]\right\}\right] = \frac{1}{\sigma^2}. \tag{A.9a}$$

This is a scalar because we only have one observation of one parameter. The CRLB is bounded by the inverse of the FIM:

$$CRLB \geq \mathbf{F}^{-1} = \sigma^2. \tag{A.9b}$$

This simple example states that if we have a single observation of a parameter in zero mean Gaussian noise, with variance of σ^2, the variance of the estimation of the measurement is at most the same as variance of the noise. If we use ML or MMSE estimates, we achieve this bound. This result makes sense intuitively, because ML and MMSE estimates of observation of a parameter in noise are the same and both estimates use the observed value of measurement as the estimate of the value of the parameter. The variance of estimate in this case is the same as the variance of the observation noise.

A.3.2 CRLB of Observation of Function of a Parameter

Modeling of single observation of a function of a parameter in noise and the log likelihood function for this model are given by Equations (A.2a and b). The FIM is then calculated from:

$$
\mathbf{F} = E\left[\frac{\partial \Lambda(O \ \alpha)}{\partial \alpha}\right]^2 = E\left[\frac{\partial}{\partial \alpha}\left\{\ln\frac{1}{\sqrt{2\pi}\sigma} - \frac{[O - \mathrm{g}(\alpha)]^2}{2\sigma^2}\right\}\right]^2
$$

$$
= E\left[-\frac{2\left[-\mathrm{g}'(\alpha)\right]\left[O - \mathrm{g}(\alpha)\right]}{2\sigma^2}\right]^2
$$

$$
= \frac{[\mathrm{g}'(\alpha)]^2}{\sigma^4}E\left[O - \mathrm{g}(\alpha)\right]^2. \tag{A.10a}
$$

Since $E\left[O - \mathrm{g}(x)\right]^2 = \sigma^2$, the FIM and the CRLB are given by:

$$
\begin{cases}
\mathbf{F} = \frac{[\mathrm{g}'(\alpha)]^2}{\sigma^4}\sigma^2 = \frac{[\mathrm{g}'(\alpha)]^2}{\sigma^2} \\
CRLB \geq \mathbf{F}^{-1} = \frac{\sigma^2}{[\mathrm{g}'(\alpha)]^2}
\end{cases}. \tag{A.10b}
$$

Equation (A.10b) is useful for calculation of CRLB for RSS ranging, where we measure the power from, $O = P_0 - 10\alpha\log(r) + \eta$ and $g(r) = P_0 - 10\log(r)$. Since $g'(r) = 10\alpha \ [\ln(10)d]$, the CRLB is:

$$
CRLB \geq \mathbf{F}^{-1} = \frac{(\ln 10)^2}{100}\frac{\sigma^2}{\alpha^2}r^2.
$$

A.3.3 CRLB of N-Observation of a Parameter

Modeling of N-observation of a single parameter in noise and the log likelihood function for this model are given in Equations (A.3a and c). The FIM is then given by:

$$\mathbf{F} = -E\left[\frac{\partial^2 \Lambda(O\ \alpha)}{\partial \alpha^2}\right]$$

$$= E\left[\frac{\partial^2}{\partial \alpha^2}\left\{\ln\left[\prod_{i=1}^{N}\frac{1}{\sqrt{2\pi}\sigma_i}\right] - \sum_{i=1}^{N}\frac{(O_i - \alpha)^2}{2\sigma_i^2}\right\}\right]$$

$$= E\left[\frac{\partial^2}{\partial \alpha^2}\left\{\ln\left[\prod_{i=1}^{N}\frac{1}{\sqrt{2\pi}\sigma_i}\right] - \frac{[\mathbf{O} - \alpha\mathbf{I}_N]^T \Sigma^{-1}[\mathbf{O} - \alpha\mathbf{I}_N]}{2}\right\}\right]$$

$$= \sum_{i=1}^{N}\frac{1}{\sigma_i^2} = \mathbf{I}_N^T \Sigma^{-1}\mathbf{I}_N = Trace\left(\Sigma^{-1}\right). \tag{A.11a}$$

Since CRLB is the inverse of the FIM, we have:

$$CRLB \geq \mathbf{F}^{-1} = \frac{1}{\displaystyle\sum_{i=1}^{N}\frac{1}{\sigma_i^2}} = \left[\mathbf{I}_N^T \Sigma^{-1}\mathbf{I}_N\right]^{-1} = \frac{1}{Trace\left(\Sigma^{-1}\right)} \tag{A.11b}$$

In this derivation, variances of the noise for different observations are not the same. This may occur when we combine observations by different quality sensors. But, in most popular applications in ranging, variances of noise from different observations are the same, $\sigma_i^2 = \sigma^2$, and the FIM and CRLB become:

$$\begin{cases} \mathbf{F} = \frac{N}{\sigma^2} \\ CRLB \geq \mathbf{F}^{-1} = \frac{\sigma^2}{N} \end{cases} \tag{A.11c}$$

This result indicates that using N sample observations can help us reducing the variance of the estimate by N times.

A.3.4 CRLB of N-Observation of Function of a Parameter

In vector notation, the general multiple observations through different functions of a parameter becomes:

$$\begin{cases} O_i = g_i(\alpha) + \eta_i \\ \mathbf{O} = \mathbf{G}(\alpha) + \boldsymbol{\eta} \end{cases}, \tag{A.12a}$$

where:

$$\begin{cases} \mathbf{O} = \begin{bmatrix} O_1 & O_2 & . & . & O_N \end{bmatrix}^T \\ \mathbf{G}(\alpha) = \begin{bmatrix} g_1(\alpha) & g_2(\alpha) & . & . & g_N(\alpha) \end{bmatrix}^T . \\ \boldsymbol{\eta} = \begin{bmatrix} \eta_1(\sigma_1) & \eta_2(\sigma_2) & . & . & \eta_N(\sigma_N) \end{bmatrix}^T \end{cases}$$

Then, the likelihood and log likelihood are given by:

$$
\begin{cases}
f(\mathbf{O} \mid \alpha) = \prod_{i=1}^{N} \frac{1}{\sqrt{2\pi}\sigma_i} e^{-\frac{[\mathbf{O}-\mathbf{G}(\alpha)]^T \Sigma^{-1} [\mathbf{O}-\mathbf{G}(\alpha)]}{2}} \\
\Lambda(\mathbf{O} \mid \alpha) = \ln\left(f(\mathbf{O} \mid \alpha)\right) = \frac{1}{\sqrt{(2\pi)^N} \prod_{i=1}^{N} \sigma_i} \\
\qquad -\frac{1}{2}\left\{[\mathbf{O} - \mathbf{G}(\alpha)]^T \Sigma^{-1} [\mathbf{O} - \mathbf{G}(\alpha)]\right\}
\end{cases}
\tag{A.12b}
$$

Then, the FIM becomes:

$$
\begin{aligned}
\mathbf{F} &= -E\left[\frac{\partial^2 \Lambda(\mathbf{O} \mid \alpha)}{\partial \alpha^2}\right] = \frac{1}{2}E\left[\frac{\partial^2}{\partial \alpha^2}\left\{[\mathbf{O} - \mathbf{G}(\alpha)]^T \Sigma^{-1} [\mathbf{O} - \mathbf{G}(\alpha)]\right\}\right] \\
&= -\frac{1}{2}E\left[\frac{\partial}{\partial \alpha}\left\{G'^T(\alpha)\Sigma^{-1}[\mathbf{O} - \mathbf{G}(\alpha)] + [\mathbf{O} - \mathbf{G}(\alpha)]^T \Sigma^{-1}\mathbf{G}'(\alpha)\right\}\right] \\
&= -\frac{1}{2}E\left[\mathbf{G}''^T(\alpha)\Sigma^{-1}[\mathbf{O} - \mathbf{G}(\alpha)] - 2\mathbf{G}'^T(\alpha)\Sigma^{-1}\mathbf{G}'(\alpha)\right. \\
&\qquad \left. + [\mathbf{O} - \mathbf{G}(\alpha)]^T \Sigma^{-1}\mathbf{G}''(\alpha)\right] \\
&= \mathbf{G}'^T(\alpha)\Sigma^{-1}\mathbf{G}'(\alpha) = \sum_{i=1}^{N} \frac{\left[g_i'(\alpha)\right]^2}{\sigma_i^2}
\end{aligned}
\tag{A.12c}
$$

The CRLB is the inverse of the FIM given by:

$$
CRLB \geq \mathbf{F}^{-1} = \frac{1}{\displaystyle\sum_{i=1}^{N} \frac{[g_i'(\alpha)]^2}{\sigma_i^2}}.
\tag{A.12d}
$$

Assuming all functions are the same, $g_i(\alpha) = g(\alpha)$, then $\mathbf{G}'(\alpha) = g'(\alpha)\mathbf{I}_N$, the CRLB becomes:

$$
\begin{cases}
\mathbf{F} = \mathbf{G}'^T(\alpha)\Sigma^{-1}\mathbf{G}'(\alpha) = [g'(\alpha)]^2 \mathbf{I}_N^T\Sigma^{-1}\mathbf{I}_N = \dfrac{[g'(\alpha)]^2}{\displaystyle\sum_{i=1}^{N} 1/\sigma_i^2} \\
CRLB \geq \mathbf{F}^{-1} = \dfrac{\displaystyle\sum_{i=1}^{N} 1/\sigma_i^2}{[g'(\alpha)]^2}
\end{cases}
.
\tag{A.12e}
$$

Assuming same noise for all observations, $\Sigma = E\left[\boldsymbol{\eta}^T\boldsymbol{\eta}\right] = \sigma^2\mathbf{I}$, FIM will be given by:

$$\left\{ \begin{array}{l} \mathbf{F} = \dfrac{|\mathbf{G}'(\alpha)|^2}{\sigma^2} = \dfrac{\sum\limits_{i=1}^{N}\left[g_i'(\alpha)\right]^2}{\sigma^2} \\[4mm] CRLB \geq \mathbf{F}^{-1} = \dfrac{\sigma^2}{\sum\limits_{i=1}^{N}\left[g_i'(\alpha)\right]^2} \end{array} \right. \qquad (A.12d)$$

Assuming that all functions and variance of the noises are the same:

$$\left\{ \begin{array}{l} \mathbf{F} = \dfrac{|\mathbf{G}'(\alpha)|^2}{\sigma^2} = \dfrac{N\left[g_i'(\alpha)\right]^2}{\sigma^2} \\[4mm] CRLB \geq \mathbf{F}^{-1} = \dfrac{\sigma^2}{N\left[g_i'(\alpha)\right]^2} \end{array} \right. \qquad (A.12e)$$

A.4 CRLB of Continuous Waveforms

In Section A.3, we derived the CRLB for single and multiple discrete observations of a single parameter and function of a single parameter. In this section, we derive CRLB for continuous observation of a waveform in additive Gaussian noise. This derivation in classical estimation theory is used for calculation of variance of measurement noise for TOA-based ranging.

The observation used for estimation of the TOA is a continuous time waveform distorted by a continuous time thermal noise:

$$o(t) = s(t - \tau) + \eta(t), \qquad (A.13a)$$

which is a zero mean Gaussian process with variance:

$$E\left\{|\eta(t)|^2\right\} = \sigma^2. \qquad (A.13b)$$

This observation is a function of time and delay; therefore, it can be interpreted as infinite number of samples of a waveform observed in additive noise at different instances of time. To find the CRLB for this example, we take two steps: first, we calculate the discrete time CRLB for samples of the observed signal, and then, we extend that to continuous time observations.

A.4.1 CRLB of Observation of Samples of a Waveform

To calculate the CRLB for continuous observation of a waveform given by Equation (A.13a), we first take N samples of the observation function every T seconds:

$$O_i = \left[s(t - \tau) + \eta(t)\right]|_{t=iT} = s_i(\tau) + \eta_i; \quad i = 1, 2, \ldots N, \qquad (A.13c)$$

where $s_i(\tau) = s(t - \tau)|_{t=iT}$ and $\eta_i = \eta(t)|_{t=iT}$ are samples of the received signal and the observation noise, respectively. We can formulate this problem in classical estimation theory as observation of N-samples of N functions of a parameter in zero mean Gaussian noise with a variance of σ^2, where

$$\begin{cases} g_i(\tau) = s_i(\tau) = s(t - \tau)|_{t=iT} \\ E\left\{|\eta_i|^2\right\} = \sigma^2 \end{cases}. \tag{A.13d}$$

Since noise is an ergodic process, $E\left\{|\eta(t)|^2\right\} = E\left\{|\eta_i|^2\right\} = \sigma^2$. For this classical N-Observations, of N-functions of a parameter in noise, the likelihood functions is:

$$f(\mathbf{O}\ \tau) = \prod_{i=1}^{N} \frac{1}{\sqrt{(2\pi)}\sigma} e^{-\frac{[O_i - g_i(\tau)]^2}{2\sigma^2}}$$

$$= \frac{1}{\left[\sqrt{(2\pi)}\sigma\right]^N} \exp\left\{-\frac{1}{2\sigma^2}\sum_{i=1}^{N}[O_i - g_i(\tau)]^2\right\}, \tag{A.14a}$$

and the log likelihood function becomes:

$$\Lambda(\mathbf{O}\ \tau) = \ln \frac{1}{\left[\sqrt{(2\pi)}\sigma\right]^N} - \frac{1}{2\sigma^2}\sum_{i=1}^{N}[O_i - g_i(\tau)]^2. \tag{A.14b}$$

Then, the FIM matrix is found to be:

$$\mathbf{F} = -E\left[\frac{\partial^2\Lambda(\mathbf{O}\ \tau)}{\partial\tau^2}\right] = -\frac{1}{2\sigma^2}E\left[\frac{\partial}{\partial\tau}\left\{-\sum_{i=1}^{N}2\left[-g_i'(\tau)\right][O_i - g_i(\tau)]\right\}\right]$$

$$= \frac{1}{\sigma^2}E\left[\sum_{i=1}^{N}[g_i''(\tau)][O_i - g_i(\tau)] + [g_i'(\tau)]^2\right] = \frac{1}{\sigma^2}\sum_{i=1}^{N}[g_i'(\tau)]^2. \tag{A.14c}$$

The CRLB is then the inverse of the FIM given by:

$$CRLB \geq \mathbf{F}^{-1} = \frac{\sigma^2}{\sum_{i=1}^{N}[g_i'(\tau)]^2} = \frac{\sigma^2}{\sum_{i=1}^{N}[s_i'(\tau)]^2} = \frac{\sigma^2}{\sum_{i=1}^{N}[s'(iT - \tau)]^2} \tag{A.14d}$$

Knowing the waveform used for localization, $s(t)$, and the variance of the noise, σ^2, we can calculate the derivative of the waveform, $s'(t)$, and use that in (A.14d) to calculate the CRLB for discrete observation of a waveform in additive Gaussian noise.

A.4.2 CRLB of Observation of a Continuous Waveform

Continuous time waveforms are limits of the discrete time waveforms as sampling time approaches zero and the number of samples approaches infinity. Therefore, for the observation of a continuous waveform defined by Equation (A.13a), the continuous time likelihood function is:

$$
f(O \ \tau) = \frac{1}{\left[\sqrt{(2\pi)}\sigma\right]^N} \exp\left\{-\frac{1}{2\sigma^2}\sum_{i=1}^{N}[O_i - g_i(\tau)]^2\right\}\Bigg|_{\substack{T \to 0 \\ N \to \infty}},
$$

where $g_i(\tau) = s_i(\tau) = s(t-\tau)|_{t=iT}$. As we approach the limit $O_i \to o(t)$, $g_i(\tau) = s(t-\tau)|_{t=iT} \to s(t-\tau)$ and we have the continuous time likelihood function:

$$
f(o|\tau) \ \propto \ \exp\left\{\frac{2}{\sigma^2}\int_{T_0}[o(t) - s(t-\tau)]^2 dt\right\}, \tag{A.15a}
$$

and the continuous log likelihood function is:

$$
\begin{aligned}
\Lambda(\tau) &\propto \frac{2}{\sigma^2}\int_{T_0}[o(t) - s(t-\tau)]^2 dt \\
&= \frac{2}{\sigma^2}\int_{T_0}[o^2(t) - 2o(t)s(t-\tau) + s^2(t-\tau)]dt
\end{aligned} \tag{A.15b}
$$

Then, the FIM for the continuous time observation of function of a parameter becomes:

$$
\begin{aligned}
\mathbf{F} &= -E\left\{\frac{d^2}{d\tau^2}[\Lambda(\tau)]\right\} \\
&= -E\left\{\frac{2}{\sigma^2}\frac{d^2}{d\tau^2}\int_{T_0}[o^2(t) - 2o(t)s(t-\tau) + s^2(t-\tau)]dt\right\} \\
&= -\frac{2}{\sigma^2}\frac{d^2}{d\tau^2}\int_{T_0}\{E[o^2(t)]dt - 2E[o(t)]s(t-\tau) + s^2(t-\tau)\}dt
\end{aligned}
$$

The first and third terms in the bracket are constant values, because the first one is the second moment of the observation and the third term after the integral is the energy of the signal. Therefore, when we apply the derivative operator, only the middle term survives and we have:

$$\mathbf{F} = \frac{1}{\sigma^2} \int_{T_0} \frac{d^2}{d\tau^2} E\left[o(t)\right] s(t-\tau) dt = \frac{1}{\sigma^2} \int_{T_0} \frac{d^2}{d\tau^2} s^2(t-\tau) dt. \qquad \text{(A.15b)}$$

Note that from (A.13a), we have $E\left[o(t)\right] = s(t-\tau)$. We can use Parseval's theorem to calculate this integral in frequency domain. Since derivative in time is equivalent to multiplication with $j\omega$ in frequency domain, we have:

$$\mathbf{F} = \frac{1}{\sigma^2} \int_{T_0} \frac{d^2}{d\tau^2} s^2(t-\tau) dt = \frac{1}{2\pi\sigma^2} \int_{-\infty}^{+\infty} |j\omega|^2 |S(\omega)|^2 \, d\omega, \qquad \text{(A.16a)}$$

where $S(\omega)$ is the Fourier Transform of $s(t)$. Since the CRLB is the inverse of the FIM:

$$CRLB \geq \mathbf{F}^{-1} = \frac{2\pi\sigma^2}{\int\limits_{-\infty}^{+\infty} \omega^2 |S(\omega)|^2 \, d\omega}. \qquad \text{(A.16b)}$$

A.5 Generalization for Positioning

All of our derivations in this appendix were for single parameter, r, estimation that relates to ranging an estimation of the distance. Positioning draws on ranging, but for 2D or 3D coordination, we have two, $(x \ y)$, or three parameters, $(x \ y \ z)$, parameters to estimate. In this section, we generalize our derivations for single-parameter estimation for α to two-parameter estimation for $(\alpha \ \beta)$. In vector notation, the general formulation of multiple observations through different functions of a parameter becomes:

$$O_i = g_i(\alpha, \beta) + \eta_i \Rightarrow \mathbf{O} = \mathbf{G}(\alpha) + \boldsymbol{\eta}, \qquad \text{(A.17a)}$$

where:

$$\begin{cases} \mathbf{O} = \begin{bmatrix} O_1 & O_2 & . & . & O_N \end{bmatrix}^T \\ \mathbf{G}(\alpha, \beta) = \begin{bmatrix} g_1(\alpha, \beta) & g_2(\alpha, \beta) & . & . & g_N(\alpha, \beta) \end{bmatrix}^T \\ \boldsymbol{\eta} = \begin{bmatrix} \eta_1(\sigma_1) & \eta_2(\sigma_2) & . & . & \eta_N(\sigma_N) \end{bmatrix}^T \end{cases} \qquad \text{(A.17b)}$$

Then, the likelihood and log likelihood functions are given by:

$$
\begin{cases}
f(\mathbf{O}\ \alpha, \beta) = \prod_{i=1}^{N} \frac{1}{\sqrt{(2\pi)^N} \sigma_i} e^{-\frac{[\mathbf{O} - \mathbf{G}(\alpha,\beta)]^T \Sigma^{-1} [\mathbf{O} - \mathbf{G}(\alpha,\beta)]}{2}} \\
\Lambda(\mathbf{O}\ \alpha, \beta) = \ln\left(f(O\ \alpha, \beta)\right) = \dfrac{1}{\sqrt{(2\pi)^N} \prod\limits_{i=1}^{N} \sigma_i} \\
\qquad - \frac{1}{2}\left\{[\mathbf{O} - \mathbf{G}(\alpha,\beta)]^T \Sigma^{-1} [\mathbf{O} - \mathbf{G}(\alpha,\beta)]\right\}
\end{cases}
\quad . \quad \text{(A.17c)}
$$

The ML estimate is obtained by setting the gradient of the log likelihood function to zero:

$$
\nabla_{\alpha,\beta}\left[\Lambda(\mathbf{O}\ \alpha)\right] = \frac{1}{2}\left\{[\nabla_{\alpha,\beta}\mathbf{G}(\alpha,\beta)]^T \Sigma^{-1} [\mathbf{O} - \mathbf{G}(\alpha,\beta)] \right.
$$
$$
\left. + [\mathbf{O} - \mathbf{G}(\alpha,\beta)]^T \Sigma^{-1} [\nabla_{\alpha,\beta}\mathbf{G}(\alpha,\beta)]\right\} = 0,
$$

which holds when

$$
\mathbf{O} = \mathbf{G}(\alpha, \beta). \tag{A.17d}
$$

We have N-equations and two unknowns, and we need a minimum of two observations to be able to look for a solution and find the ML estimate of the parameters.

Then, the FIM becomes:

$$
\mathbf{F} = -E\left\{\nabla^2_{\alpha,\beta}\left[\Lambda(\mathbf{O}\ \alpha)\right]\right\}
$$
$$
= \frac{1}{2}E\left[\nabla^2_{\alpha,\beta}\left\{[\mathbf{O} - \mathbf{G}]^T \Sigma^{-1} [\mathbf{O} - \mathbf{G}]\right\}\right]
$$
$$
= -\frac{1}{2}E\left[\nabla_{\alpha,\beta}\left\{[\nabla_{\alpha,\beta}G]^T \Sigma^{-1} [\mathbf{O} - \mathbf{G}] + [\mathbf{O} - \mathbf{G}]^T \Sigma^{-1} [\nabla_{\alpha,\beta}G]\right\}\right]
$$
$$
= -\frac{1}{2}E\left[(\nabla^2_{\alpha,\beta}\mathbf{G})^T \Sigma^{-1} [\mathbf{O} - \mathbf{G}] - 2[\nabla_{\alpha,\beta}G]^T \Sigma^{-1} [\nabla_{\alpha,\beta}G]\right.
$$
$$
\left. + [\mathbf{O} - \mathbf{G}]^T \Sigma^{-1} (\nabla^2_{\alpha,\beta}\mathbf{G})\right]
$$
$$
= [\nabla_{\alpha,\beta}G]^T \Sigma^{-1} [\nabla_{\alpha,\beta}G] \tag{A.18a}
$$

Defining $\mathbf{H} = \nabla_{\alpha,\beta}\mathbf{G}$, the FIM and the CRLB are given by:

$$
\begin{cases}
\mathbf{F} = \mathbf{H}^T \Sigma^{-1} \mathbf{H} \\
CRLB \geq Trace\left(\mathbf{F}^{-1}\right) = Trace\left(\mathbf{H}^T \Sigma^{-1} \mathbf{H}\right)^{-1}
\end{cases}
\quad . \quad \text{(A.18b)}
$$

Assuming same and noise for all observations, $\boldsymbol{\Sigma} = \mathrm{E}\left[\boldsymbol{\eta}^T\boldsymbol{\eta}\right] = \sigma^2\mathbf{I}$, FIM and the CRLB are given by:

$$\begin{cases} \mathbf{F} = \frac{\mathbf{H}^T\mathbf{H}}{\sigma^2} \\ CRLB \geq Trace\left(\mathbf{F}^{-1}\right) = \sigma^2 Trace\left(\mathbf{H^T H}\right)^{-1} \end{cases} \tag{A.19}$$

These results can be easily extended to 3D with the same frame formulation.

List of Abbreviations

1D, 2D, 3D	1, 2, and 3-dimensional
4G	Fourth-generation cellular networks
5G	Fifth-generation cellular networks
ACF	Autocorrelation function
AP	Access points
ASIFT	Affine Scale Invariant Feature Transform
BAN	Body Area Network
B-UDP	Blocked Undetectable Direct Path
CAT-Scan	Computer-Aided Tomography Scan
CDF	Cumulative Distribution Function
CDMA	Code Division Multiple Access
CLOQ	Cooperative Localization for Optimum Quality
CNLS	Closest Neighbor Least Square
CNPD	Closest Neighbor Power Difference
CPS	Cell-tower Positioning System
CRLB	Cramer–Rao Lower Bound
CWINS	Center for wireless information network studies
CZT	Chrip Z-transform
dB	Decibel, 10 log of relative power
dBm	10 log of power in milliwatt
DDP	Detectable Direct Path
DFIR	DiFfused InfRared
DME	Distance Measurement Error
DML	D by M by L matrix
DOA	Direction Of Arrival
DSL	Digital Subscriber Line
DSSS	Direct Sequence Spread Spectrum
DS-UWB	Direct Sequence Ultra-WideBand
E-911	Emergency-911
ECEF	Earth-Centered, Earth-Fixed
EIRP	Equivalent Isotropic Radio Propagation

EKF	Extended Kalman Filtering
EV/FBCM	Eigenvector Forward Backward Correction Matrix
FCC	Federal Communications Commission
FDMA	Frequency Division Multiple Access
FDTD	Finite Difference Time Domain
FFT	Fast Fourier Transform
FHSS	Frequency Hopping Spread Spectrum
FIM	Fisher Information Matrix
GI	GastroIntestinal
GPS	Global Positioning System
HF	High-Frequency band
HFSS	High-Frequency Structure Simulation
IEEE	Institute of Electrical and Electronics Engineering
IFT	Inverse Fourier Transform
IMU	Inertia Motion Unit
IoT	Internet of Things
ISM	Industrial, Scientific, and Medical
K-CNPD	K-Closest Neighbor Power Difference
LEB	Low-Energy Bluetooth
LiDAR	Light Detection and Ranging
LOS	Line of Sight
LS	Least Square
MAC	Medium Access Control
MB-OFDM	Multi-Band Orthogonal Frequency Division Multiplexing
MEMS	Micro-Electro-Mechanical Systems
MIMO	Multiple Input Multiple Output
MLGT	Maximum Likelihood Grid Triangulation
mmWave	millimeter Wave
MRP	Mobile Reference Point
MUSIC	MUltiple SIgnal Classification
NB	NarrowBand
NC	No Coverage
NIST	National Institute of Standards and Technology
NOAA	National Oceanic and Atmospheric Administration
N-UDP	Natural Undetectable Direct Path
OFDM	Orthogonal Frequency Division Multiplexing
OLOS	Obstructed Line of Sight
PGCN	Physical Grid Closest Neighbor

PN-sequence	Pseudo-Noise sequence
PRGCN	PRobabilistic Grid Closest Neighbor
QoL	Quality of the Link
RLS	Recursive Least Square
RF	Radio Frequency
RFID	Radio-Frequency IDentification
RP	Reference Point
RPT	Receiver Processing Time
RSS	Received Signal Strength
RTLS	Real-Time Location Services
RTTOA	Round Trip Time Of Arrival
RW-RLS	Residual Weighted Recursive Least Square
SLAM	Simultaneous Localization And Mapping
SNR	Signal-to-Noise Ratio
SPF	Spatial Periodogram Filter
TDMA	Time Division Multiple Access
TDOA	Time Difference Of Arrival
TDR	Time Domain Response
TG	TarGet
TLS-ESPRIT	Total Least-Square Estimation of Signal Parameters by Rotational Invariance Techniques
TOA	Time Of Arrival
TV	TeleVision
Rx	Receiver
Tx	Transmitter
UAV	Unmanned Automotive Vehicles
UDP	Undetected Direct Path
UHF	Ultra High Frequency
U-TDOA	Uplink Time Difference of Arrival
UWB	Ultra Wideband
VNA	Vector Network Analyzer
WB	Wideband
WPS	Wireless Positioning System
WLAN	Wireless Local Area Network
WPAN	Wireless Personal Area Network
WPI	Worcester Polytechnic Institute
WVCE	Wireless Video Capsule Endoscope

List of Parameters

Chapter 2

P_r	RSS (the average received power)
r	Distance between transmitter and the receiver
r_0	An arbitrary known range as a reference distance
P_0	RSS at the reference distance r_0
α	Distance-power gradient
X	A Gaussian random variable representing shadow fading
σ	Standard deviation of zero mean shadow fading
$\widehat{P_i}$	The average of the expected received power at a distance
N	Number of samples
$\varepsilon(P_0, \alpha)$	A cost function to estimate parameters using LS algorithm
L_P	Path Loss
P_t	Transmitted Power
P_r	Received Power
h_b	BS antenna height
h_m	Height of mobile antenna
f_c	Center frequency
r_{bp}	Distance at break point
O	Observation
$g(\alpha)$	Function of the desired parameter
$\eta(\sigma)$	Zero mean Gaussian noise with standard deviation of σ
$f(O/\alpha)$	Likelihood function
$\mathbf{F}$	Fisher Information Matrix
σ_r	Standard deviation of DME
P_M	Probability that device measure the RSS
F_σ	Fade margin
γ	Certainty of coverage
P_s	Receiver Sensitivity

Chapter 3

$\mathbf{O}$	Observation Matrix
$\mathbf{G}(x, y)$	Function Matrix
$\mathbf{X}$	Gaussian Noise Matrix
$\mathbf{H}$	Jacobian Matrix
$\mathbf{I}_N$	Identity Matrix

Chapter 4

τ	Time of Flight
c	Speed of Light
T_M	Measurement Duration
W	Bandwidth of a sinc pulse
ω	Angular frequency of the transmitted siganl
ϕ	Phase of the received signal
λ	Wavelength of the carrier frequency
T_s	Transmission period
R	Symbol transmission rate
P_s	Received Power
N_0	Background Noise
$S(f)$	Power spectral density of the received signal
E_s	Energy per symbol
ρ^2	Normalized SNR
β^2	Normalized Bandwidth
σ_r	CRLB of the ranging error
σ_τ	CRLB of estimated time of flight
l	Distance between two receiver antennas
α	Angle of arrival of paths
$\Delta\tau$	TDOA between two paths measured at the receiver

Chapter 5

A	Peak amplitude of the pulse
τ	Constant determining the width of the pulse
t	Time variable
f_c	Center frequency of the pulse
T_b	Transmitted information bits duration
T_c	Sequence duration
$Y(k)$	Normalized received signal

ϕ_k	Measured phase by each carrier
Δf	Separation of frequencies
T_p	Period of signal
τ_s	Delay of the reference signal
v	Radio propagation velocity
ε_r	Permittivity
ω	Radial frequency of operation

Chapter 6

β_i	Amplitude of ith path
Φ_i	Phase of ith path
τ_i	Delay of ith path
$\zeta(d)$	Binary Variable
$D_{G0}(\theta)$	Diffraction coefficient function
$F(S)$	Transition Fresnel Function

Chapter 7

$\sum$	Covariance Matrix
$\mathbf{U}$	Matrix used in Newton-Gauss method

Chapter 8

K	Grid points
P_k^x	Weights of averaging along X
P_k^y	Weights of averaging along Y
$l_i(x, y)$	Location of i-th signature

Chapter 9

$p(t)$	Envelop
R_{ss}	Autocorrelation function of PN=sequence
$\mathbf{q}_k$	Noise eigenvectors
$\mathbf{W}$	Spatial Filter
$\mathbf{V}$	Frequency Response Vector
$\mathbf{U}_m$	Measured Calibration Matrix
$\mathbf{D}$	Diagonal time delay matrix
$\mathbf{J}$	Cost Function

Chapter 10

θ	Direction of compass heading
p	Air pressure
h	Height
φ	Angular depths of two points
$D(\lambda)$	Doppler Spectrum

References

[Akg09] Akgul, Ferit Ozan, and Kaveh Pahlavan. "Location awareness for everyday smart computing." In Telecommunications, 2009. ICT'09. International Conference on, pp. 2–7. IEEE, 2009.

[Ala03] Alavi, Bardia, and Kaveh Pahlavan. "Bandwidth effect on distance error modeling for indoor geolocation." In Personal, Indoor and Mobile Radio Communications, 2003. PIMRC'03. 14th IEEE International Symposium on, vol. 3, pp. 2198–2202. IEEE, 2003.

[Ala06] Alavi, Bardia, and Kaveh Pahlavan. "Modeling of the TOA-based distance measurement error using UWB indoor radio measurements." IEEE communications letters 10.4 (2006): 275–277.

[Ali09] Alizadeh-Shabdiz, Farshid, Kaveh Pahlavan, and Edward J. Morgan. "Estimation of speed and direction of travel in a WLAN positioning system using multiple position estimations." U.S. Patent 7,551,929, issued June 23, 2009.

[Ali09A] Alizadeh-Shabdiz, Farshid, Kaveh Pahlavan, and Nicolas Brachet. "Calculation of quality of WLAN access point characterization for use in a WLAN positioning system." U.S. Patent 7,551,579, issued June 23, 2009.

[Ali09B] Alizadeh-Shabdiz, Farshid, and Kaveh Pahlavan. "Estimation of position using WLAN access point radio propagation characteristics in a WLAN positioning system." U.S. Patent 7,515,578, issued April 7, 2009.

[Als06] Alsindi, Nayef A., Kaveh Pahlavan, Bardia Alavi, and Xinrong Li. "A novel cooperative localization algorithm for indoor sensor networks." In Personal, Indoor and Mobile Radio Communications, 2006. PIMRC'06, IEEE 17th International Symposium on, pp. 1–6. IEEE, 2006.

[Als07] Alsindi, Nayef, Xinrong Li, and Kaveh Pahlavan. "Analysis of time of arrival estimation using wideband measurements of indoor radio propagations." IEEE Transactions on Instrumentation and Measurement 56, no. 5 (2007): 1537–1545.

449

[Als08] Alsindi, Nayef, and Kaveh Pahlavan. "Cooperative localization bounds for indoor ultra-wideband wireless sensor networks." EURASIP Journal on Advances in Signal Processing 2008 (2008): 125.

[Als09] Alsindi, Nayef A., Bardia Alavi, and Kaveh Pahlavan. "Measurement and modeling of ultrawideband TOA-based ranging in indoor multipath environments." IEEE Transactions on Vehicular Technology 58, no. 3 (2009): 1046–1058.

[Aoy09] Aoyagi, Takahiro, Kenichi Takizawa, Takehiko Kobayashi, Jun-ichi Takada, and Ryuji Kohno. "Development of a WBAN channel model for capsule endoscopy." In Antennas and Propagation Society International Symposium, 2009. APSURSI'09, IEEE, pp. 1–4. IEEE, 2009.

[Ask11] Askarzadeh, Fardad, Yunxing Ye, Kaveh Ghaboosi, Sergey Makarov, and Kaveh Pahlavan. "A new perspective on the impact of diffraction in proximity of micro-metals for indoor geolocation." In Personal Indoor and Mobile Radio Communications, 2011, PIMRC'11, 22nd IEEE International Symposium on, pp. 1177–1181. IEEE, 2011.

[Ask16] Askarzadeh, Fardad, Kaveh Pahlavan, Sergey Makarov, Yunxing Ye, and Umair Khan. "Analyzing the effect of human body and metallic objects for indoor geolocation." In Medical Information and Communication Technology (ISMICT), 2016 10th International Symposium on, pp. 1–5. IEEE, 2016.

[Ask17a] Askarzadeh, Fardad, Kaveh Pahlavan, Yishuang Geng, Sergey N. Makarov, Yunxing Ye, and Umair Khan. "Modeling the Effect of Human Body on ToA Ranging Using Ray Theory." International Journal of Wireless Information Networks 24, no. 2 (2017): 140–152.

[Ask17b] Askarzadeh, Fardad. "Diffraction Analysis with UWB Validation for ToA Ranging in the Proximity of Human Body and Metallic Objects." (2017).

[Ass07] Assad, Muhammad A., Mohammad Heidari, and Kaveh Pahlavan. "Effects of channel modeling on performance evaluation of WiFi RFID localization using a laboratory testbed." In Global Telecommunications Conference, 2007. GLOBECOM'07. IEEE, pp. 366–370. IEEE, 2007.

[Au13] Au, Anthea Wain Sy, Chen Feng, Shahrokh Valaee, Sophia Reyes, Sameh Sorour, Samuel N. Markowitz, Deborah Gold, Keith Gordon, and Moshe Eizenman. "Indoor tracking and navigation using received signal strength and compressive sensing on a mobile device." IEEE Transactions on Mobile Computing 12, no. 10 (2013): 2050–2062.

[Bae07] Baeg, Seung-Ho, Jae-Han Park, Jaehan Koh, Kyung-Wook Park, and Moon-Hong Baeg. "Building a smart home environment for service robots based on RFID and sensor networks." In Control, Automation and Systems, 2007. ICCAS'07. International Conference on, pp. 1078–1082. IEEE, 2007.

[Bah00] Bahl, Paramvir, and Venkata N. Padmanabhan. "RADAR: An in-building RF-based user location and tracking system." In INFOCOM 2000. Nineteenth Annual Joint Conference of the IEEE Computer and Communications Societies. Proceedings. IEEE, vol. 2, pp. 775–784. Ieee, 2000.

[Bai06] Bailey, Tim, and Hugh Durrant-Whyte. "Simultaneous localization and mapping (SLAM): Part II." IEEE Robotics & Automation Magazine 13, no. 3 (2006): 108–117.

[Bao12] Guanqun Bao, 3D Localization of the Endoscopy Capsule, Animation at YoutTube (https://www.youtube.com/wat ch?v=h-zqFyWAZ9c), May 29, 2012.

[Bao14] Bao, Guanqun, Liang Mi, Yishuang Geng, and Kaveh Pahlavan. "A computer vision based speed estimation technique for localizing the wireless capsule endoscope inside small intestine." In 36th Annual International Conference of the IEEE Engineering in Medicine and Biology Society (EMBC), vol. 123. 2014.

[Bao15] Bao, Guanqun, Kaveh Pahlavan, and Liang Mi. "Hybrid localization of microrobotic endoscopic capsule inside small intestine by data fusion of vision and RF sensors." IEEE Sensors Journal 15, no. 5 (2015): 2669–2678.

[Bar03] Alavi, Bardia, and Kaveh Pahlavan. "Bandwidth effect on distance error modeling for indoor geolocation." In Personal, Indoor and Mobile Radio Communications, 2003. PIMRC'03. 14th IEEE International Symposium on, vol. 3, pp. 2198–2202. IEEE, 2003.

[Bar15] Bargshady, Nader, Kaveh Pahlavan, and Nayef A. Alsindi. "Hybrid WiFi/UWB, cooperative localization using particle filter." In 2015 International Conference on Computing, Networking and Communications (ICNC), pp. 1055–1060. IEEE, 2015.

[Ben99] Beneat, Jacques, Kaveh Pahlavan, and Prashant Krishnamurthy. "Radio channel characterization for geolocation at 1 GHz, 500 MHz, 90 MHz and 60 MHz in SUO/SAS." In Military Communications Conference Proceedings, 1999. MILCOM 1999. IEEE, vol. 2, pp. 1060–1063. IEEE, 1999.

[Ber94] Bertoni, Henry L., Walter Honcharenko, L. R. Macel, and Howard H. Xia. "UHF propagation prediction for wireless personal communications." Proceedings of the IEEE 82, no. 9 (1994): 1333–1359.

[Bir11] Bird, Jeff, and Dale Arden. "Indoor navigation with foot-mounted strapdown inertial navigation and magnetic sensors [emerging opportunities for localization and tracking]." IEEE Wireless Communications 18, no. 2 (2011): 28–35.

[Bis12] Biswas, Joydeep, and Manuela Veloso. "Depth camera based indoor mobile robot localization and navigation." In Robotics and Automation (ICRA), 2012 IEEE International Conference on, pp. 1697–1702. IEEE, 2012.

[Bru13] Bruno, Luigi, and Patrick Robertson. "Observability of path loss parameters in WLAN-based Simultaneous Localization and Mapping." In IPIN, pp. 1–10. 2013.

[Cas12a] Castellanos, Jose A., and Juan D. Tardos. Mobile robot localization and map building: A multisensor fusion approach. Springer Science & Business Media, 2012.

[Cas12b] Castro, Pablo Samuel, Daqing Zhang, and Shijian Li. "Urban traffic modelling and prediction using large scale taxi GPS traces." In International Conference on Pervasive Computing, pp. 57–72. Springer, Berlin, Heidelberg, 2012.

[Che02] Chen, Yongguang, and Hisashi Kobayashi. "Signal strength based indoor geolocation." In Communications, 2002. ICC 2002. IEEE International Conference on, vol. 1, pp. 436–439. IEEE, 2002.

[Che12] Chen, S. Y. "Kalman filter for robot vision: a survey." IEEE Transactions on Industrial Electronics 59, no. 11 (2012): 4409–4420.

[Che99] Chen, Pi-Chun. "A non-line-of-sight error mitigation algorithm in location estimation." In Wireless Communications and Networking Conference, 1999. WCNC. 1999 IEEE, vol. 1, pp. 316–320. IEEE, 1999.

[Coc91] Coco, David S., Clayton Coker, Scott R. Dahlke, and James R. Clynch. "Variability of GPS satellite differential group delay biases." IEEE Transactions on Aerospace and Electronic Systems 27, no. 6 (1991): 931–938.

[Col13] Collotta, Mario, Arcengelo Lo Cascio, Giovanni Pau, and Gianfranco Scatá. "Smart localization platform for IEEE 802.11 industrial networks." In Industrial Embedded Systems (SIES), 2013 8th IEEE International Symposium on, pp. 69–72. IEEE, 2013.

[Col15] Collotta, Mario, Giovanni Pau, Giovanni Tesoriere, and Salvatore Tirrito. "Intelligent shoe system: A self-powered wearable device for personal localization." In AIP Conference Proceedings, vol. 1648, no. 1, p. 780004. AIP Publishing, 2015.

[Dar08] Dardari, Davide, Andrea Conti, Jaime Lien, and Moe Z. Win. "The effect of cooperation on UWB-based positioning systems using experimental data." EURASIP Journal on Advances in Signal Processing 2008 (2008): 124.

[Dav07] Davison, Andrew J., Ian D. Reid, Nicholas D. Molton, and Olivier Stasse. "MonoSLAM: Real-time single camera SLAM." IEEE Transactions on Pattern Analysis & Machine Intelligence 6 (2007): 1052–1067.

[Dav13] David Schneider, New Indoor Navigation Technologies Work Where GPS Can't, IEEE Spectrum, 2013.

[Dea12] Deak, Gabriel, Kevin Curran, and Joan Condell. "A survey of active and passive indoor localisation systems." Computer Communications 35, no. 16 (2012): 1939–1954.

[Dum94] Dumont, L., M. Fattouche, and G. Morrison. "Super-resolution of multipath channels in a spread spectrum location system." Electronics Letters 30, no. 19 (1994): 1583–1584.

[Duv99] Duvall, William R. "Vehicles tracking transponder system and transponding method." U.S. Patent 5,917,423, issued June 29, 1999.

[Est08] Esteban, I., O. Booij, Z. Zivkovic, and B. Krose. "Mapping large environments with an omnivideo camera." In ASCII, Heijen, The Netherlands, 2008. meters-1. 2008.

[Fan14] Fang, Xi, Satyajayant Misra, Guoliang Xue, and Dejun Yang. "Smart grid—The new and improved power grid: A survey." IEEE communications surveys & tutorials 14, no. 4 (2012): 944–980.

[For06] Fort, Andrew, Claude Desset, Philippe De Doncker, Piet Wambacq, and Leo Van Biesen. "An ultra-wideband body area propagation channel model-from statistics to implementation." IEEE Transactions on Microwave Theory and Techniques 54, no. 4 (2006): 1820–1826.

[Fu11] Fu, Ruijun, Yunxing Ye, Ning Yang, and Kaveh Pahlavan. "Doppler spread analysis of human motions for body area network applications." In Personal Indoor and Mobile Radio Communications, 2011. PIMRC'11, IEEE 22nd International Symposium on, pp. 2209–2213. IEEE, 2011.

[Fue15] Fuentes-Pacheco, Jorge, José Ruiz-Ascencio, and Juan Manuel Rendón-Mancha. "Visual simultaneous localization and mapping: a survey." Artificial Intelligence Review 43, no. 1 (2015): 55–81.

[Gan13] Pan, Gang, Guande Qi, Wangsheng Zhang, Shijian Li, Zhaohui Wu, and Laurence Tianruo Yang. "Trace analysis and mining for smart cities: issues, methods, and applications." IEEE Communications Magazine 51, no. 6 (2013): 120–126.

[Gav04] Ghavami, Mohammad, Lachlan Michael, and Ryuji Kohno. Ultra wideband signals and systems in communication engineering. John Wiley & Sons, 2007.

[Gen13] Geng, Yishuang, Jie He, and Kaveh Pahlavan. "Modeling the effect of human body on TOA based indoor human tracking." International Journal of Wireless Information Networks 20, no. 4 (2013): 306–317.

[Gen15] Geng, Yishuang, and Kaveh Pahlavan. "On the accuracy of rf and image processing based hybrid localization for wireless capsule endoscopy." In Wireless Communications and Networking Conference (WCNC), 2015 IEEE, pp. 452–457. IEEE, 2015.

[Gen16] Geng, Yishuang, Jin Chen, Ruijun Fu, Guanqun Bao, and Kaveh Pahlavan. "Enlighten wearable physiological monitoring systems: On-body rf characteristics based human motion classification using a support vector machine." IEEE Transactions on Mobile Computing 15, no. 3 (2016): 656–671.

[Gez05] Gezici, Sinan, Zhi Tian, Georgios B. Giannakis, Hisashi Kobayashi, Andreas F. Molisch, H. Vincent Poor, and Zafer Sahinoglu. "Localization via ultra-wideband radios: a look at positioning aspects for future sensor networks." IEEE signal processing magazine 22, no. 4 (2005): 70–84.

[Gre01] GREWAL, M.S. and ANDREWS, A.P., 2001. Kalman Filtering: Theory and Practice Using MATLAB. John Wiley & Sons.

[Gu15] Gu, Yu, and Fuji Ren. "Energy-efficient indoor localization of smart hand-held devices using Bluetooth." IEEE Access 3 (2015): 1450–1461.

[Gud91] Gudmundson, Mikael. "Correlation model for shadow fading in mobile radio systems." Electronics letters 27, no. 23 (1991): 2145–2146.

[Gus02] Gustafsson, Fredrik, Fredrik Gunnarsson, Niclas Bergman, Urban Forssell, Jonas Jansson, Rickard Karlsson, and P-J. Nordlund. "Particle filters for positioning, navigation, and tracking." IEEE Transactions on Signal Processing 50, no. 2 (2002): 425–437.

[Hac14] Hackmann, Gregory, Weijun Guo, Guirong Yan, Zhuoxiong Sun, Chenyang Lu, and Shirley Dyke. "Cyber-physical codesign of distributed structural health monitoring with wireless sensor networks." IEEE Transactions on Parallel and Distributed Systems 25, no. 1 (2014): 63–72.

[Hag12] Hagenauer, Julian, and Marco Helbich. "Mining urban land-use patterns from volunteered geographic information by means of genetic algorithms and artificial neural networks." International Journal of Geographical Information Science 26, no. 6 (2012): 963–982.

References 455

[Har99] Har, Dongsoo, Howard H. Xia, and Henry L. Bertoni. "Path-loss prediction model for microcells." IEEE Transactions on Vehicular Technology 48, no. 5 (1999): 1453–1462.

[Hat05] Hatami, Ahmad, and Kaveh Pahlavan. "A comparative performance evaluation of RSS-based positioning algorithms used in WLAN networks." In Wireless Communications and Networking Conference, 2005 IEEE, vol. 4, pp. 2331–2337. IEEE, 2005.

[Hat06] Hatami, Ahmad. "Application of Channel Modeling for Indoor Localization Using TOA and RSS." PhD diss., Worcester Polytechnic Institute, 2006.

[Hat06a] Hatami, Ahmad, and Kaveh Pahlavan. "Comparative statistical analysis of indoor positioning using empirical data and indoor radio channel models." In Consumer Communications and Networking Conference, 2006. CCNC 2006. 3rd IEEE, vol. 2, pp. 1018–1022. IEEE, 2006.

[Hat06b] Hatami, A., 2006. Application of Channel Modeling for Indoor Localization Using TOA and RSS (Doctoral dissertation, Worcester Polytechnic Institute).

[Hay06] Haykin, Simon, ed. Nonlinear methods of spectral analysis. Vol. 34. Springer Science & Business Media, 2006.

[He11] He, Miao, and Junshan Zhang. "A dependency graph approach for fault detection and localization towards secure smart grid." IEEE Transactions on Smart Grid 2, no. 2 (2011): 342–351.

[Hei06] Heidari, Mohammad, and Kaveh Pahlavan. "Markov model for dynamic behavior of ranging errors in indoor geolocation systems." IEEE Communications Letters 11, no. 12 (2007).

[Hei07] M. Heidari and K. Pahlavan, "Performance Evaluation of RFID Localization Techniques", Chapter 10, RFID Technology and Applications, Cambridge University Press, 2007.

[Hei09] Heidari, Mohammad, Nayef Ali Alsindi, and Kaveh Pahlavan. "UDP identification and error mitigation in ToA-based indoor localization systems using neural network architecture." IEEE Transactions on Wireless Communications 8, no. 7 (2009).

[Hel05] Helal, Sumi, William Mann, Hicham El-Zabadani, Jeffrey King, Youssef Kaddoura, and Erwin Jansen. "The gator tech smart house: A programmable pervasive space." Computer 38, no. 3 (2005): 50–60.

[Hel97] Hellebrandt, Martin, Rudolf Mathar, and Markus Scheibenbogen. "Estimating position and velocity of mobiles in a cellular radio network." IEEE Transactions on Vehicular Technology 46, no. 1 (1997): 65–71.

[Hen12] Henry, Peter, Michael Krainin, Evan Herbst, Xiaofeng Ren, and Dieter Fox. "RGB-D mapping: Using Kinect-style depth cameras for dense 3D modeling of indoor environments." The International Journal of Robotics Research 31, no. 5 (2012): 647–663.

[Hil08] Hile, Harlan, and Gaetano Borriello. "Positioning and orientation in indoor environments using camera phones." IEEE Computer Graphics and Applications 28, no. 4 (2008).

[HLS] Homeland Security, Critical Infrastructure Sectors, http://www.dhs.gov/critical-infrastructure-sectors.

[Hol09] Holm, Sverre. "Hybrid ultrasound-RFID indoor positioning: Combining the best of both worlds." In RFID, 2009 IEEE International Conference on, pp. 155–162. IEEE, 2009.

[Hol92] Holt, T., Kaveh Pahlavan, and Jin-Fa Lee. "A graphical indoor radio channel simulator using 2D ray tracing." In Personal, Indoor and Mobile Radio Communications, 1992. PIMRC'92, 3rd IEEE International Symposium on, pp. 411–416. IEEE, 1992.

[How90] Howard, Steven J., and Kaveh Pahlavan. "Measurement and analysis of the indoor radio channel in the frequency domain." IEEE Transactions on Instrumentation and Measurement 39, no. 5 (1990): 751–755.

[How92] Howard, Steven J., and Kaveh Pahlavan. "Autoregressive modeling of wide-band indoor radio propagation." IEEE Transactions on Communications 40, no. 9 (1992): 1540–1552.

[Hua11] Huang, Joseph, David Millman, Morgan Quigley, David Stavens, Sebastian Thrun, and Alok Aggarwal. "Efficient, generalized indoor wifi graphslam." In Robotics and Automation (ICRA), 2011 IEEE International Conference on, pp. 1038–1043. IEEE, 2011.

[Idd00] Iddan, Gavriel, Gavriel Meron, Arkady Glukhovsky, and Paul Swain. "Wireless capsule endoscopy." Nature 405, no. 6785 (2000): 417.

[Kan04A] Kanaan, Muzaffer, and Kaveh Pahlavan. "A comparison of wireless geolocation algorithms in the indoor environment." In Wireless Communications and Networking Conference, 2004. WCNC. 2004 IEEE, vol. 1, pp. 177–182. IEEE, 2004.

[Kan04B] Kanaan, M., and K. Pahlavan. "Algorithm for TOA-based indoor geolocation." Electronics Letters 40, no. 22 (2004): 1.

[Kar12] Karaliopoulos, Merkourios, and Christian Rohner. "Trace-based performance analysis of opportunistic forwarding under imperfect node cooperation." In INFOCOM, 2012 Proceedings IEEE, pp. 2651–2655. IEEE, 2012.

[Kay13] Kay, Steven M. Fundamentals of statistical signal processing: Practical algorithm development. Vol. 3. Pearson Education, 2013.

[Kel11] Kelly, Jonathan, and Gaurav S. Sukhatme. "Visual-inertial sensor fusion: Localization, mapping and sensor-to-sensor self-calibration." The International Journal of Robotics Research 30, no. 1 (2011): 56–79.

[Kha11] Pahlavan, Kaveh, Guanqun Bao, Yunxing Ye, Sergey Makarov, Umair Khan, Pranay Swar, D. Cave, Andrew Karellas, Prashant Krishnamurthy, and Kamran Sayrafian. "Rf localization for wireless video capsule endoscopy." International Journal of Wireless Information Networks 19, no. 4 (2012): 326–340.

[Kha18] Khan, Umair Ishaq, Sergey Makarov, Yunxing Ye, Ruijun Fu, Pranay Swar, and Kaveh Pahlavan. "Review of Computational Techniques for Performance Evaluation of RF Localization inside the Human Body." IEEE Reviews in Biomedical Engineering (2018).

[Kha18b] Khan, Umair. "Statistical Modelling and Performance Evaluation of TOA for Localization inside the Human Body using Computational Techniques." (2018).

[Kha18c] Khan, Umair, Yunxing Ye, Ain-Ul Aisha, Pranay Swar, and Kaveh Pahlavan. "Precision of EM Simulation Based Wireless Location Estimation in Multi-Sensor Capsule Endoscopy." IEEE journal of translational engineering in health and medicine 6 (2018): 1–11.

[Kle06] https://www.mathworks.com/matlabcentral/fileexchange/7941-convert-cartesian–ecef–coordinates-to-lat–lon–alt?focused=5062924&tab=function

[Ko12] Ko, Nak Yong, and Tae Gyun Kim. "Comparison of Kalman filter and particle filter used for localization of an underwater vehicle." In Ubiquitous Robots and Ambient Intelligence (URAI), 2012 9th International Conference on, pp. 350–352. IEEE, 2012.

[Koh04] Kohno, Ryuji. "Project: IEEE P802. 15 Working Group for Wireless Personal Area Networks (WPANs)." doc.: IEEE (2004): 802–15.

[Kur09] Kurup, Divya, Wout Joseph, Günter Vermeeren, and Luc Martens. "Path loss model for in-body communication in homogeneous human muscle tissue." Electronics letters 45, no. 9 (2009): 453–454.

[Kut06] R. Kuth, J. Reinschke and R. Rockelein, U.S. Patent Application 11/481,935, 2006.

[Lee12] Lee, En-Shiun Annie, Franky Kin-Wai Yeung, and Tzu-Yang Yu. "Variable categorization and modelling: A novel adversarial approach to mobile location-based advertising." In AAAI Workshops. 2012.

[Li03] Xinrong Li, Super-resolution TOA estimation with diversity for indoor geolocation, PhD Dissertation, WPI, May 2003

[Li04] Li, Xinrong, and Kaveh Pahlavan. "Super-resolution TOA estimation with diversity for indoor geolocation." IEEE Transactions on Wireless Communications 3, no. 1 (2004): 224–234.

[Li12] Li, Shen, Yishuang Geng, Jie He, and Kaveh Pahlavan. "Analysis of three-dimensional maximum likelihood algorithm for capsule endoscopy localization." In Biomedical Engineering and Informatics (BMEI), 2012 5th International Conference on, pp. 721–725. IEEE, 2012.

[Li16] Li, Zhouchi, Yang Yang, and Kaveh Pahlavan. "Using iBeacon for newborns localization in hospitals." In Medical Information and Communication Technology (ISMICT), 2016 10th International Symposium on, pp. 1–5. IEEE, 2016.

[Liu15] Liu, Guanxiong, Yishuang Geng, and Kaveh Pahlavan. "Effects of calibration RFID tags on performance of inertial navigation in indoor environment." In Computing, Networking and Communications (ICNC), 2015 International Conference on, pp. 945–949. IEEE, 2015.

[Ma14] Ma, Yongtao, Kaveh Pahlavan, and Yishuang Geng. "Comparison of POA and TOA based ranging behavior for RFID application." In Personal, Indoor, and Mobile Radio Communication, 2014. PIMRC'14, 25th IEEE International Symposium on, pp. 1722–1726. IEEE, 2014.

[Mak11] Makarov, Sergey N., Umair I. Khan, Md Monirul Islam, Reinhold Ludwig, and Kaveh Pahlavan. "On accuracy of simple FDTD models for the simulation of human body path loss." In Sensors Applications Symposium (SAS), 2011 IEEE, pp. 18–23. IEEE, 2011.

[Man00] D. Manolakis, V. Ingle, and S. Kogon, Statistical and Adaptive Signal Processing. New York: McGraw-Hill, 2000.

[Mar12] Martel, Sylvain. "Journey to the center of a tumor." IEEE Spectrum 49, no. 10 (2012).

[Mar14] Marya, Neil, Andrew Karellas, Anne Foley, Abhijit Roychowdhury, and David Cave. "Computerized 3-dimensional localization of a video capsule in the abdominal cavity: validation by digital radiography." Gastrointestinal endoscopy 79, no. 4 (2014): 669–674.

[Mar85] Marcus, Michael J. "Recent US regulatory decisions on civil uses of spread spectrum." In IEEE GLOBECOM, vol. 16, pp. 1–16. 1985.

[Mi14] Liang Mi, Guanqun Bao, and Kaveh Pahlavan, "Analysis of the Impact of Intestinal Motility on the Speed Estimation of Video Capsule Endoscope," The 36th Annual International Conference of the IEEE Engineering in Medicine and Biology Society (EMBC'14), Chicago, USA, August 26–30, 2014.

[Mil08] Stephen, B. MILES, E. SARMA Sanjay, and R. WILLIAMS JOHN. "RFID Technology and Applications." Cambridge Univ Pr (2008).

[Mis06] Misra, Pratap, and Per Enge. "Global Positioning System: signals, measurements and performance second edition." Massachusetts: Ganga-Jamuna Press (2006).

[Moa11] N. Moayeri, J. Mapar, S. Tompkins and K. Pahlavan (Editors), "Localization and Tracking for Emerging Wireless Systems", IEEE Wireless Communications Magazine, vol 18 (2), pp. 8–9, April 2011.

[Moe09] Morel, Jean-Michel, and Guoshen Yu. "ASIFT: A new framework for fully affine invariant image comparison." SIAM journal on imaging sciences 2, no. 2 (2009): 438–469.

[Moh14] Mohedano, Raul, Andrea Cavallaro, and Narciso Garcia. "Camera localization using trajectories and maps." IEEE Transactions on Pattern Analysis and Machine Intelligence 36, no. 4 (2014): 684–697.

[Mol12] Molisch, Andreas F. Wireless communications. Vol. 34. John Wiley & Sons, 2012.

[Mul09] Mulloni, Alessandro, Daniel Wagner, Istvan Barakonyi, and Dieter Schmalstieg. "Indoor positioning and navigation with camera phones." IEEE Pervasive Computing 2 (2009): 22–31.

[Nad12] Nadimi, Esmaeil S., and Vahid Tarokh. "Bayesian source localization in networks with heterogeneous transmission medium." Navigation: Journal of The Institute of Navigation 59, no. 3 (2012): 163–175.

[Niu16] Niu, Luyao, Yingyue Fan, Kaveh Pahlavan, Guanxiong Liu, and Yishuang Geng. "On the accuracy of Wi-Fi localization using robot and human collected signatures." In Consumer Electronics (ICCE), 2016 IEEE International Conference on, pp. 375–378. IEEE, 2016.

[Opp04] Oppermann, Ian, Matti Hämäläinen, and Jari Iinatti, eds. UWB: theory and applications. John Wiley & Sons, 2005.

[Pah02] Pahlavan, Kaveh, Xinrong Li, and Juha-Pekka Makela. "Indoor geolocation science and technology." IEEE Communications Magazine 40, no. 2 (2002): 112–118.

[Pah05] Pahlavan, Kaveh, and Allen H. Levesque. Wireless information networks. Vol. 93. John Wiley & Sons, 2005.

[PAh06] Pahlavan, Kaveh, Ferit O. Akgul, Mohammad Heidari, Ahmad Hatami, John M. Elwell, and Robert D. Tingley. "Indoor geolocation in the absence of direct path." IEEE Wireless Communications 13, no. 6 (2006): 50–58.

[Pah10] Pahlavan, K., F. Akgul, Y. Ye, T. Morgan, F. Alizadeh-Shabdiz, M. Heidari, and C. Steger. "Taking positioning indoors Wi-Fi localization and GNSS." Inside GNSS 5, no. 3 (2010): 40–47.

[Pah11] Pahlavan, Kaveh, Nayef Al-Sindi, and Bardia Alavi. "Precise node localization in sensor ad-hoc networks." U.S. Patent 8,005,486, issued August 23, 2011.

[Pah11] Pahlavan, Kaveh, Yunxing Ye, Umair Khan, and Ruijun Fu. "RF localization inside human body: Enabling micro-robotic navigation for medical applications." In Localization and GNSS (ICL-GNSS), 2011 International Conference on, pp. 133–139. IEEE, 2011.

[Pah12] Pahlavan, Kaveh, Guanqun Bao, Yunxing Ye, Sergey Makarov, Umair Khan, Pranay Swar, D. Cave, Andrew Karellas, Prashant Krishnamurthy, and Kamran Sayrafian. "Rf localization for wireless video capsule endoscopy." International Journal of Wireless Information Networks 19, no. 4 (2012): 326–340.

[Pah12] Pahlavan, Kaveh, Yunxing Ye, Ruijun Fu, and Umair Khan. "Challenges in channel measurement and modeling for RF localization inside the human body." International Journal of Embedded and Real-Time Communication Systems (IJERTCS) 3, no. 3 (2012): 18–37.

[Pah13] Pahlavan, Kaveh, and Prashant Krishnamurthy. Principles of wireless access and localization. John Wiley & Sons, 2013.

[Pah88] Pahlavan, Kaveh, and Jerry L. Holsinger. "Voice-band data communication modems-a historical review: 1919–1988." IEEE Communications Magazine 26, no. 1 (1988): 16–27.

[Pah89] Pahlavan, K., and S. J. Howard. "Frequency domain measurements of indoor radio channels." Electronics Letters 25, no. 24 (1989): 1645–1647.

[Pah95] K. Pahlavan and A. Levesque, Wireless Information Networks, John Wiley and Sons, 1995.

[Pah98] Pahlavan, Kaveh, Prashant Krishnamurthy, and A. Beneat. "Wideband radio propagation modeling for indoor geolocation applications." IEEE Communications Magazine 36, no. 4 (1998): 60–65.

[Pal91] Pallas, M-A., and Genevieve Jourdain. "Active high resolution time delay estimation for large BT signals." IEEE Transactions on Signal Processing 39, no. 4 (1991): 781–788.

[Poo13] Poor, H. Vincent. An introduction to signal detection and estimation. Springer Science & Business Media, 2013.

[Pu13] Pu, Qifan, Sidhant Gupta, Shyamnath Gollakota, and Shwetak Patel. "Whole-home gesture recognition using wireless signals." In Proceedings of the 19th annual international conference on Mobile computing & networking, pp. 27–38. ACM, 2013.

[Rap02] Theodore, S. "Rappaport, wireless communications principles and practice (pp. 387–390)." (2002).

[Rob09] Roberts, Brian, and Kaveh Pahlavan. "Site-specific RSS signature modeling for WiFi localization." In Global Telecommunications Conference, 2009. GLOBECOM 2009. IEEE, pp. 1–6. IEEE, 2009.

[Roo02] Roos, Teemu, Petri Myllymäki, Henry Tirri, Pauli Misikangas, and Juha Sievänen. "A probabilistic approach to WLAN user location estimation." International Journal of Wireless Information Networks 9, no. 3 (2002): 155–164.

[Roo02b] Roos, Teemu, Petri Myllymaki, and Henry Tirri. "A statistical modeling approach to location estimation." IEEE Transactions on Mobile Computing 99, no. 1 (2002): 59–69.

[Ruf09] T. Ruff, "RFID Tracking Systems – Conventional and Reverse RFID," Mine Communications and Tracking Workshop, Lakewood, CO, May 2009.

[Saa97] Saarnisaari, Harri. "TLS-ESPRIT in a time delay estimation." In Vehicular Technology Conference, 1997, IEEE 47th, vol. 3, pp. 1619–1623. IEEE, 1997.

[Say09] Sayrafian-Pour, Kamran, Wen-Bin Yang, John Hagedorn, Judith Terrill, and Kamya Yekeh Yazdandoost. "A statistical path loss model for medical implant communication channels." In Personal, Indoor and Mobile Radio Communications, 2009 IEEE 20th International Symposium on, pp. 2995–2999. IEEE, 2009.

[Say10] Sayrafian-Pour, Kamran, Wen-Bin Yang, John Hagedorn, Judith Terrill, Kamya Yekeh Yazdandoost, and Kiyoshi Hamaguchi. "Channel models for medical implant communication." International Journal of Wireless Information Networks 17, no. 3–4 (2010): 105–112.

[Shi10] Zhong, Billy, Cengiz Karakoyunlu, Ramsey D. Abouzahra, and Yuan Shi. "Hybrid Indoor Geolocation for Robotic Applications." (2010).

[Siw03] Siwiak, K., H. Bertoni, and S. M. Yano. "Relation between multipath and wave propagation attenuation." Electronics Letters 39, no. 1 (2003): 142–143.

[Spe00] Spencer, Quentin H., Brian D. Jeffs, Michael A. Jensen, and A. Lee Swindlehurst. "Modeling the statistical time and angle of arrival characteristics of an indoor multipath channel." IEEE Journal on Selected areas in communications 18, no. 3 (2000): 347–360.

[Spi10] Spinella, Silverio C., Antonio Iera, and Antonella Molinaro. "On potentials and limitations of a hybrid WLAN-RFID indoor positioning technique." International Journal of Navigation and Observation 2010 (2010).

[Swa12] Swar, Pranay, Kaveh Pahlavan, and Umair Khan. "Accuracy of localization system inside human body using a fast FDTD simulation technique." In Medical Information and Communication Technology (ISMICT), 2012 6th International Symposium on, pp. 1–6. IEEE, 2012.

[Tar11] Tarzia, Stephen P., Peter A. Dinda, Robert P. Dick, and Gokhan Memik. "Indoor localization without infrastructure using the acoustic background spectrum." In Proceedings of the 9th international conference on Mobile systems, applications, and services, pp. 155–168. ACM, 2011.

[Tay01] Taylor, James D. Ultra-wideband radar technology. CRC press, 2000.

[Tha12] Than, Trung Duc, Gürsel Alici, Hao Zhou, and Weihua Li. "A review of localization systems for robotic endoscopic capsules." IEEE Transaction on Biomedical Engineering 59, no. 9 (2012): 2387–2399.

[Tin01] Tingley, Robert D., and Kaveh Pahlavan. "Space-time measurement of indoor radio propagation." IEEE Transactions on Instrumentation and Measurement 50, no. 1 (2001): 22–31.

[Van04] Van Trees, Harry L. Detection, estimation, and modulation theory, part I: detection, estimation, and linear modulation theory. John Wiley & Sons, 2004.

[V-Est15] Estimote Indoor Location and Nearables, Published on Sep 2, 2015 Estimote, https://www.youtube.com/watch?v=9cH44pE1Fks&t=6s

[V-Est16] Estimote Mirror - the world's first video-enabled beacon, Published on Sep 14, 2016, Estimote. https://www.youtube.com/watch?v=FoVvPZRFd1I.

[V-Gyr09] Gyroscope, Youtube, published Jan 2, 2009, https://www.youtube.com/watch?v=cquvA_IpEsA

[Wan11] Wang, Dashun, Dino Pedreschi, Chaoming Song, Fosca Giannotti, and Albert-Laszlo Barabasi. "Human mobility, social ties, and link prediction." In Proceedings of the 17th ACM SIGKDD international conference on Knowledge discovery and data mining, pp. 1100–1108. Acm, 2011.

[Wan11] Wang, Yi, Ruijun Fu, Yunxing Ye, Umair Khan, and Kaveh Pahlavan. "Performance bounds for RF positioning of endoscopy camera capsules." In Biomedical Wireless Technologies, Networks, and Sensing Systems (BioWireleSS), 2011 IEEE Topical Conference on, pp. 71–74. IEEE, 2011.

[Win00] Win, Moe Z., and Robert A. Scholtz. "Ultra-wide bandwidth time-hopping spread-spectrum impulse radio for wireless multiple-access communications." IEEE Transactions on Communications 48, no. 4 (2000): 679–689.

[Xu18] Xu, Liyuan, Jie He, Julang Ying, Peng Wang, Kaveh Pahlavan, and Qin Wang. "UWB body motion assisted indoor geolocation with a single reference point." In Position, Location and Navigation Symposium (PLANS), 2018 IEEE/ION, pp. 1509–1514. IEEE, 2018.

[Yan94] Yang, Ganning, Kaveh Pahlavan, and Timothy J. Holt. "Sector antenna and DFE modems for high speed indoor radio communications." IEEE Transactions on Vehicular Technology 43, no. 4 (1994): 925–933.

[Ye10] Yunxing Ye, Ferit Ozan Akgul, Nader Bargshady, Kaveh Pahlavan, "Performance of Hybrid WiFi Localization for Cooperative Robotics Applications", IEEE/ION PLANS 2010.

[Ye12] Ye, Haibo, Tao Gu, Xiaorui Zhu, Jinwei Xu, Xianping Tao, Jian Lu, and Ning Jin. "FTrack: Infrastructure-free floor localization via mobile phone sensing." In Pervasive Computing and Communications (PerCom), 2012 IEEE International Conference on, pp. 2–10. IEEE, 2012.

[Ye12] Ye, Yunxing, Pranay Swar, Kaveh Pahlavan, and Kaveh Ghaboosi. "Accuracy of RSS-based RF localization in multi-capsule endoscopy." International Journal of Wireless Information Networks 19, no. 3 (2012): 229–238.

[Yim13] Yim, Sehyuk, and Metin Sitti. "3-D localization method for a magnetically actuated soft capsule endoscope and its applications." IEEE Transactions on Robotics 29, no. 5 (2013): 1139–1151.

[Yin15] Ying, Julang, Chao Ren, and Kaveh Pahlavan. "On automated map selection problem in indoor navigation for smart devices." In Proc. IEEE Telecommun. Submit. 2015.

[Yin17] Ying, Julang, Kaveh Pahlavan, and Xinrong Li. "Precision of RSS-based indoor geolocation in IoT applications." In Personal, Indoor, and Mobile Radio Communications, 2017. PIMRC'17, 28th IEEE International Symposium on, pp. 1–5. IEEE, 2017.

[Zan09] Zandbergen, Paul A. "Accuracy of iPhone locations: A comparison of assisted GPS, WiFi and cellular positioning." Transactions in GIS 13 (2009): 5–25.

[Zha90] Zhang, Ker, and Kaveh Pahlavan. "An integrated voice/data system for mobile indoor radio networks." IEEE Transactions on Vehicular Technology 39, no. 1 (1990): 75–82.

[Zhe16] Zheng, Yang, Yuzhang Zang, and Kaveh Pahlavan. "UWB localization modeling for electronic gaming." In Consumer Electronics (ICCE), 2016 IEEE International Conference on, pp. 170–173. IEEE, 2016.

Index

About the Author

Kaveh Pahlavan is a Professor of ECE, a Professor of CS, and Director of the Center for Wireless Information Network Studies, Worcester Polytechnic Institute, Worcester, MA. Since the inception of Skyhook Wireless, Boston, MA (2005), the world's pioneer in Wi-Fi localization for smart devices, he has been the chief technical advisor of the company. From 1995 to 2007, he had a long-term and productive cooperation with the University of Oulu and Nokia in Finland (1995–2007). He is renowned for his pioneering research in Wi-Fi and indoor geolocation and has contributed to numerous seminal visionary papers and key patents related to these areas. His current area of research is localization techniques and location-based security for body area networks. He is the author of several pioneering textbooks translated and taught around the world in several languages. He is the founding Editor-in-Chief of the International Journal Wireless Information Networks, Springer, which was established in 1994 as the first journal in wireless networks. He has founded, chaired and organized a number of pioneering international events in wireless access and localization, which includes Workshops on Opportunistic RF Localization for Emerging Smart Devices, (2008, 2010, 2012). For his pioneering entrepreneurship activities in the growth of wireless networking industry, he has been selected as a member of the Committee on Evolution of Untethered Communication, US National Research Council (1997) and has led the US team for the review of the Finnish National R&D Programs (2000 and 2003). For his contributions in research and scholarship, he was the Westin Hadden Professor of Electrical and Computer Engineering at WPI (1993–1996), elected as a fellow of the IEEE (1996), became the first non-Finn fellow of the Nokia (1999), received the first Fulbright-Nokia fellowship (2000), and received WPI board of trustee's award for Outstanding Research and Creative Scholarship (2011). Recently, he has received an "overseas famous scholar award" from R.I. China to serve as a visiting professor at University of Science and Technology of Beijing (2019–2021).